Introduction to Quantum Groups

Introduction to Quantum Groups

M. Chaichian

High Energy Physics Division
University of Helsinki
and
Physics Research Institute
Helsinki

A. Demichev

Nuclear Physics Institute
Moscow State University
and
Centro Brasileiro de Pesquisas Fisicas
Rio de Janeiro

Published by

World Scientific Publishing Co. Pte. Ltd.

P O Box 128, Farrer Road, Singapore 912805

USA office: Suite 1B, 1060 Main Street, River Edge, NJ 07661

UK office: 57 Shelton Street, Covent Garden, London WC2H 9HE

Library of Congress Cataloging-in-Publication Data

Chaichian, M. (Masud), 1941–
Introduction to quantum groups / by M. Chaichian and A. P. Demichev.
356 p. 22 cm.
Includes bibliographical references and index.
ISBN 9810226233
1. Quantum groups. I. Demichev, A. P. (Andrei Pavlovich)
II. Title.
QC20.7.G76C47 1996
530.1'5255--dc20 96-25942
CIP

British Library Cataloguing-in-Publication Data

A catalogue record for this book is available from the British Library.

This book is printed on acid-free paper.

Printed in Singapore by Uto-Print

To our parents

Preface

The importance of the theory of Lie groups and algebras for theoretical physics can hardly be disputed. Its applications in most branches of modern physics have proved to be extremely fruitful not only for the solution of already existing problems but also as a guide for the formulation and development of essentially new ideas and approaches for the description of physical phenomena. It turns out, however, that the present level of theoretical problems requires more general algebraic objects called *quantum groups* which contain Lie groups and Lie algebras as particular and, in a sense, degenerate cases. The mathematical theory of these new objects arose as an abstraction from constructions developed in the frame of the inverse scattering method of solution of quantum integrable models. This origin implied, from the very beginning, the importance of this mathematical theory for physics. On the other hand, the theory has proved to be so rich and powerful that it seems natural to apply it to different problems far beyond the original area, generalizing such basic physical notions as space-time symmetries, gauge symmetries, different types of dynamical symmetries and so on.

At present the theory being highly developed, is still far from full completeness. Its mathematical foundation are summarized in a number of books [171, 64, 61, 137, 174]. But many results, especially concerning physical applications, are scattered in a variety of original papers and reviews, often rather difficult for a first reading.

In this book we have aimed at giving an introduction to some of the principal ideas of the quantum group theory accompanied by examples of its different applications, assuming that the reader is familiar with the theory of Lie groups, Lie algebras and quantum mechanics. A certain acquaintance with some basic facts from the theory of supergroups and quantum field theory is also assumed. Actually, the latter is important only for a better understanding of some applications of the new mathematics. Though the most natural applications of quantum groups are in the theory of integrable systems, we have not concentrated on this subject. The main reason is that we have intended to show a variety and diversity of possible applications of quantum groups, including the theory of elementary particles and field theory. Besides, the inverse scattering method is a very extensive subject which cannot be reduced only to quantum groups. Its self-contained exposition would require an essential increase of the volume of the book and would lead us away from the main topic.

By now quantum groups and, in a more general sense, non-commutative geometry have become an immense and versatile subject. We have selected certain topics for discussion and we are well aware of the fact that many important topics have been either left out or are only mentioned briefly. We hope that partially this is compensated by references in our book to the original papers and appropriate reviews. Also we have tried to use a minimum of mathematical tools. Thus proofs of a number of theorems and details of applications are either briefly sketched or omitted, adequate references being given to enable the interested reader to study the subject in full completeness.

In writing this book we have benefited from discussions on various topics with many of our colleagues, in particular with Peter Prešnajder. We thank all of them for useful discussions and for their advice. Especially, it is a pleasure to express our gratitude to Peter Kulish and to acknowledge stimulating discussions, suggestions and criticism. We are deeply indebted for all what he has offered to us. We are also grateful to Wen-Feng Chen for making many improvements in the manuscript. A.D. is much grateful to José Helayel-Neto for constant support and to the Brazilian Research Council and INTAS for financial support.

Helsinki, Rio de Janeiro
March 1996

M. Chaichian, A. Demichev

Contents

Preface **vii**

Introduction **1**

Notational Conventions 5

1 Mathematical Aspects of Quantum Group Theory and Non-Commutative Geometry **9**

1.1 Non-commutative algebras, differential calculi, transformations and all that 9

1.2 Hopf algebra and Poisson structure of classical Lie groups and algebras 22

1.3 Deformation of co-Poisson structures 41

1.4 Quasi-triangular Hopf algebras and quantum double construction 52

1.5 Quantum matrix groups 60

1.5.1 Quantum groups $GL_q(n)$ and $SL_q(n)$ 62

1.5.2 Quantum groups $SO_q(N)$ and $Sp_q(n)$ 66

1.5.3 Twists of quantum groups and multiparametric deformations 70

1.6 Quantum deformation of differential and integral calculi 77

1.6.1 Differential calculus on q-groups 78

1.6.2 Differential calculus on quantum spaces 81

1.6.3 q-Deformation of integral calculus 85

1.7 Elements of quantum group representations 92

1.7.1 Corepresentations of quantum groups 94

1.7.2 Representations of quantum universal enveloping algebras 99

1.7.3 Representations of quantized algebras of functions . . . 107

2 q-Deformation of Harmonic Oscillators, Quantum Symmetry and All That **111**

2.1 q-Deformation of single harmonic oscillator 112
2.2 Bargmann-Fock representation for q-oscillator algebra in terms of operators on quantum planes 126
2.3 Quasi-classical limit of q-oscillators and q-deformed path integrals 132
2.3.1 Quasi-classical limit of q-oscillators (with real parameter of deformation) 133
2.3.2 Path integral for q-oscillators (real q) 143
2.3.3 Path integral for q-oscillators with root of unity value of deformation parameter 146
2.4 q-Oscillators and representations of QUEA 151
2.4.1 q-Deformed Jordan-Schwinger realization 151
2.4.2 Quantum Clebsch-Gordan coefficients and Wigner-Eckart theorem 153
2.4.3 Covariant systems of q-oscillators 158
2.5 q-Deformation of supergroups and conception of braided groups 166
2.5.1 q-Supergroups, q-superalgebras and q-superoscillators . 166
2.5.2 Braided groups and spaces 171
2.6 Quantum symmetries and q-deformed algebras in physical systems 177
2.6.1 Integrable one-dimensional spin-chain model 177
2.6.2 A model in quantum optics 182
2.6.3 Magnetic translations and the algebra $sl_q(2)$ 183
2.6.4 Pseudoeuclidian quantum algebra as symmetry of phonons 185
2.6.5 q-Oscillators and regularization of quantum field theory 188
2.6.6 q-Deformed statistics and the ideal q-gas 192
2.6.7 Nonlinear Regge trajectory and quantum dual string theory 198
2.6.8 q-Deformation of the Virasoro algebra 203

3 q-Deformation of Space-Time Symmetries **209**

3.1 One-dimensional lattice and q- deformation of differential calculus 210
3.2 Multidimensional Jackson calculus and particle on two-dimensional quantum space 214
3.3 Projective construction of quantum inhomogeneous groups . . 224

3.4 Twisted Poincaré group and geometry of q-deformed Minkowski space 229
3.4.1 Quantum deformation of the Poincaré group 230
3.4.2 Quantum Minkowski geometry 234
3.4.3 q-tetrades and transformation to commuting coordinates 237
3.4.4 Twisted Poincaré algebra and induced representations of the q-group 239
3.4.5 Twisted Minkowski space in the case of related q and $\hbar$ constants 244
3.5 Jordan-Schwinger construction for q-algebras of space-time symmetries and contraction of quantum groups 246
3.5.1 Fock space representation of the q-Lorentz algebra . . . 247
3.5.2 q-Deformed anti-de Sitter algebra and its contraction . 249
3.5.3 Quantum inhomogeneous groups from contraction of q-deformed simple groups 256
3.6 Elements of general theory of q-inhomogeneous groups and classification of q-Poincaré groups and q-Minkowski spaces 261
3.6.1 Classification of q-Lorentz groups and q-Minkowski spaces 261
3.6.2 General definition and properties of inhomogeneous quantum groups 268
3.6.3 Classification of quantum Poincaré groups 271

4 Non-commutative Geometry and Internal Symmetries of Field Theoretical Models 281
4.1 Non-commutative geometry of Yang-Mills-Higgs models 283
4.2 Posets, discrete differential calculus and Connes-Lott-like models 294
4.2.1 Yang-Mills-Higgs theory from dimensional reduction . . 294
4.2.2 Finite approximation of topological spaces 296
4.3 Basic elements of quantum fibre bundle theory 307

Appendix: Short Glossary of Selected Notions from the Theory of Classical Groups 315

Bibliography 323

Index 339

Introduction

Over the few past decades the development in theoretical physics have tended to be formulated in terms of different types of symmetries and patterns and types of their breaking. Thus a discovery and development of any new type of symmetry transformations and corresponding mathematical structures leads to an essential increase of our abilities to describe and explain complicated physical phenomena. Symmetry transformations based on Lie groups and Lie algebras are most known and still most exploited in all branches of physics and other fields of science. But complicated problems of fundamental physics have required different generalizations and further development of the conception of symmetry: there have appeared local transformations (gauge groups), supertransformations (Lie supergroups and Lie superalgebras) and infinite dimensional Lie algebras.

Theory of quantum integrable systems [100, 91, 152, 101, 105] has initiated [151, 102] a new type of symmetry and mathematical objects called *quantum groups*. Though the name "quantum" can sometimes cause confusion with usual *physical* quantization, it is not accidental and quite meaningful. The point is that these new objects are related to usual Lie groups as quantum mechanics is related to its classical limit. And mathematically the procedure of derivation of quantum groups out of classical Lie groups is quite analogous to the well known method of quantization of classical systems (see, e.g., [19]). Actually, this procedure is a highly nontrivial generalization of usual quantization which takes into account geometrical and topological properties of a group manifold and group structure on it. In a sense, the role of non-commutative geometry and quantum groups in quantum mechanics seems to be analogous to that of usual differential geometry and Lie groups in the formulation of Einstein's General Relativity: both mathematical conceptions provide so natural formalism for description of corresponding physical phenomena that, in fact, distinction between "geometry" and "physics" disappears and one speaks about "geometric dynamics" or "geometric physics". In the General Relativity this is well known since the time of its creation. Contrary to this, the essential step to comprehensive geometrization of quantum phenomena, the invention of quantum groups, has been done quite recently, long after the formulation of basic principles of quantum mechanics.

However, the physical meaning of a group quantization can be absolutely different and corresponding parameter of the quantization which is usually denoted by the letter q, *in general* has nothing common with the Planck constant $\hbar$. For example, physically a group quantization can lead to a kind of discretization of the group manifold or of a homogeneous space, somehow related to a physical (classical or quantum) system. In fact, the situation when a classical object corresponds to some limiting value of a parameter other than the Planck constant is not new: it is well known, that classical limit of a quantum system on a sphere (quantum top) corresponds to infinite value of a spin [19], but not to zero value of $\hbar$. To stress this physical difference one often uses the term "q-deformation" along with "quantization" or simply the addition of the letter "q" to a corresponding "classical" object, e.g., q-group, q-algebra etc.

Initially, a quantum analog (q-analog) of Lie algebra $su(2)$ was discovered by P. P. Kulish and N. Yu. Reshetikhin [151] as an unusual algebra of observables in the quantum version of the sine-Gordon model. The new property of this algebra is that commutators of the observables are regular functions (infinite series) of the latter but not just linear in Lie algebra generators. Such objects have proved to be described adequately [86] in terms of the so-called *Hopf algebras* [1].

The mathematical formulation with important contributions by V. V. Drinfel'd, L. D. Faddeev, S. L. Woronowicz, Yu. I. Manin, M. Jimbo, N. Yu. Reshetikhin, L. A. Takhtajan, P. P. Kulish, E. Sklyanin, M. A. Semenov-Tian-Shansky, J. Wess, B. Zumino, Y. S. Soibelman and many other mathematicians and physicists have resulted in the very rich and beautiful theory of quantum (q-deformed) objects: q-spaces, q-differential calculi, q-supergroups, braided groups, etc.

There exist different approaches to the construction of quantum groups: Drinfel'd's approach [86] heavily based on deformation of a Lie group Poisson structure; R-matrix Faddeev-Reshetikhin-Takhtajan's approach [104] which is in a sense dual to the Drinfel'd's one; Manin's approach [177, 178] with the initial object being a quadratic algebra on quantum linear spaces and Woronowicz's approach [244, 245, 246] with an essentially different background of $\mathbb{C}^*$-algebra theory.

The last approach is close in its origin to Connes' general considerations

[63, 64] of non-commutative geometry, the starting point for both approaches being Gel'fand-Naimark theorem [9], which states that any commutative $\mathbb{C}^*$-algebra with unit element is isomorphic to an algebra of all continuous functions on some compact topological manifold.

A $\mathbb{C}^*$-algebra $\mathcal{A}$, commutative or otherwise, is an algebra with a norm $\| \cdot \|$ and an antilinear involution * such that $\| a \| = \| a^* \|$, $\| a^* a \| = \| a^* \| \| a \|$ and $(ab)^* = b^* a^*$ for $a, b \in \mathcal{A}$. $\mathcal{A}$ is also assumed to be complete in the given norm. Examples of $\mathbb{C}^*$-algebras are: 1) The algebra of $n \times n$ matrices T with T^* equal to the hermitian conjugate of T, and $\| T \|^2$ equal to the largest eigenvalue of $T^* T$; 2) The algebra $Fun(\mathcal{X})$ of continuous functions on a manifold $\mathcal{X}$, with * equal to complex conjugation and the norm given by the supremum norm, $\| f \| = \sup_{x \in M} |f(x)|$.

Let $\mathcal{Y}$ denote the space of (equivalence classes of) irreducible representations of a given commutative C^*-algebra $\mathcal{C}$, called the structure space of $\mathcal{C}$. As $\mathcal{C}$ is commutative, every irreducible representation is one-dimensional. Hence if $x \in \mathcal{Y}$ and $f \in \mathcal{C}$, the image $x(f)$ of f in the representations defined by x is a complex number. Writing $x(f)$ as $f(x)$, we can therefore regard f as a complex-valued function on $\mathcal{Y}$ with the value $f(x)$ at $x \in \mathcal{Y}$. We thus get the interpretation of $\mathcal{C}$ as $\mathbb{C}$-valued functions on $\mathcal{Y}$ or, in other words, a reconstruction of a topological space $\mathcal{Y}$ from $\mathcal{C}$ such that the algebra of smooth function on $\mathcal{Y}$ is isomorphic to $\mathcal{C}$.

A generalization to non-commutative $\mathbb{C}^*$-algebras leads to disappearance of the underlying manifold but one can develop the theory postulating that all geometrical properties are put into the $\mathbb{C}^*$-algebra language similarly to that in the commutative case. If a manifold is a topological group then the space of functions on it carries extra structures which can be generalized to the case of non-commutative $\mathbb{C}^*$-algebras and lead to Woronowicz's definition and the development of quantum group theory. This approach requires rather involved mathematics and we do not give its detailed exposition in this book, restricting ourselves with some remarks and appropriate references.

The three first abovementioned approaches are close and complementary to each other and we will follow them in our description of the mathematical structure of quantum groups and algebras.

We present the theory without attempting to supply rigorous proofs, to which, however, adequate references are given. Our aim is to give a guided tour through the subject, illustrating the various points with examples. Clearly it is hopeless, in a reasonable space of an introductory book to attempt an exhaustive treatment, taking into account the full diversity of the subject. Our book naturally reflects personal preferences, or even prejudices.

Chapter 1 is of central importance and gives a general introduction to the mathematical theory of quantum groups and related constructions. The exposition assumes that the reader is somewhat familiar with the standard theory of Lie groups and algebras. Next in Chapter 2 we give in some details a basic example of a q-deformed dynamical systems: q-deformed harmonic oscillators, and consider their quantum symmetries and relations with q-deformed Lie algebras. Two last sections of the Chapter are devoted to a brief description of q-superalgebras, corresponding q-superoscillator systems, braided groups and statistics and to a few examples of more involved dynamical systems which exhibit quantum symmetries. Chapter 3 deals with the problem of quantization of non-semisimple (inhomogeneous) Lie groups. This is an important topic in its own right, but it also provides, hopefully, a background for the construction of self-consistent (without ultraviolet divergencies) quantum field theory on the q-deformed space-time. This may be important especially for the quantum theory of gravitation. An attempt to explain one of the most important phenomenon of modern unified theory of elementary particle interactions on the basis of non-commutative geometry is presented in the last Chapter 4. The approach has emerged from the Connes' general development of the idea of non-commutative geometry and in its original version does not use the tools of the quantum group theory. However, we have tried to argue that the latter is at least useful if not absolutely necessary for further development of the approach.

For the reader's convenience we give the definitions and explanations of the most important mathematical objects right in the main text. Some other notions and facts from the theory of Lie groups and algebras which we use in the book essentially, are collected in the separated Short Glossary.

Notational Conventions

a^*	*-algebra involution of an element a
$\bar{c}$	complex conjugation of $c \in \mathbb{C}$
$A^\dagger$	hermitian conjugation
$T^\top$	matrix transposition
lhs	left hand side of a relation
rhs	right hand side of a relation
:=	rhs is a definition of lhs
=:	lhs is a definition of rhs
BF	Bargmann-Fock
BFR	Bargmann-Fock representation
CR	commutation relations
CG	Clebsch-Gordan
CYBE	classical Yang-Baxter equation
HP	Holstein-Primakoff
MCYBE	modified classical Yang-Baxter equation
QISM	quantum inverse scattering method
QUEA	quantum universal enveloping algebra
QYBE	quantum Yang-Baxter equation
UEA	universal enveloping algebra
YBE	Yang-Baxter equation

We assume usual summation convention for repeated indices *if the opposite is not indicated explicitly*; in ambiguous cases we use the explicit sign of summation.

Signs and page numbers where the signs appear first time:

$\otimes$	16, 20	$\textcircled{\scriptsize T}$	262
$\hat{\otimes}$	15, 56	$\preceq$	297
$\star$	18	$\mathbf{I}$	9
$\diamond$	37		
$\triangleright$	176	$[x]_q$	21

$[x;q]$	59
$\begin{bmatrix} n \\ k \end{bmatrix}_q$	97
$\lVert \cdot \rVert$	3
$\{\cdot,\cdot\}_m$	19
$\{\cdot,\cdot\}_p$	19
$\langle\langle\cdot,\cdot\rangle\rangle$	25
$\langle\cdot,\cdot\rangle$	41
$[\cdot,\cdot]_q$	50
$[\cdot,\cdot\}$	166
$(r,r)_S$	38
(α_i,α_j)	47
$(p;q)_m$	98
$[A,B]_{p,r}$	153
$(x+y)^{[n]}$	155
$a\hat{*}\rho$	269
$\rho\hat{*}a$	269
$\rho\hat{*}\rho'$	269
A_H	9
A^0	31
$A\otimes A$	24
$\mathrm{ad}_X^{\pm}$	48
$\mathrm{ad}^{(q)}_{\tilde{X}_i^{\pm}}$	50
$BGL_q(2)$	173
$BSL_q(2)$	174
$\mathbb{C}^n_{f,R}$	61
$\mathbb{C}^n_q$	64
$\wedge\mathbb{C}^n_q$	64
D	68
D_q	83
$D^{(q)}_s$	154
$\mathcal{D}$	63
$\det_q$	63
$\det_B$	174
$\dim_q$	105
$\partial_R, \bar{\partial}_R$	140
$End\,V$	52
$\exp_q x,\ e_q^x$	60, 129
$Fun(\mathcal{X})$	3
Fun_R	60
$Fun(G)$	19, 25
G	25
$G(2,4)$	230
$\mathcal{H}$	18
$\mathcal{I}$	11, 29

L 25
l_q 69

$\mathcal{M}_q$ 235

$O_q^N(\mathbb{C})$ 68
$\wedge O_q^N$ 69

P 61
$p(X)$ 167
$P_q^{\pm}$ 64
$Pol(G)$ 95
$\mathcal{P}_q$ 232

q_i 49

$\hat{R}$ 61
$\mathbb{R}_q^n$ 66

S 27
$SL_q(2)$ 17

Tr 18
Tr_q 81
$\mathcal{T}_4$ 229

$\mathcal{U}(L)$ 25
$\mathcal{U}^*(L)$ 25

$\mathcal{U}(su(2))$ 20
$\mathcal{U}_q(s\ell(2))$ 21

$\mathcal{W}_A$ 18

$Z_{(ij)}$ 230

Δ 20
Δ' 31
$\Psi_{V,W}$ 176
$\Omega^m(A_H)$ 10
ω_{ij} 18
σ 29

Chapter 1

Mathematical Aspects of Quantum Group Theory and Non-Commutative Geometry

The aim of this chapter is to present general mathematical structure of quantum groups and related constructions. In Section 1 we consider examples of non-commutative algebras and quantum deformation of commutative algebras, partially known from the standard quantum mechanics, to give the reader an idea about the main objects of the book - quantum group and algebra, their duality, non-commutative geometry, comultiplication, R-matrix etc. Starting from Section 2 we set to systematic description of mathematical theory of quantum deformations.

1.1 Non-commutative algebras, differential calculi, transformations and all that

The aim of this section is to give the reader a preliminary feeling about main objects, tools and goals of the subject. As follows from the very term "non-commutative geometry" we are going to define structures analogous to that in differential geometry of manifolds on sets of non-commutative elements. To get accustomed to new constructions more easily, it is reasonable to use the experience in quantum mechanical formalism. No doubt that the most known to physicists non-commutative object is the Heisenberg algebra A_H of the position operator x, momentum operator p and unity $\mathbf{I}$ satisfying the

canonical commutation relation

$$[x, p] = i\hbar \mathbf{I} \ . \tag{1.1.1}$$

A basis for A_H consists of all ordered monomials in p and x (including zero-degree monomials generated by the algebra unity $\mathbf{I}$) and one says that the algebra A_H is *generated* by x, p and $\mathbf{I}$. Let us now consider them as a non-commutative analog of phase space coordinates. Classical (commutative) geometries can be classified by the groups of transformations which they admit. The simplest but very important class of transformations consists of linear transformations of coordinates. In the non-commutative case we may try to proceed in an analogous way and raise the question about groups of (linear) transformations preserving the defining relations for an object which we want to consider as a non-commutative geometry.

In the case of A_H such a group is well known: the defining relation (1.1.1) is invariant with respect to $SL(2, \mathbb{R}) \approx Sp(1, \mathbb{R})$ group of transformations of phase "coordinates" x, p. So in this case non-commutative algebra is invariant with respect to usual (classical, commutative) group.

If we want to define some differential geometry on the algebra A_H, for example an analog of symplectic geometry which plays so important role for classical Hamiltonian systems, we have to construct, first of all, an analog of a differential calculus on the algebra. To this aim let us associate with the elements x and p their "differentials" dx and dp which are considered as generators of a space $\Omega^1(A_H)$ of 1-forms. Natural condition $d\mathbf{I} = 0$, eliminates the differential of unity. We allow a multiplication of the differentials dx, dp by elements of A_H from the left and from the right and by definition a result of the multiplications belongs to $\Omega^1(A_H)$ again. On mathematical language, this means that $\Omega^1(A_H)$ is an A_H-*bimodule*.

A differential calculus on an arbitrary algebra A is a prescription on how to multiply 1-forms to create m-forms $\Omega^m(A_H)$ for arbitrary $m \in \mathbb{Z}_+$ and then also how to multiply m-forms

$$\wedge : \ \Omega^m(A) \otimes \Omega^{m'}(A) \longrightarrow \Omega^{m+m'}(A) \ ,$$

together with a $\mathbb{C}$-linear *exterior differential operator* d which defines the map

$$d : \ \Omega^m(A) \longrightarrow \Omega^{m+1}(A) \ ,$$

where Ω^m is a space of m-form and A-bimodule, A is arbitrary associative (in general, non-commutative) algebra. Natural generalization of a usual exterior differential calculus leads to the rules [64, 247]

$$\begin{aligned} &d^2 = 0\,, \\ &d(\omega \wedge \omega') = (d\omega) \wedge \omega' + (-1)^m \omega \wedge d\omega'\,, \\ &\omega \in \Omega^m(A)\,, \qquad \omega' \in \Omega^{m'}(A)\,. \end{aligned} \tag{1.1.2}$$

If one considers an algebra with *-involution (in the case of A_H the involution is the usual hermitian conjugation $\dagger$), one has to define properties of $\wedge$ and d under the involution. In particular, one can postulate

$$\begin{aligned} (\omega \wedge \omega')^* &= (-1)^{mm'}(\omega')^* \wedge \omega^*\,, \\ (df)^* &= d(f^*)\,, \end{aligned} \tag{1.1.3}$$

$f \in A$ (i.e., f is 0-form). Note, however, that there are other possibilities for definition of properties of $\wedge$ under an involution, for example, one could choose

$$(\omega \wedge \omega')^* = (\omega')^* \wedge \omega^*\,. \tag{1.1.4}$$

For a non-commutative algebra A we cannot expect that differentials commute with the algebra elements and assume

$$f d\xi^i = d\xi^k \Xi^i_k(f)\,, \tag{1.1.5}$$

where ξ^i, $i \in \mathcal{I}$ ($\mathcal{I}$ is some set of indices) generate A, $d\xi^i$ generate $\Omega^1(A)$ (differentials of ξ^i), f is an arbitrary element of A and Ξ^i_k is a map $A \to A$.

Application of the exterior differential d to (1.1.5) gives commutation relations (CR) for the differentials $d\xi^i$.

Self-consistency of (1.1.5) requires

$$\begin{aligned} \Xi(fh)^i_j &= \Xi^k_j(f)\Xi^i_k(h)\,, \quad \forall\, f,h \in A\,, \\ \Xi(\mathbf{I})^i_j &= \delta^i_j\,, \end{aligned} \tag{1.1.6}$$

($\mathbf{I}$ is the unity of A). To obtain further consistency conditions for Ξ [247, 239], one has to

- apply d to commutation relations of an algebra A, use (1.1.5) to move all differentials to the left and compare coefficients at independent differentials;
- apply an involution to (1.1.5) and use (1.1.3);

- commute the differentials $d\xi^i$ through the CR of an algebra A;
- commute the algebra generators ξ^i through CR for the differentials.

Let us come back to the Heisenberg algebra A_H and consider possible differential calculi on it with *linear* map Ξ

$$\Xi(x) = c_1 x + c_2 p \ ,$$

$$\Xi(p) = c_3 x + c_4 p \ .$$

To define the coefficients c_i one has to go through the steps indicated above and as a result, one obtains [78] $SL(2,\mathbb{R})$ invariant differential calculus with defining relations

$$dx \wedge dx = 0 \ , \quad dp \wedge dp = 0 \ , \quad dx \wedge dp = -dp \wedge dx \ ,$$

$$\begin{aligned} x dx &= (dx)x \ , & p dx &= (dx)p \ , \\ x dp &= (dp)x \ , & p dp &= (dp)p + ic(dx)x \ , \end{aligned} \qquad (1.1.7)$$

where $c \in \mathbb{R}$ is an arbitrary constant. For $c = 0$ the differential calculus looks like the ordinary calculus on a classical phase space (but we must remember, of course, that x and p are non-commuting operators). If nonlinear transformations are also admitted, then the transformation

$$\begin{pmatrix} x \\ p \end{pmatrix} \longrightarrow \begin{pmatrix} x \\ p + \frac{c}{6\hbar} x^3 \end{pmatrix} ,$$

which preserve the CR (1.1.1) and hermiticity, convert the calculus (1.1.7) to ordinary one even for $c \neq 0$. Now we can introduce the next ingredient of a differential calculus, partial derivatives ∂_x, ∂_p, through the natural relation

$$d =: dp\partial_p + dx\partial_x \ ,$$

and derive its properties which also do not differ essentially from the usual classical case (for details of the calculations and discussion of possible applications see [78]).

Consider now the operators $v^1 := \exp\{isx\}$, $v^2 := \exp\{itp\}$ which satisfy the Weyl relation

$$v^1 v^2 = e^{-i\hbar st} v^2 v^1 \ , \qquad (1.1.8)$$

and again raise the question about *linear* transformations which leave this relation invariant. It is easy to see that there are no such transformations *mixing* the operators v^1, v^2 with *c-number* coefficients (simple rescaling $v^i \to c^i v^i$ (no summation), $c^i \in \mathbb{C}$, obviously preserve (1.1.8)). One of the basic ideas of quantum group theory is to consider transformations of more general type: with matrix elements of transformations being also non-commutative with each other.

Let us now drop the connection with Heisenberg or Weyl algebra and consider the possibility from more general point of view [177, 178, 104, 98]. For an introductory example, we restrict ourselves to a two dimensional linear space with non-commuting variables x^1 and x^2. Classically, for a two dimensional vector, one has three independent quadratic combinations of x^1, x^2. So to define non-commutative space as close as possible to the usual space, one has to impose one linear relation between quadratic combinations

$$E_{ij} x^i x^j = 0 \,, \qquad E_{ij} \in \mathbb{C} \,. \tag{1.1.9}$$

It is clear that (1.1.8) is the particular case of (1.1.9) Analogously, one can introduce covectors with coordinates u_1, u_2 and the relation

$$F^{ij} u_i u_j = 0 \,, \qquad F^{ij} \in \mathbb{C} \,. \tag{1.1.10}$$

If the matrices E and F are non-degenerate, the number of independent monomials in x^1, x^2 or u_1, u_2 of any degree are the same as in the usual commutative case. This can be shown by direct inspection. We want to find transformation matrices with generally non-commuting entries which preserve the relations (1.1.9) and (1.1.10). First of all, we have to agree about CR between matrix elements and the non-commutative coordinates. In quantum group theory one postulates that entries of transformation matrices *commute* with space coordinates. Note that this prevents us from considering the quantum group transformations as a generalization of supertransformations of superspace (cf., e.g., [111]). Thus one requires that the transformed coordinates

$$x'^{\,i} = T^i{}_j x^j \,,$$

$$u'_i = u_j T^j{}_i \,,$$

satisfy the same relations (1.1.9),(1.1.10) as x^i and u_j, respectively. This gives a set of quadratic relations for the entries of T

$$E_{ij} T^i{}_l x^l T^j{}_k x^k = 0 \,,$$

$$F^{ij} u_l T^l{}_i u_k T^k{}_j = 0 \,.$$

Due to commutativity of the matrix entries with the coordinates these equations can be rewritten in the equivalent form

$$E_{ij}T^i{}_k T^j{}_l = DE_{kl} \;, \tag{1.1.11}$$

$$F^{ij}T^m_i T^n_j = \tilde{D}F^{mn} \tag{1.1.12}$$

for some, possibly non-commuting elements $D, \tilde{D}$.

In the classical case there are 10 independent quadratic combinations of $T^i{}_j$. To make the transformations compatible with the properties of quantum space the same has to be held for non-commutative case. One can show [98] that if $\tau := E_{ij}F^{ij} \neq 0$, the system (1.1.11),(1.1.12) allows to express $D, \tilde{D}$ (actually it is easy to see that $D = \tilde{D}$) and six quadratic combination of $T^i{}_j$ in terms of ten other independent quadratic monomials of the matrix entries.

Multiplying (1.1.11) by $F^{lm}T^s{}_m$ from the right and by $F^{nk}T^t{}_n$ from the left, one obtains two equality

$$DTM = MTD \;,$$

$$NDT = TDN \;,$$

where $M^k_i := E_{ij}F^{jk}$, $N^k_i := F^{kj}E_{ji}$ and matrix multiplication is understood. Therefore,

$$D(TMN - MNT) = 0$$

and if D is invertible element of the algebra, one must put

$$MN = \lambda \mathbf{1} \;, \tag{1.1.13}$$

since otherwise there would exist linear dependence between the elements of the matrix T. The invertibility of the element D is necessary for the invertibility of quantum space transformations. Indeed, it is easy to see, using (1.1.11), that $(T^{-1})^l{}_j := D^{-1}E_{ij}T^i{}_k E^{lk}$ (here E^{lk} is the inverse matrix: $E_{ij}E^{jk} = \delta_i{}^k$) is left inverse for the matrix T. Thus, from (1.1.13) one obtains the constraints for matrices E_{ij}, F^{kl}

$$E_{ab}F^{bk}F^{ac}E_{cd} = \lambda\delta^k_d \;. \tag{1.1.14}$$

It is important that the same constraints provide a coincidence of numbers of independent cubic and higher monomials for quantum and for classical matrix entries $T^i{}_j$ [98].

Enumeration of all solutions of the constraint (1.1.14) leads to classification of all possible two-dimensional quantum spaces and corresponding quantum groups of transformations. They fall into two families: two-parametric quantum general linear groups $GL_{r,q}(2)$ and one parametric family $GL_{\alpha}^{J}(2)$. Consider the $GL_{r,q}(2)$ group with $r = q^2 \in \mathbb{C}$. In this case the matrices E and F coincide

$$E = F = \begin{pmatrix} 0 & 1 \\ -q & 0 \end{pmatrix} \tag{1.1.15}$$

and it is easy to check that the constraint (1.1.14) is indeed satisfied. The relations (1.1.9) and (1.1.10) become

$$\begin{aligned} x^1x^2 &= qx^2x^1 \ , \\ u_1u_2 &= qu_2u_1 \ , \end{aligned} \tag{1.1.16}$$

and the relations (1.1.11),(1.1.12) are equivalent to

$$\begin{aligned} ab &= qba \ , \qquad & bd &= qdb \\ cd &= qdc \ , \qquad & bc &= cb \ , \\ ac &= qca \ , \qquad & ad &= da + (q - q^{-1})bc \ , \end{aligned} \tag{1.1.17}$$

where a, b, c, d are entries of the matrix T

$$T =: \begin{pmatrix} a & b \\ c & d \end{pmatrix} . \tag{1.1.18}$$

The commutation relations (CR) (1.1.17) can be written in a very convenient form with help of the so-called *R-matrix*, which in the case under consideration is

$$R^{ab}_{\ \ cd} = \delta^b_c \delta^a_d - q^{-2} F^{ba} E_{cd} \ , \tag{1.1.19}$$

the condition (1.1.14) being equivalent to the famous Yang-Baxter equation (YBE)

$$R_{12}R_{13}R_{23} = R_{23}R_{13}R_{12} \ . \tag{1.1.20}$$

The subscripts in this equation mean the following. R-matrix entries carry four indices and so it acts in the tensor product $\mathbb{C}^n \hat{\otimes} \mathbb{C}^n$ of two copies of vector spaces (in the case under consideration $\mathbb{C}^2 \hat{\otimes} \mathbb{C}^2$). The subscripts 12, 13 and 23 in (1.1.20) indicate different embeddings of a space of matrices acting in $\mathbb{C}^n \hat{\otimes} \mathbb{C}^n$ into a space of matrices acting in a triple tensor product $\mathbb{C}^n \hat{\otimes} \mathbb{C}^n \hat{\otimes} \mathbb{C}^n$, according to the singling out of a pair of factors in the latter. Thus, for example, R_{12} acts like R in the tensor product of the first two factors and like unity matrix $\mathbb{I}$ in the last one. In components this looks as follows

$$\begin{aligned} (R_{12})^{abc}_{\ \ \ def} &= R^{ab}_{\ \ de} \delta^c_{\ f} \ , \\ (R_{13})^{abc}_{\ \ \ def} &= R^{ac}_{\ \ df} \delta^b_{\ e} \ , \end{aligned} \tag{1.1.21}$$

$$(R_{12})^{abc}_{\ \ def} = \delta^a_{\ d} R^{bc}_{\ \ ef} \ .$$

Using analogous notations, the CR (1.1.17) can be written in the matrix form

$$RT_1T_2 = T_2T_1R \ , \tag{1.1.22}$$

where $T_1 = T \otimes \mathbf{I}$ and $T_2 = \mathbf{I} \otimes T$, i.e.

$$\begin{aligned} (T_1)^{ij}_{\ \ kl} &= T^i_{\ k}\delta^j_{\ l} \ , \\ (T_2)^{ij}_{\ \ kl} &= \delta^i_{\ k}T^j_{\ l} \ . \end{aligned} \tag{1.1.23}$$

In (1.1.22) R acts in a tensor product of *two* spaces, so we use no subscripts. In explicit form, (1.1.22) reads as

$$\sum_{p,s} R^{mn}_{\ \ ps} T^p_{\ u} T^s_{\ v} = \sum_{s,r} T^n_{\ s} T^m_{\ r} R^{rs}_{\ \ uv} \ . \tag{1.1.24}$$

The equivalence of (1.1.17) and (1.1.22) is easy to check using the explicit form of the R-matrix, which is derived from (1.1.19) and (1.1.15)

$$R = R^{ik}_{\ \ jl} = \begin{pmatrix} 1 & 0 & 0 & 0 \\ 0 & q^{-1} & 0 & 0 \\ 0 & 1-q^{-2} & q^{-1} & 0 \\ 0 & 0 & 0 & 1 \end{pmatrix} \ . \tag{1.1.25}$$

The R-matrix formalism of the commutation relations emerges from the inverse scattering method and appears to be a very convenient tool for the quantum group theory (the R-matrix or FRT-formalism)[104]. In particular, one has the same form (1.1.22) of the CR (*TT-relations*) for the q-analog of any simple Lie group, but with different R-matrices (see Section 1.5).

To illustrate the advantage of the compact R-matrix form of CR, we show that the product of two copies T and T' of the q-matrix with mutually commuting elements has the same CR as the initial ones. Take the identity

$$T_2T_1R_{12}T_1'T_2' = T_2T_1R_{12}T_1'T_2' \ .$$

Due to (1.1.22) this is equivalent to

$$R_{12}T_1T_2T_1'T_2' = T_2T_1T_2'T_1'R_{12} \ ,$$

or, because of commutativity of $T^i_{\ j}$ with $(T')^k_{\ l}$

$$R_{12}T_1T_1'T_2T_2' = T_2T_2'T_1T_1'R_{12} \ .$$

This gives the desirable identity

$$R_{12}T_1''T_2'' = T_2''T_1''R_{12} ,$$

where

$$\begin{aligned} T_1'' &= T'' \otimes 1 = TT' \otimes \mathbf{1} = (T \otimes \mathbf{1})(T' \otimes \mathbf{1}) = T_1T_1' , \\ T_2'' &= \mathbf{1} \otimes T'' = T_2T_2' . \end{aligned}$$

Thus the matrix product has the same properties as the initial matrix T. Of course, this is a necessary condition for the matrices to form some generalization of a group. Another essential property of a group is the existence of an inverse element. Fortunately, in the non-commutative case one can also define an appropriate analog. The eqs. (1.1.11), (1.1.12), (1.1.15) define the element D

$$D = ad - qbc , \tag{1.1.26}$$

which in the considered case prove to be a central element (i.e. commuting with any one) of the algebra, generated by matrix entries. This permits to put the condition

$$D = 1 \tag{1.1.27}$$

and to obtain the quantum unimodular group $SL_q(2)$ with the inverse matrix

$$S(T) = \begin{pmatrix} d & -q^{-1}b \\ -qc & a \end{pmatrix} . \tag{1.1.28}$$

It is easy to see by direct calculations with use of (1.1.26), (1.1.27) that

$$TS(T) = S(T)T = \mathbf{1} .$$

Thus we have obtained an object - *a quantum group* $SL_q(2)$ - which can be considered as an appropriate generalization of the corresponding Lie group. Of course, the complete structure of quantum groups is much richer and will be discussed in the subsequent sections.

In the limit $q \to 1$ one recovers usual $SL(2)$ group. This means that the quantum group $SL_q(2)$ is a smooth deformation of the corresponding classical one. Later we will clarify the explicit construction of this deformation. Here we would like to illustrate the notion of deformation using again the experience in quantum mechanics.

Let us recall that quantum mechanics can be considered as a deformation of classical mechanics (with the Planck constant $\hbar$ being the deformation parameter) [182, 14, 17, 20]. It is possible to set up a linear one-to-one map among operators $A, B, \dots$ in some Hilbert space $\mathcal{H}$ and complex valued functions $\mathcal{S}_A, \mathcal{S}_B, \dots \in Fun(\mathcal{X})$, where $Fun(\mathcal{X})$ is a space of functions defined on an appropriate finite-dimensional manifold $\mathcal{X}$. The space $Fun(\mathcal{X})$ can be converted to an algebra if we define some multiplication for its elements. It is obvious that an algebra of operators in Hilbert space $\mathcal{H}$ is not homomorphic to the algebra $Fun(\mathcal{X})$ with *usual* (commutative) pointwise multiplication of functions. So we may look for another multiplication, called *star-product* $\star$, which satisfies the homomorphic property

$$\mathcal{S}_{AB} = \mathcal{S}_A \star \mathcal{S}_B \ .$$

The well known ordering problem in quantum mechanics leads to a variety of possibilities of associating an operator A to a functions $\mathcal{S} \in Fun(\mathcal{X})$.

Assume, for simplicity, that canonical coordinates

$$\xi^i := (\mathcal{S}_{p^1}, \dots, \mathcal{S}_{p^n}, \mathcal{S}_{x^1}, \dots, \mathcal{S}_{x^n}) \ , \quad i = 1, \dots, 2n \ ,$$

on a space $\mathcal{X}_{2n}$ can be introduced globally. As a particular type of the symbol map one can use a *Weyl symbol* $\mathcal{W}_A$ which has the property that if $\mathcal{W}_A(\mathcal{S}_p, \mathcal{S}_x)$ is a polynomial in $\mathcal{S}_p$ and $\mathcal{S}_x$, the operator $A(p, x)$ is the symmetrically ordered polynomial in p and x. The symbol is given by the formula

$$\mathcal{W}_A(\xi^i) = \int \frac{d^{2n}\xi_0}{(2\pi\hbar)^n} \exp\left\{\frac{i}{\hbar}\xi_0^i \omega_{ij} \xi^j\right\} \mathrm{Tr}\left(\exp\left\{\frac{i}{\hbar}\xi_0^i \omega_{ij} \hat{\xi}^j\right\} A\right) \ ,$$

where $\hat{\xi}^i = (p^1, \dots, p^n, x^1, \dots, x^n)$ denote the quantum canonical operators, Tr denotes an operator trace and $\omega = \omega_{ij} d\xi^i \wedge d\xi^j$ is a symplectic 2-form on $\mathcal{X}_{2n}$.

The inverse map reads

$$A = \int \frac{d^{2n}\xi_0 d^{2n}\xi}{(2\pi\hbar)^n} \mathcal{W}_A(\xi) \exp\left\{\frac{i}{\hbar}\xi^i \omega_{ij} \xi_0^j\right\} \exp\left\{\frac{i}{\hbar}\xi_0^i \omega_{ij} \hat{\xi}^j\right\} \ .$$

In Section 2.3 we will use another well known symbol, namely, the normal one. The star product which makes the algebra of Weyl symbols isomorphic to the operator algebra is defined by the expression

$$\begin{aligned}
(\mathcal{W}_A \star \mathcal{W}_B)(\xi) &= \mathcal{W}_A(\xi) \exp\left\{i\frac{\hbar}{2} \overleftarrow{\partial}_i \, \omega^{ij} \overrightarrow{\partial}_j\right\} \mathcal{W}_B(\xi) \\
&= \sum_{m=0}^{\infty} \frac{1}{m!} \left(\frac{i\hbar}{2}\right)^m \omega^{i_1 j_1} \cdots \omega^{i_m j_m} (\partial_{i_1} \cdots \partial_{i_m} \mathcal{W}_A)(\partial_{j_1} \cdots \partial_{j_m} \mathcal{W}_B) \\
&= \mathcal{W}_A(\xi) \mathcal{W}_B(\xi) + \mathcal{O}(\hbar) \ ,
\end{aligned} \tag{1.1.29}$$

where ω^{ij} is the inverse symplectic matrix: $\omega^{ij}\omega_{jk} = \delta^i{}_k$. This expression shows that the star-product is indeed *a smooth deformation* of the usual product $\mathcal{W}_A(\xi)\mathcal{W}_B(\xi)$ and in the classical limit $\hbar \to 0$ reduces to the latter. The corresponding *Moyal bracket*

$$\begin{aligned}\{\mathcal{S}_A, \mathcal{S}_B\}_m &:= \frac{1}{i\hbar}(\mathcal{S}_A \star \mathcal{S}_B - \mathcal{S}_B \star \mathcal{S}_A) \\ &= \mathcal{S}_{[A,B]/i\hbar}\,, \end{aligned} \tag{1.1.30}$$

reduces in the classical limit $\hbar \to 0$ to the Poisson bracket $\{\cdot,\cdot\}_p$

$$\{\mathcal{S}_1, \mathcal{S}_2\}_m = \{\mathcal{S}_1, \mathcal{S}_2\}_p + \mathcal{O}(\hbar^2)\,, \qquad \forall\, \mathcal{S}_1, \mathcal{S}_2 \in Fun(\mathcal{X}_{2n})\,,$$

$$\{\mathcal{S}_1, \mathcal{S}_2\}_p := \partial_i \mathcal{S}_1 \omega^{ij} \partial_j \mathcal{S}_2\,.$$

This symbol calculus permits to consider a quantization process as a smooth deformation of an algebra of classical observables

$$\mathcal{S}_1\mathcal{S}_2 \overset{\text{deformation}}{\longrightarrow} \mathcal{S}_1 \star \mathcal{S}_2 \overset{\text{contraction: } \hbar\to 0}{\longrightarrow} \mathcal{S}_1\mathcal{S}_2 + \mathcal{O}(\hbar)\,,$$

$$\{\mathcal{S}_1, \mathcal{S}_2\}_p \overset{\text{deformation}}{\longrightarrow} \{\mathcal{S}_1, \mathcal{S}_2\}_m \overset{\text{contraction: } \hbar\to 0}{\longrightarrow} \{\mathcal{S}_1, \mathcal{S}_2\}_p + \mathcal{O}(\hbar^2)\,,$$

rather than a sharp change in the nature of observables:

$$\boxed{\mathbb{C}\text{-valued functions}} \longrightarrow \boxed{\text{operators in a Hilbert space}}$$

The transition from classical to quantum groups also can be considered as a process of a deformation. One starts from a classical matrix group G and takes matrix entries as generators (in general, with some constraints, e.g. of the form (1.1.27)) of an algebra of functions $Fun(G)$ on the group. Then one can use operator or pure algebraic formalism to introduce "quantum" analog of the group as we did above on the example of $SL_q(2)$. But one may also look for a deformation of usual multiplication in $Fun(G)$ of the star-product type. Of course, in this case, in distinction from usual quantum mechanics, one must take care about preservation of *group properties* in an appropriate form. This is indeed a possible approach for the group quantization [227, 28].

However, if one considers a group manifold there is still one more way of deformation. Recall that in the case of a usual Lie group there is another closely related object, namely its Lie algebra. It is natural to expect the existence of such an object in the non-commutative case also. Indeed, this can be defined but the relation between q-groups and q-Lie algebras is not so straightforward as in the classical case: there is no analog of the exponential map from

Lie algebras to the corresponding Lie groups. Nevertheless, groups and algebras (more precisely, algebra of functions on a group and *universal enveloping algebra* (UEA) of the corresponding Lie algebra) still keep the duality relation after quantization (duality is not used widely in Lie group theory as there exist stronger relation provided by the exponential map). The duality permits to derive properties of (quantized) algebra of functions on a group from those of (quantized) UEA. So one can start from the quantization of UEA and then derive the multiplication and other structures for quantized groups. We will proceed in this way.

A general theory of Lie (actually universal enveloping) algebra quantization was developed by V.I.Drinfel'd [86] and M.Jimbo [129] and will be discussed later. Of course, these algebras are non-commutative already in the classical case. The most important new feature of q-Lie algebras in comparison with the classical ones is their action in a tensor product $V_{j_1}\hat{\otimes}V_{j_2}$, of the representation spaces V_j. In the case of classical $su(2)$ algebra this action is defined according to the well known addition of quantum mechanical angular momentum, $\vec{J}_{tot} = \vec{J}_1 + \vec{J}_2$, and is realized with help of the Leibniz rule

$$\vec{J}_{tot}|\psi\rangle_{tot} = \vec{J}_1|\varphi\rangle_1\hat{\otimes}|\chi\rangle_2 + |\varphi\rangle_1\hat{\otimes}\vec{J}_2|\chi\rangle_2 \,, \tag{1.1.31}$$

where

$$|\psi\rangle_{tot} = |\varphi\rangle_1\hat{\otimes}|\chi\rangle_2 \ \in V_{j_1}\hat{\otimes}V_{j_2} \,, \tag{1.1.32}$$

$\vec{J}_1, \vec{J}_2$ are the operators of angular momenta for the states $|\varphi\rangle \in V_{j_1}$ and $|\chi\rangle \in V_{j_2}$, respectively.

Equation (1.1.31) can be written in the abstract algebraic form

$$\Delta(\vec{J}) = \vec{J}\otimes\mathbf{1} + \mathbf{1}\otimes\vec{J} \,, \tag{1.1.33}$$

where Δ is a homomorphism

$$\Delta : \mathcal{U}(su(2)) \to \mathcal{U}(su(2))\otimes\mathcal{U}(su(2)) \,,$$

$\mathcal{U}(su(2))$ is universal enveloping algebra of $su(2)$, $\otimes$ is tensor product of algebras. This homomorphism is called *a comultiplication* (or *a coproduct*). Note that in the case of $su(2)$ the element $\Delta(\vec{J})$ is symmetric with respect to the permutation of two copies of the algebra in the tensor product (rhs of (1.1.33)). Such a comultiplication is called *cocommutative*. It is this map which receives essentially new feature after the deformation (quantization) and becomes *non-cocommutative*, i.e. non-symmetric under a permutation of factors in tensor product. For the algebra $\mathcal{U}_q(s\ell(2))$ it has the form

$$\begin{aligned} \Delta(H) &= H\otimes\mathbf{1} + \mathbf{1}\otimes H \,, \\ \Delta(X^{\pm}) &= X^{\pm}\otimes q^{-H/2} + q^{H/2}\otimes X^{\pm} \,. \end{aligned} \tag{1.1.34}$$

To stress that the generators form now a *deformed* algebra $\mathcal{U}_q(s\ell(2))$ we have changed the notations for them with the obvious correspondence: $J_\pm \longrightarrow X^\pm$, $J_0 \longrightarrow H$ (mathematically more consistent would be to change the notations for the multiplication etc. in the UEA, but this leads to too bulky expressions). As is seen from (1.1.34), the comultiplication for $U_q(s\ell_q(2))$ is not symmetric and hence is *non-cocommutative.*

The appearance of non-trivial operators $q^{H/2}$, $q^{-H/2}$ in the comultiplication for $X_\pm$ (instead of unity in the classical case) can be interpreted as a kind of non-commutativity of the generators with factors in the vector space tensor product (1.1.32). Actually, the structure of the quantum algebra $s\ell_q(2)$ can be derived in a way similar to that described above for the quantum group $SL_q(2)$. One can make an ansatz for an action of the generators on non-commutative coordinates of representation spaces and derive CR between them and the form of comultiplication by requiring preservation of defining CR and space invariants under the generator actions. The calculations are rather cumbersome [213] and we shall not repeat them here, moreover that in the subsequent sections we will discuss general procedure of Lie algebra quantum deformation. As is seen, the rhs of (1.1.34) contains the infinite series in the generator H (as usual, exponent type operators $q^{H/2}$, $q^{-H/2}$ are understood as series expansion), so that actually one obtains the deformation of the universal enveloping algebra (UEA) of $s\ell_q(2)$ which is called *quantum universal enveloping algebra* (QUEA).

As Δ is a homomorphism, the rhs of (1.1.34) must satisfy the CR of $s\ell_q(2)$ itself. But it is easy to check that q-deformed "sum of the angular momenta" does not satisfy the usual CR for the algebra $sl(2)$ (or $su(2)$ if appropriate reality conditions are imposed)

$$[J_0, J_\pm] = \pm 2J_\pm\,, \qquad [J_+, J_-] = J_0\,, \tag{1.1.35}$$

but the q-deformed CR

$$\begin{aligned} [H, X^\pm] &= \pm 2X^\pm \\ [X^+, X^-] &= \frac{q^H - q^{-H}}{q - q^{-1}} = \frac{\sinh(\chi H)}{\sinh(\chi)} = [H]_q\,. \end{aligned} \tag{1.1.36}$$

Here $q \equiv e^\chi$, and $[x]_q$ denotes the so-called *q-square bracket* (notice that sometime the term "q-number" is also used for this object. However, this term is widely used in theoretical physics after Dirac to denote operators as distinct from usual commutative "c-numbers")

$$[x]_q := \frac{q^x - q^{-x}}{q - q^{-1}}\,, \tag{1.1.37}$$

with the property

$$[x]_q \underset{q\to 1}{\longrightarrow} x \ , \tag{1.1.38}$$

so that in this limit the CR (1.1.36) coincides with (1.1.35). The deformation of the commutator means that along with comultiplication, the multiplication in the universal enveloping algebra of $s\ell(2)$ is changed too. This reflects strong correlation between different structures of a quantum algebra.

With help of the q-square brackets many formulas of q-group theory can be written in a form similar to the classical non-deformed case. For example, the Casimir operator of $s\ell(2)$ Lie algebra

$$C_2 = J_\mp J_\pm + \left(\frac{1}{2}(J_0 \pm 1)\right)^2$$

becomes after the deformation

$$C_2^q = X^\mp X^\pm + \left[\frac{1}{2}(H \pm 1)\right]_q^2 . \tag{1.1.39}$$

If q is not a root of unity, representations of $s\ell_q(2)$ have the same dimensions and structure as those in the classical case

$$\begin{aligned} H|j,m\rangle &= 2m|j,m\rangle \ , \\ X^\pm|j,m\rangle &= \sqrt{[j \mp m]_q [j \pm m + 1]_q}\, |j, m \pm 1\rangle \ , \end{aligned} \tag{1.1.40}$$

(j is integer or half-integer number, $m = -j, -j+1, \ldots, j$). Using (1.1.38) it is easy to see that the q-deformed expressions have correct classical limit $q \to 1$.

It is the algebra (1.1.36) that had been discovered by Kulish and Reshetikhin [151] in their study of quantum sine-Gordon equation as the first example of q-deformed UEA.

General pattern of the q-deformation (quantization) of Lie algebras and groups is depicted on Fig.1.1.

1.2 Hopf algebra and Poisson structure of classical Lie groups and algebras

As we discussed, quantum groups and algebras are smooth deformation of corresponding classical Lie groups and algebras. This means, in particular,

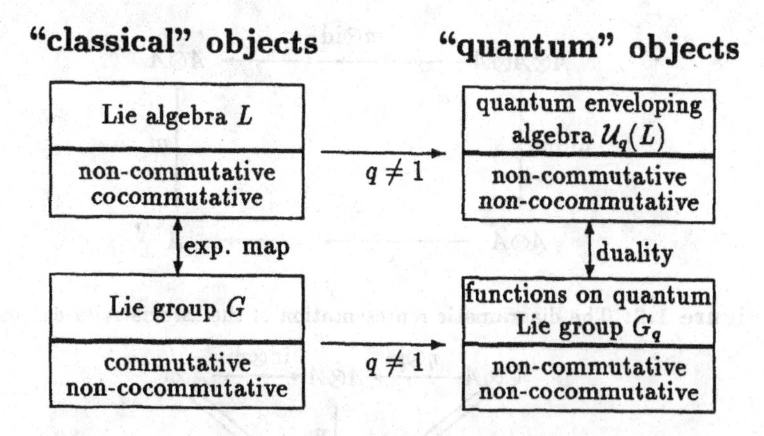

Figure 1.1: General pattern of q-deformation of Lie algebras and groups

that the latter contain in degenerate form all structures of the former. In this section we will describe these structures which exist in the classical objects but not used generally because of the degeneracy and possibility to make use of another language (manifolds, points, etc.) for a description of group properties. The structures introduced in this section, give more general group description and admit their smooth deformation.

We mentioned already that the first step in the construction of non-commutative geometry is the description of usual manifolds in terms of algebras of smooth functions on them. If one wants to describe a group manifold, these algebras have to possess a set of properties reflecting the group structure. An algebra with such a set of properties are known in mathematics as *a Hopf algebra* (see, e.g.[1]).

We start with reminding the definition of the very notion of an algebra, more precisely, unital associative algebra.

Definition 1.1 *A unital associative algebra is a linear space A together with two maps*

$$m : A \otimes A \to A\,, \quad \textit{(multiplication)} \tag{1.2.1}$$

$$\eta : \mathbb{C} \to A\,, \quad \textit{(unity map)} \tag{1.2.2}$$

such that

1. *m and η are linear;*
2. *$m(m \otimes \mathbf{I}) = m(\mathbf{I} \otimes m)$ (associativity);*
3. *$m(\mathbf{I} \otimes \eta) = m(\eta \otimes \mathbf{I}) = \mathrm{id}$ (existence of unit element);*

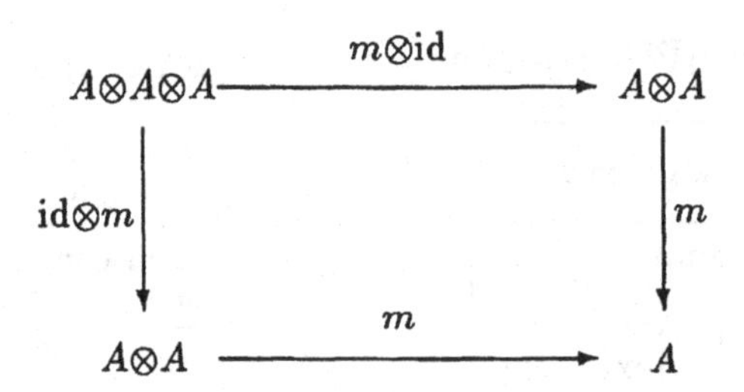

Figure 1.2: The diagramatic representation of the associativity axiom 2.

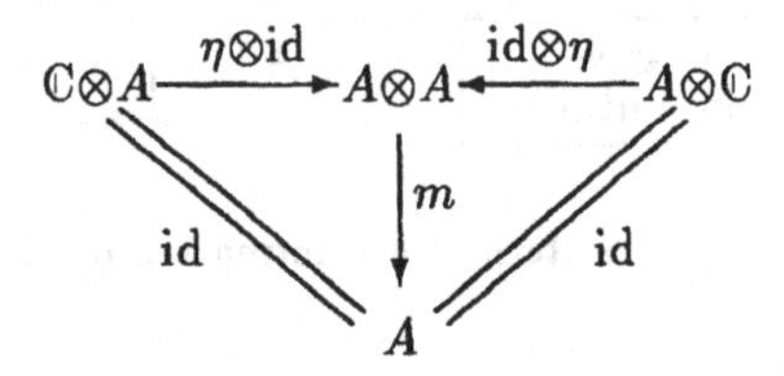

Figure 1.3: The diagramatic representation of the unity axiom 3.

Here $A \otimes A$ means tensor product of two algebras, $\mathbf{1}$ is the unit element of A and "id" means the identity map.

Let $\lambda \in \mathbb{C}$, $a \in A$. Then

$$m(\eta\otimes\mathbf{1})(\lambda\otimes a) = \eta(\lambda)a \ .$$

From the other hand, the property 3 gives

$$m(\eta \otimes \mathbf{1})(\lambda \otimes a) = \mathrm{id}(\lambda\otimes a) = \lambda\otimes a = \lambda a \ ,$$

where we have used the obvious isomorphism $\mathbb{C} \otimes A \sim A$. Comparison of these equalities leads to the conclusion $\eta(\lambda) = \lambda\mathbf{1}$, so that the property 3 is a formal way to express that A has a unit element. As will be clear, this rather unusual formulation prove to be convenient for the algebraic description of group structure. The axioms 2 and 3 can be expressed with the help of the diagrams Fig.1.2 and Fig.1.3, requiring their commutativity (in other words, an equivalence of the different ways through the diagrams along the arrows).

The usual notation for an algebra multiplication is simply

$$ab := m(a \otimes b)$$

and we will use such notation in all cases where it will not cause a confusion. Also we will use short term *algebra* instead of *unital associative algebra*.

With help of algebras we want to describe Lie groups and algebras and then to deform them. Using their well known properties one can realize which additional structures must be added to a general associative algebra for such a description.

Consider the algebra $Fun(G)$ of complex valued smooth functions on compact topological group G with multiplication

$$m(\phi \otimes \psi)(g) \equiv \phi\psi(g) := \phi(g)\psi(g)\,, \quad g \in G\,; \quad \phi, \psi \in Fun(G) \tag{1.2.3}$$

and unity map

$$\eta(\alpha) = \alpha u\,, \tag{1.2.4}$$

where $u \in Fun(G): \ u(g) = 1 \, \forall \, g \in G\,; \, \alpha \in \mathbb{C}$. The algebra $Fun(G)$ is the starting object for the algebraic formulation of group theory. On the other hand, to any Lie algebra L one associates its universal enveloping algebra $\mathcal{U}$ with ordinary associative multiplication and natural unity map

$$\eta(\alpha) = \alpha \mathbf{I}\,, \tag{1.2.5}$$

where $\mathbf{I}$ is unity in $\mathcal{U}(L)$.

First of all we would like to express, on the algebraic level, the well known relation between Lie groups and Lie algebras.

We hardly can expect to construct an analog of usual exponential map from a Lie algebra L to a Lie group G in the non-commutative case. But we can use the exponential map as a starting point to establish more general relation between L and G. Consider the action of $X \in L$ in the algebra $Fun(G)$ of smooth (infinitely differentiable) functions on G

$$(\rho(X)\phi)(g) \equiv X\,|_g\, \phi := \frac{d}{dt}(\phi(e^{tX}g))\,|_{t=0}$$

(the derivative of a function ϕ at $g \in G$ and in the direction X). This, in turn, defines the action $\rho(a)$ of any element $a \in \mathcal{U}(L)$ in $Fun(G)$ and permits to introduce the pairing

$$\langle\langle \cdot, \cdot \rangle\rangle : \ Fun(G) \times \mathcal{U}(L) \to \mathbb{C}$$

by the definition

$$\langle\langle \phi, a \rangle\rangle := (\rho(a)\phi)(e) \in \mathbb{C}\,, \tag{1.2.6}$$

where $a \in \mathcal{U}(L)$, $\phi \in Fun(G)$ and e is the unit element of G. So there is a correspondence between a space of functions $Fun(G)$ and a space $\mathcal{U}^*(L)$ dual to $\mathcal{U}(L)$. Moreover, if $\phi_1 \neq \phi_2$ the corresponding pairing $\langle\langle\phi_1, a\rangle\rangle$ cannot be equal to $\langle\langle\phi_2, a\rangle\rangle$ for all $a \in \mathcal{U}(L)$. Otherwise,

$$\langle\langle\phi_1 - \phi_2, a\rangle\rangle = (\rho(a)(\phi_1 - \phi_2))(e) = 0$$

and as $\rho(a)$ is a derivative of arbitrary order, the function $\phi_1 - \phi_2$ must be zero (a smooth function is determined by all its derivatives at arbitrary point).

This means that the pairing (1.2.6) defines the embedding of $Fun(G)$ in $\mathcal{U}^*(L)$ and so there is duality between $\mathcal{U}(L)$ and $Fun(G)$.

We can hope to preserve the duality relation after quantization procedure (of course, with another explicit definition of the pairing). As one can endow an algebra with a structure which is induced by the one existing in a dual algebra, general algebraic structures defined on the spaces $Fun(G)$ and $\mathcal{U}(L)$ must be related.

From the discussion of $s\ell(2)$ example in the previous section we know already that an algebra action in a tensor product of representation spaces (analog of angular momenta addition) is defined by a homomorphism

$$\Delta : A \to A \otimes A \tag{1.2.7}$$

(here A is an abstract algebra and can be, in particular, either $Fun(G)$ or $\mathcal{U}(L)$). The homomorphism Δ must be linear and the representations $(V_1 \otimes V_2) \otimes V_3$ and $V_1 \otimes (V_2 \otimes V_3)$ must be equal. The latter property reminds the requirement of associativity and is called *coassociativity* of Δ. Thus the map Δ satisfies the following conditions:

1. Δ is linear;
2. $$\begin{aligned} &\Delta(ab) = \Delta(a)\Delta(b) \,, \qquad a, b \in A \,; \\ &(\Delta \otimes \mathrm{id})\Delta = (\mathrm{id} \otimes \Delta)\Delta \,. \end{aligned} \tag{1.2.8}$$

The homomorphism Δ is a counterpart of multiplication m (1.2.1) and thus is called *comultiplication.* The coassociativity can be expressed by commutativity of the diagram on Fig.1.4. One can note that it is quite similar to that on Fig.1.2 but all arrows have opposite direction. This corresponds to opposite directions of the arrows in the very definitions ((1.2.1) and (1.2.7)) of the maps. It is natural to add a counterpart for the unity map η reversing the arrow in (1.2.2) as in the case of m and Δ. So we introduce *counity map* ε

$$\varepsilon : A \to \mathbb{C}$$

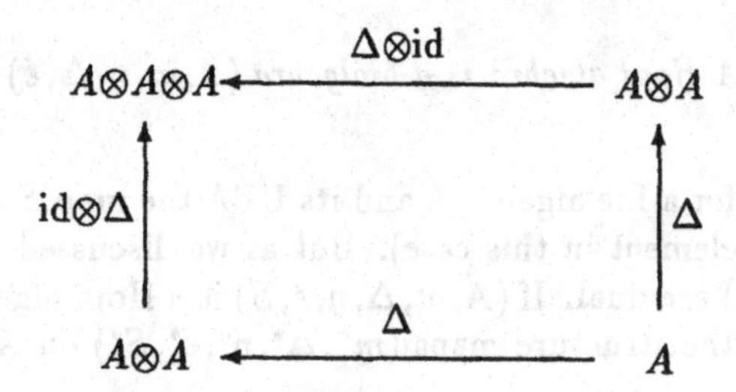

Figure 1.4: The diagramatic representation of the coassociativity of a comultiplication map.

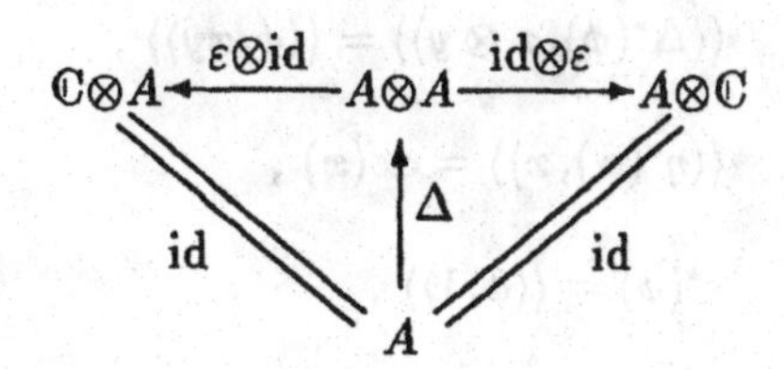

Figure 1.5: The diagramatic representation of the counity axiom

with the property

$$(\mathrm{id}\otimes\varepsilon)\circ\Delta = (\varepsilon\otimes\mathrm{id})\circ\Delta = \mathrm{id}$$

This property can be expressed by commutativity of the diagram in Fig.1.5.

Again this diagram can be obtained from that for the map η by reversing all the arrows.

Definition 1.2 *A bialgebra* $(A, m, \Delta, \eta, \varepsilon)$ *is a linear space* A *with maps* $m, \Delta, \eta, \varepsilon$ *which satisfy all the above properties.*

This object is still unable to describe all group properties. One must add a new map S

$$S : A \to A\ ,$$

which is assumed to be an antihomomorphism

$$S(ab) = S(b)S(a)\ , \qquad a, b \in A\ ,$$

is called an *antipode* (or coinverse) and is related to the fact that every element of a group has an inverse one. This map has the defining property

$$m(S\otimes\mathrm{id})\circ\Delta = m(\mathrm{id}\otimes S)\circ\Delta = \eta\circ\varepsilon\ . \qquad (1.2.9)$$

Definition 1.3 *A Hopf algebra is a bialgebra* $(A, m, \eta, \Delta, \varepsilon)$ *together with an antipode map* S.

At first sight for a Lie algebra L and its UEA the map S may seem redundant (no inverse element in this case). But as we discussed above the spaces $Fun(G)$ and $\mathcal{U}(L)$ are dual. If $(A, m, \Delta, \eta, \varepsilon, S)$ is a Hopf algebra and A^* is its dual space, then the structure maps $(m^*, \Delta^*, \eta^*, \varepsilon^*, S^*)$ on A^* can be defined by the relations

$$
\begin{aligned}
&\langle\langle m^*(\phi\otimes\psi), x\rangle\rangle = \langle\langle \phi\otimes\psi, \Delta(x)\rangle\rangle \ , \\
&\langle\langle \Delta^*(\phi), x\otimes y\rangle\rangle = \langle\langle \phi, xy\rangle\rangle \ , \\
&\langle\langle \eta^*(\alpha), x\rangle\rangle = \alpha\varepsilon(x) \ , \\
&\varepsilon^*(\phi) = \langle\langle \phi, 1\rangle\rangle \ , \\
&\langle\langle S^*(\phi), x\rangle\rangle = \langle\langle \phi, S(x)\rangle\rangle \ ,
\end{aligned}
\tag{1.2.10}
$$

$\forall \phi, \psi \in A^*$ and $x, y \in A$.
It is easy to verify that A^* is a Hopf algebra if A is a Hopf algebra.

Thus we expect that both spaces $Fun(G)$ and $\mathcal{U}(L)$ must be dual Hopf algebras. This is indeed the case. The explicit form of maps defining the Hopf algebra structure are the following:

1. <u>For a space of continuous functions $Fun(G)$:</u>

- $(\phi\psi)(g) = \phi(g)\psi(g)$;
- $\Delta(\phi)(g_1\otimes g_2) = \phi(g_1 g_2)$
  (note that $\phi(g_1 g_2)$ is considered as a function of *two* variables and $Fun(G)\otimes Fun(G)$ is identified with $Fun(G\times G)$);
- $\eta(x) = \alpha u, \quad u(g) = 1 \quad \forall\, g \in G,\ u \in Fun(G),\ \alpha \in \mathbb{C}$;
- $\varepsilon(\phi) = \phi(e)$, e is unity in G
- $S(\phi)(g) = \phi(g^{-1})$.

2. <u>For an universal enveloping algebra $\mathcal{U}(L)$</u> the maps are defined for the subset $L \in \mathcal{U}(L)$ but are extended uniquely to all $\mathcal{U}(L)$:

- the map m is the usual associative multiplication in $\mathcal{U}(L)$;

- $\Delta(X) = X \otimes \mathbf{I} + \mathbf{I} \otimes X, \quad X \in L,$ $\mathbf{I}$ is unity of $\mathcal{U}(L)$;
- $\eta(\alpha) = \alpha\mathbf{I}, \quad \alpha \in \mathbb{C}$;
- $\varepsilon(\mathbf{I}) = 1,\ \varepsilon(X) = 0 \quad \forall\, X \neq \mathbf{I}, \quad X \in \mathcal{U}(L)$;
- $S(X) = -X, \quad X \in L.$

A Lie algebra itself is not a Hopf algebra, of course, as it is not associative.

We leave to the reader as an exercise to check that all above introduced maps for $Fun(G)$ and $\mathcal{U}(L)$ have the properties of the Hopf algebra structure maps and that they are indeed dual in the sense of the relations (1.2.10).

Note that the multiplication of dual space is defined by the comultiplication in the initial one and vice versa. As a result, one can see that Hopf algebra of functions on a group is commutative but not cocommutative (this means that $\Delta \neq \sigma \circ \Delta$ for it, where σ is an operator of permutation of two factors in a direct product: $\sigma(a \otimes b) := b \otimes a$) and the Hopf algebra based on $\mathcal{U}(L)$ is non-commutative but cocommutative ($\Delta = \sigma \circ \Delta$).

To get accustomed to the Hopf algebra structure maps, it is useful to rewrite them in terms of structure constants. This form is more customary for physicists (e.g. from Lie algebra theory).

In chosen linear basis $\{e_i\}$ ($i \in \mathcal{I}$, where $\mathcal{I}$ is a set of indices, e.g. set of positive integers $\mathbb{Z}_+$ or finite set $\mathbb{Z}^k$) the structure maps are defined as follows

$$\begin{aligned} m(e_i \otimes e_j) &= e_i e_j = m_{ij}^p e_p\ , \\ \Delta(e_i) &= \Delta_i^{jk} e_j \otimes e_k\ , \\ S(e_i) &= S_i^j e_j\ , \\ \varepsilon(e_i) &= \varepsilon_i\ . \end{aligned} \tag{1.2.11}$$

The unit element $\mathbf{I}$ of a Hopf algebra can be written in the form

$$\mathbf{I} = \varepsilon^i e_i\ .$$

To avoid any confusion, we remind that $\{e_i\}$ is the basis of a Hopf algebra, and in the case of $\mathcal{U}(L)$ this can be infinite basis of ordered monomials in Lie algebra generators, but not a Lie algebra basis.

The properties of the Hopf algebra structure maps lead to many relations for the structure constants:

1. unity element $\mathbf{1}$

$$m(a\otimes\mathbf{1}) = m(\mathbf{1}\otimes a) = a \quad \Rightarrow \quad m_{ij}^{k}\varepsilon^{i} = m_{ji}^{k}\varepsilon^{i} = \delta_{j}^{k} \; ; \qquad (1.2.12)$$

2. associativity

$$m(m\otimes\mathrm{id}) = m(\mathrm{id}\otimes m) \quad \Rightarrow \quad m_{ij}^{k}m_{pt}^{j} = m_{ip}^{j}m_{jt}^{k} \; ; \qquad (1.2.13)$$

3. coassociativity

$$(\Delta\otimes\mathrm{id})\circ\Delta = (\mathrm{id}\otimes\Delta)\circ\Delta \quad \Rightarrow \quad \Delta_{k}^{ij}\Delta_{i}^{pt} = \Delta_{k}^{pi}\Delta_{i}^{tj} \; ; \qquad (1.2.14)$$

4. counity

$$(\varepsilon\otimes\mathrm{id})\circ\Delta = (\mathrm{id}\otimes\varepsilon)\circ\Delta = \mathrm{id} \quad \Rightarrow \quad \Delta_{k}^{ij}\varepsilon_{j} = \Delta_{k}^{ji}\varepsilon_{j} = \delta_{k}^{i} \; ; \qquad (1.2.15)$$

$$\varepsilon(ab) = \varepsilon(a)\varepsilon(b) \quad \Rightarrow \quad m_{ij}^{k}\varepsilon_{k} = \varepsilon_{i}\varepsilon_{j} \; ; \qquad (1.2.16)$$

5. Δ-homomorphism of multiplication

$$\Delta(a)\Delta(b) = \Delta(ab) \quad \Rightarrow \quad \Delta_{p}^{ij}\Delta_{n}^{uv}m_{iu}^{r}m_{jv}^{t} = m_{pn}^{s}\Delta_{s}^{rt} \; ; \qquad (1.2.17)$$

$$\Delta(\mathbf{1}) = \mathbf{1}\otimes\mathbf{1} \quad \Rightarrow \quad \varepsilon^{k}\Delta_{k}^{ij} = \varepsilon^{i}\varepsilon^{j} \; ; \qquad (1.2.18)$$

6. antipode

$$S(ab) = S(b)S(a) \quad \Rightarrow \quad m_{ij}^{k}S_{k}^{l} = S_{j}^{p}S_{i}^{n}m_{pn}^{l} \; ; \qquad (1.2.19)$$

$$\Delta\circ S = (S\otimes S)\circ\sigma\circ\Delta \quad \Rightarrow \quad \Delta_{k}^{ij}S_{l}^{k} = S_{p}^{i}S_{n}^{j}\Delta_{l}^{np} \; ; \qquad (1.2.20)$$

$$m(S\otimes\mathrm{id})\circ\Delta(a) = m(\mathrm{id}\otimes S)\circ\Delta(a) = \varepsilon(a)\mathbf{1}$$

$$\Rightarrow \quad m_{ij}^{k}S_{l}^{j}\Delta_{p}^{il} = m_{ij}^{k}S_{l}^{i}\Delta_{p}^{lj} = \varepsilon_{p}\varepsilon^{k} \; . \qquad (1.2.21)$$

Structure constants of a dual Hopf algebra A^{*} in a basis $\{e^{i}\}$, $i \in \mathcal{I}$ are expressed through the structure constants of algebra A by the use of the pairing conditions (1.2.10)

$$\begin{aligned} m^{*}(e^{i}\otimes e^{j}) &= \Delta_{k}^{ij}e^{k} \, , \\ \Delta^{*}(e^{i}) &= m_{kl}^{i}(e^{k}\otimes e^{l}) \, , \\ S^{*}(e^{i}) &= S_{j}^{i}e^{j} \, , \\ \varepsilon^{*}(e^{i}) &= \varepsilon^{i} \, . \end{aligned} \qquad (1.2.22)$$

Because of the properties of a Hopf algebra map, the structure constants are strongly related to each other. As a useful exercise, let us show that the antipode S' of the Hopf algebra A^0 which has structure maps m^*, ε^* but the comultiplication Δ' with additional permutation

$$\Delta^* \longrightarrow \Delta' := \sigma \circ \Delta^* , \tag{1.2.23}$$

can be expressed in terms of the inverse matrix of the structure constants

$$(S')^i{}_j = (S^{-1})^i{}_j . \tag{1.2.24}$$

Indeed, as other structure constants of A^0 have the form

$$\begin{aligned} m^{*ij}_k &= \Delta^{ij}_k , \\ \Delta^{lk}_{ij} &= m^k_{ji} , \\ \varepsilon^{*i} &= \varepsilon^i , \end{aligned} \tag{1.2.25}$$

it is easy to see that $S^i{}_j$ does not satisfy the antipode axioms (1.2.19)-(1.2.21). To prove that (1.2.24) satisfies the antipode properties, we notice that (1.2.19) and (1.2.20) can be rewritten in the form

$$\begin{aligned} m^l_{pn}(S^{-1})^k_l &= (S^{-1})^i_n (s^{-1})^j_p m^k_{ij} , \\ \Delta^{pn}_l (S^{-1})^l_k &= (S^{-1})^n_i (S^{-1})^p_j \Delta^{ij}_k , \end{aligned} \tag{1.2.26}$$

which just correspond to the relations for Δ', m^* and S'. Multiplying (1.2.21) by S^{-1} from the right, using (1.2.26) and the relation

$$S^i_j \varepsilon^j = \varepsilon^i ,$$

which follows from (1.2.19) (take the algebra unity $\mathbf{I}$ as $a \in A$ in this relation), one has

$$m^k_{ij}(S^{-1})^i_l \Delta^{jl}_p = \varepsilon^k \varepsilon_p . \tag{1.2.27}$$

Finally, (1.2.27) and the definitions (1.2.24),(1.2.25) give the relation (1.2.21) for the structure constants of a Hopf algebra A^0

$$\Delta'^{lk}_{ji} S'^{i}_l m^{*jl}_p = \varepsilon^k \varepsilon_p \equiv \varepsilon^{*k} \varepsilon^*_p .$$

The second equality (1.2.21)

$$\Delta'^{lk}_{ji} S'^{j}_l m^{*li}_p \equiv m^k_{ij}(S^{-1})^j_l \Delta^{li}_p = \varepsilon^{*k} \varepsilon^*_p , \tag{1.2.28}$$

can be proven analogously. The algebra A^0 and the relations for its structure constants will be used in Section 1.4 for a very important quantum double construction.

Now from pure algebraic point of view, it is clear how to generalize the above considered example of Lie groups and algebras: one has to consider Hopf algebras which are both *non-commutative* and *non-cocommutative*. But there is the obvious problem how to construct the corresponding (generalized) structure maps explicitly. As we want to generalize (to deform), in particular the multiplication in the algebra of function $Fun(G)$ and make it non-commutative ("quantize" it) one can use the experience from usual quantum mechanics and try to carry out a similar procedure of "quantization". As we discussed, a commutator of algebra elements (Moyal bracket), defined in terms of deformed product, reduces in classical limit to a Poisson bracket (the "correspondence principle"). So if one starts from a classical object and looks for a possible quantization, one has to define a Poisson bracket as a first step. From mathematical point of view this means that one must define on a smooth manifold a closed non-degenerate (symplectic) two-form (see, e.g., [8]), which defines a Poisson bracket $\{\cdot,\cdot\}_p$. The latter defines a map $Fun(\mathcal{X})\otimes Fun(\mathcal{X}) \longrightarrow Fun(\mathcal{X})$, where $Fun(\mathcal{X})$ is a commutative algebra of smooth functions on a manifold $\mathcal{X}$. The translation of this construction to the algebraic language leads to the notion of a Poisson algebra.

Definition 1.4 *A Poisson algebra is a commutative algebra* (A, m, η) *together with a map*

$$\gamma : A\otimes A \to A\,,$$

such that A is a Lie algebra with respect to γ *which means:*

1. *antisymmetry of* γ

$$\gamma \circ \sigma = -\gamma\;; \tag{1.2.29}$$

2. *Jacobi identity*

$$\gamma(1\otimes\gamma)(1\otimes 1\otimes 1 + (1\otimes\sigma)(\sigma\otimes 1) + (\sigma\otimes 1)(1\otimes\sigma)) = 0\;; \tag{1.2.30}$$

and, besides, γ *satisfies the Leibniz rule*

$$\gamma(m\otimes 1) = m(1\otimes\gamma)(1\otimes 1\otimes 1 + \sigma\otimes 1)\,. \tag{1.2.31}$$

The last relation expresses in algebraic language the well known property of a Poisson bracket

$$\{\phi\psi,\zeta\}_p = \phi\{\psi,\zeta\}_p + \{\phi,\zeta\}_p\psi\,.$$

Definition 1.5 *A smooth manifold $\mathcal{X}$ is called a Poisson manifold if the algebra of smooth functions $Fun(\mathcal{X})$ is a Poisson algebra.*

Let us recall that a manifold $\mathcal{X}$ is called *symplectic* if there is *non-degenerate* closed 2-form ω on $\mathcal{X}$. A Poisson manifold may not be a symplectic since a Poisson structure may be degenerate at some points. In this case $\mathcal{X}$ can be decomposed into a union of so-called *symplectic leaves* [238]. First introduce *a Hamiltonian curve* $\Gamma(t)$ which is defined by a Hamiltonian equation of motion for some Hamiltonian $H \in Fun(\mathcal{X})$, and the equivalence relation for points of $\mathcal{X}$: $x \in \mathcal{X}$ is equivalent to $y \in \mathcal{X}$ if there exists a piece-Hamiltonian curve Γ between x and y.

Definition 1.6 *The equivalence class is called symplectic leaf.*

There is the following proposition:

- Poisson manifold $\mathcal{X}$ is a union of symplectic leaves S_r : $\mathcal{X} = \cup_r S_r$.

An important example of symplectic leaves is group orbits of coadjoint action of a group G in the space L^* of linear functions on the Lie algebra L of G (i.e. L^* is dual to a Lie algebra L)[142]. In this case L^* plays the role of a Poisson manifold, L is the space of linear functions on L^* with so-called *Lie-Kirillov-Kostant brackets*

$$\{X,Y\}_p(z) := \langle\langle z, [X,Y]\rangle\rangle \,, \tag{1.2.32}$$

$X, Y \in L$, $z \in L^*$, $\langle\langle\cdot,\cdot\rangle\rangle$ is dual pairing of L and L^*. Using the Leibniz rule (1.2.31) the brackets can be extended to the polynomial algebra $Pol(L^*) \subset Fun(L^*)$.

If one would consider an arbitrary manifold as the phase space for some classical mechanics one could choose any symplectic form and then carry out the usual quantization procedure (see, e.g., [19]). But our present aim is to quantize (to deform) also the group structure. The obvious requirement in such a case is the group covariance of a Poisson bracket

$$\{\phi,\psi\}_p(g'g) = \{\phi(g'g), \psi(g'g)\}_p \,, \quad g', g \in G \,. \tag{1.2.33}$$

A Lie group with a covariant Poisson structure is called *Poisson-Lie group.* However, we want to use Hopf algebras as a general guide for a group quantization according to the principles:

- quantization corresponds to a transition from commutative to *non-* commutative or (for dual objects) from cocommutative to *non-* cocommutative Hopf algebras;

- relation between a Lie group and a Lie algebra survive after the quantization in the form of duality of the corresponding Hopf algebras.

So we have to rewrite (1.2.33) in terms of Hopf algebra structure maps. Explicit form in the case of Lie groups (see page 28) immediately gives that (1.2.33) is equivalent to the homomorphism condition for comultiplication

$$\{\Delta(\phi), \Delta(\psi)\}_{p,A\otimes A} = \Delta(\{\phi, \psi\}_{p,A}) . \tag{1.2.34}$$

The Poisson bracket in rhs of this relation contains as inputs, the elements of tensor product space and hence it must be defined for this case. One can check that all necessary properties are fulfilled for the following definition

$$\{\phi\otimes\psi, \zeta\otimes\xi\}_p = \{\phi, \zeta\}_p\otimes\psi\xi + \phi\zeta\otimes\{\psi, \xi\}_p . \tag{1.2.35}$$

In terms of the Poisson map γ the definition (1.2.35) takes a form

$$\gamma_{A\otimes B} = (\gamma_A\otimes m_B + m_A\otimes\gamma_B)(1\otimes\sigma\otimes 1) , \tag{1.2.36}$$

and the relation(1.2.34) becomes

$$\gamma_{A\otimes A} \circ (\Delta\otimes\Delta) = \Delta \circ \gamma_A . \tag{1.2.37}$$

Definition 1.7 *A commutative Hopf algebra $(A, m, \Delta, \eta, \varepsilon, S)$ together with a Poisson map γ, which satisfies the relation (1.2.37) is called a Poisson Hopf algebra.*

The next definition is very natural and does not need any comments.

Definition 1.8 *A Lie group G together with a Poisson map γ defined on smooth functions $Fun(G)$ is called a Poisson Lie group if $Fun(G)$ is a Poisson Hopf algebra.*

Now we can use the advantage of the algebraic language and give at once the definition of a dual object:

Definition 1.9 *A co-Poisson bialgebra is a co-commutative bi-algebra $(A, m, \Delta, \eta, \varepsilon)$ together with a map*

$$\delta : A \rightarrow A\otimes A ,$$

such that

1. *$\sigma \circ \delta = -\delta$ (co-antisymmetry);*

2.

$$(1\otimes 1\otimes 1+(1\otimes\sigma)(\sigma\otimes 1)+(\sigma\otimes 1)(1\otimes\sigma))(1\otimes\delta)\delta=0 \quad \textit{(co-Jacobi identity)}; \tag{1.2.38}$$

3.

$$(\Delta\otimes 1)\delta=(1\otimes 1\otimes 1+\sigma\otimes 1)(1\otimes\delta)\Delta \quad \textit{(co-Leibniz rule)}; \tag{1.2.39}$$

4.

$$(m\otimes m)\circ\delta_{A\otimes A}=\delta\circ m \quad \textit{(i.e., m is a co-Poisson homomorphism)}, \tag{1.2.40}$$

where

$$\delta_{A\otimes A}=(1\otimes\sigma\otimes 1)(\delta\otimes\Delta+\Delta\otimes\delta)\ . \tag{1.2.41}$$

The relations (1.2.38), (1.2.39), (1.2.40), (1.2.41) are obtained from (1.2.30), (1.2.31), (1.2.37), (1.2.36) correspondingly by substitution of co-objects and reversing of the operations.

We already have explicit realization of a Hopf algebra structure maps in the cases of $Fun(G)$ and $\mathcal{U}(L)$ (see page 28). Now we present an explicit construction for a Poisson map γ.

Let $\{X_\mu\}_{\mu=1}^{dim(G)}$ be a set of right invariant vector fields on G [145], such that $\{X_\mu|_g\}$ is a basis in tangent space T_gG for all $g\in G$. Remind that vector fields on a manifold $\mathcal{X}$ act as derivatives in an algebra $Fun(\mathcal{X})$ of functions on $\mathcal{X}$. Since a Poisson map satisfies the Leibniz rule (1.2.31) and hence is a derivation, one can write

$$\{\phi,\cdot\}_p(g)=\sum_\mu c^\mu(g)X_\mu|_g\ .$$

Applying the same arguments to $\{\cdot,\psi\}_p$ one obtains the general form of Poisson bracket on $Fun(G)$

$$\{\phi,\psi\}_p=\sum_{\mu\nu}\eta^{\mu\nu}(g)X_\mu|_g(\phi)\,X_\nu|_g(\psi)=m\circ\eta(g)(\phi\otimes\psi)\ ,\quad g\in G\ , \tag{1.2.42}$$

where $\eta(g):\ G\to L\otimes L$ reads as

$$\eta(g):=\eta^{\mu\nu}(g)(X_\mu|_g\otimes X_\nu|_g) \tag{1.2.43}$$

(right invariant vector fields form a Lie algebra of G).

Actually, this is the usual expression for Poisson brackets on a curved manifold, the $\eta^{\mu\nu}$ playing the role of inverse matrix with respect to that of symplectic two-form coefficients (cf. Section 1.1). The only peculiarity is that the right invariant derivatives (vector fields) are used instead of usual ones.

The expression (1.2.42) reduces the problem of finding of a Poisson map consistent with comultiplication (relation (1.2.37) or (1.2.34)) to the determination of the appropriate function $\eta(g)$ with values in $L\otimes L$.

To derive the equation for $\eta(g)$ from the homomorphism condition (1.2.34) it is convenient to write $\Delta(\phi)$ in the form

$$\Delta\phi(g_1, g_2) = \phi(g_1 g_2) = \sum_i \phi_i^{(1)}(g_1)\otimes\phi_i^{(2)}(g_2) .$$

Explicitly this can be done using the harmonic expansion on a group [233]

$$\phi(g_1 g_2) = \sum_i c_i(g_1) D_i(g_2) ,$$

where $\{D_i(g)\}$ is orthonormal basis of $Fun(G)$,

$$c_i(g_1) = \int d\mu_{g_2} D_i(g_2)\phi(g_1 g_2) ,$$

are coefficients of the expansion, $d\mu_{g_2}$ is the Haar measure.

Then from the definition (1.2.35) one finds

$$\begin{aligned} &\{\Delta(\phi), \Delta(\psi)\}_p(g_1, g_2) \\ &= \sum_{\mu\nu}\sum_{ij} \left[\eta^{\mu\nu}(g_1)\, X_\mu|_{g_1}(\phi_i^{(1)})\, X_\nu|_{g_1}(\psi_j^{(1)})\right] \phi_i^{(2)}(g_2)\psi_j^{(2)}(g_2) \\ &+ \phi_i^{(1)}(g_1)\psi_j^{(1)}(g_1) \left[\eta^{\mu\nu}(g_2)\, X_\mu|_{g_2}(\phi_i^{(2)})\, X_\nu|_{g_2}(\psi_j^{(2)})\right] . \end{aligned} \tag{1.2.44}$$

Also one has

$$\begin{aligned} \sum_i X_\mu|_{g_1}(\phi_i^{(1)})\, \phi_i^{(2)}(g_2) &= \frac{d}{dt}\sum_i \phi_i^{(1)}(e^{tX_\mu} g_1)\, \phi_i^{(2)}(g_2)\, |_{t=0} \\ &= \frac{d}{dt}\phi(e^{tX_\mu} g_1 g_2)\, |_{t=0} = X_\mu|_{g_1 g_2}\,(\phi) \end{aligned}$$

and

$$\begin{aligned} \sum_i \phi_i^{(1)}(g_1)\, X_\mu|_{g_2}(\phi_i^{(2)}) &= \frac{d}{dt}\sum_i \phi_i^{(1)}(g_1)\, \phi_i^{(2)}(e^{tX_\mu} g_2)\, |_{t=0} \\ &= \frac{d}{dt}\phi(g_1 e^{tX_\mu} g_2)\, |_{t=0} = \frac{d}{dt}\phi(g_1 e^{tX_\mu} g_1^{-1} g_1 g_2)\, |_{t=0} \\ &= Ad_{g_1} X_\mu\, |_{g_1 g_2}\,(\phi) . \end{aligned}$$

Thus (1.2.44) can be written in the form

$$\{\Delta(\phi),\Delta(\psi)\}(g_1,g_2)=(\eta(g_1)+g_1\diamond\eta(g_2))\phi\otimes\psi\ , \tag{1.2.45}$$

where

$$g\diamond(X\otimes Y):=Ad_g^{\otimes 2}(X\otimes Y):=Ad_gX\otimes Ad_gY\ . \tag{1.2.46}$$

Equating (1.2.45) to the rhs of (1.2.34)

$$\Delta\{\phi,\psi\}_p(g_1,g_2)=\eta(g_1g_2)\phi(g_1g_2)\otimes\psi(g_1g_2)\ ,$$

one finally obtains the condition on $\eta(g)$

$$\delta_G\eta:=g_1\diamond\eta(g_2)-\eta(g_1g_2)+\eta(g_1)=0\ . \tag{1.2.47}$$

To find a solution of this equation it is useful to consider lhs of (1.2.47) as a map from the space $C(G;L)$ of functions $\eta^{(1)}:G\to L\otimes L$ to the space $C(G^{\otimes 2};L)$ of functions $\eta^{(2)}:G\otimes G\to L\otimes L$. This map can be generalized to the map

$$C(G^{\otimes n};L)\to C(G^{\otimes(n+1)};L)\ ,$$

where $C(G^{\otimes n};L)=\{\eta^{(n)}:\underbrace{G\otimes\ldots\otimes G}_{n}\to L\otimes L\}$,

$$\begin{aligned}(\delta_G\eta^{(n)})(g_1,\ldots,g_{n+1}) &= g_1\diamond\eta^{(n)}(g_2,\ldots,g_{n+1})\\ &+\sum_{i=1}^{n}(-1)^i\eta^{(n)}(g_1,\ldots,g_ig_{i+1},\ldots,g_{n+1})\\ &+(-1)^{n+1}\eta^{(n)}(g_1,\ldots,g_n)\ .\end{aligned}$$

The gain from this generalization is such that δ_G has a property of a coboundary operator

$$\delta_G^2=0$$

(this can be shown by straightforward calculations). And hence the sequence $\{C^n(G;L)\}_{n=0}^{\infty}$ becomes a complex (for elements of cohomology theory see, e.g., [214]). This in turn permits to write immediately the particular solution of (1.2.47) in the form

$$\eta=\delta_Gr\ ,$$

for some $r\in L\otimes L$: $\delta_G\eta=\delta_G^2r=0$. Hence, from the definition of δ_G one finds

$$(\delta_Gr)(g)=r-g\diamond r\ , \tag{1.2.48}$$

and Poisson bracket (1.2.42) takes a form

$$\{\phi,\psi\} = r^{\mu\nu}(X_\mu|_g \otimes X_\nu|_g - Ad_g X_\mu|_g \otimes Ad_g X_\nu\,|_g)(\phi\otimes\psi)\,, \tag{1.2.49}$$

where $r^{\mu\nu}$ is defined by $r = r^{\mu\nu}X_\mu\otimes X_\nu$. The matrix r is called *classical r-matrix*. Note that

$$Ad_g X_\mu|_g\,\phi = \frac{d}{dt}\phi\left(ge^{tX}g^{-1}g\right)|_{t=0} = \frac{d}{dt}\phi\left(ge^{tX}\right)|_{t=0} = \widetilde{X}_\mu|_g\,\phi\,,$$

where $\widetilde{X}_\mu$ are left invariant vector fields. So the Poisson bracket is

$$\{\phi,\psi\}_P = r^{\mu\nu}(X_\mu\phi X_\nu\psi - \widetilde{X}_\mu\phi\widetilde{X}_\nu\psi)\,. \tag{1.2.50}$$

Of course, r must satisfy additional conditions which follows from Poisson bracket antisymmetry and Jacobi identity. Antisymmetry leads to the condition

$$(Ad_g\otimes Ad_g)(r^{\mu\nu}+r^{\nu\mu})X_\mu\otimes X_\nu = (r^{\mu\nu}+r^{\nu\mu})X_\mu\otimes X_\nu\,, \tag{1.2.51}$$

i.e. symmetric part of r is $Ad_g^{\otimes 2}$-invariant.

Lengthy but straightforward calculations show that the Jacobi identity gives

$$\begin{aligned} &Ad_g^{\otimes 3} r^{\mu\nu} r^{\mu'\nu'}\,(X_{\mu'}|_g \otimes X_\mu|_g \otimes [X_{\nu'}|_g, X_\nu|_g] \\ &+\; X_{\mu'}|_g \otimes [X_{\nu'}|_g, X_\mu|_g] \otimes X_\nu|_g + [X_{\mu'}|_g, X_\mu|_g] \otimes X_{\nu'}|_g \otimes X_\nu|_g) = 0 \end{aligned} \tag{1.2.52}$$

Using the notations

$$\begin{aligned} r_{12} &:= r^{\mu\nu}(X_\mu\otimes X_\nu\otimes 1)\,, \\ r_{13} &:= r^{\mu\nu}(X_\mu\otimes 1\otimes X_\nu)\,, \\ r_{23} &:= r^{\mu\nu}(1\otimes X_\mu\otimes X_\nu) \end{aligned} \tag{1.2.53}$$

the condition (1.2.52) can be written in the compact form

$$Ad_g^{\otimes 3}(r,r)_S = 0\,, \quad \forall g\in G\,, \tag{1.2.54}$$

where

$$(r,r)_S := [r_{12},r_{13}] + [r_{13},r_{23}] + [r_{12},r_{23}]\,. \tag{1.2.55}$$

$(r,r)_S$ is called *a Schouten bracket* of r with itself. The simplest way to satisfy (1.2.54) is to put $(r,r)_S = 0$. This equation for r is called *the classical Yang-Baxter equation* (CYBE). The complete equation (1.2.54) is called *modified classical Yang-Baxter equation* (MCYBE) [218].

Now we must take care about dual structure - a co-Poisson map δ introduced in the Definition 1.9. As L generates the whole UEA $\mathcal{U}(L)$, it is enough to construct the map δ for $L \subset \mathcal{U}(L)$ only. The map δ, being dual to the Poisson map γ, satisfies the relation

$$\langle\langle \phi\otimes\psi, \delta(Y)\rangle\rangle = \langle\langle \gamma(\phi\otimes\psi), Y\rangle\rangle \ . \tag{1.2.56}$$

The definition of dual pairing (1.2.6) gives for lhs of (1.2.56)

$$\langle\langle \phi\otimes\psi, \delta(Y)\rangle\rangle = \delta^{\mu\nu}(Y)\,(X_\mu|_e\phi\otimes X_\nu|_e\psi) \ ,$$

and for the rhs

$$\begin{aligned}\sum_{\mu\nu}\langle\langle \eta^{\mu\nu} X_\mu\phi X_\nu\psi, Y\rangle\rangle &= \frac{d}{dt}\eta^{\mu\nu}\left(e^{tY}\right) X_\mu\phi\left(e^{tY}\right)\otimes X_\nu\psi\left(e^{tX}\right)|_{t=0} \\ &=: \zeta^{\mu\nu}(Y)\,(X_\mu|_e\phi\, X_\nu|_e\psi) \ .\end{aligned}$$

Here we have introduced the map

$$\zeta : \ L \to L\otimes L$$

by

$$\zeta(X) := \frac{d}{dt}\eta(e^{tX})\,|_{t=0} \ , \tag{1.2.57}$$

η being the map (1.2.43).

Thus the restriction of δ to $L \subset \mathcal{U}(L)$ is equal to ζ. By definition of $\mathcal{U}(L)$ this can be expanded to the whole $\mathcal{U}(L)$.

It is clear that the map ζ is a counterpart of a Lie bracket $[\cdot,\cdot] : \ L\otimes L \to L$. Let us calculate $\zeta\left([X,Y]\right)$:

$$\begin{aligned}\zeta([X,Y]) &= \frac{d}{dt}\eta\left(e^{t[X,Y]}\right) = \frac{d}{ds}\frac{d}{dt}\eta\left(e^{sX}e^{tY}e^{-sX}\right)|_{t=s=0} \\ &= \frac{d}{ds}\frac{d}{dt}\Big[\eta\left(e^{sX}\right) + Ad^{\otimes 2}_{\exp\{sX\}}\eta\left(e^{tY}\right) \\ &\quad + Ad^{\otimes 2}_{\exp\{sX\}}Ad^{\otimes 2}_{\exp\{tY\}}\eta\left(e^{-sX}\right)\Big]\,|_{s,t=0}\end{aligned}$$

$$= \frac{d}{ds}\left(Ad^{\otimes 2}_{\exp\{sX\}}\right)|_{s=0}\,\zeta(Y) + \frac{d}{dt}\left(Ad^{\otimes 2}_{\exp\{tY\}}\right)|_{t=0}\,\zeta(-X) \ , \tag{1.2.58}$$

where

$$ad^{\otimes 2}_X(Y\otimes Z) := X.(Y\otimes Z) \equiv [X,Y]\otimes Z + Y\otimes[X,Z] \ , \quad \forall\, X,Y,Z \in L \ .$$

The third equality follows from the condition (1.2.47). The equality

$$\begin{aligned} 0 &= \partial_t \eta(e^{tX} e^{-tX}) \mid_{t=0} \\ &= \partial_t \eta(e^{tX}) \mid_{t=0} + \partial_t \left[(Ad^{\otimes 2}_{\exp\{tX\}}) \eta(e^{-tX}) \right] \mid_{t=0} \\ &= \zeta(X) + \zeta(-X) \end{aligned}$$

(here the condition (1.2.47) has been used again and the condition $\eta(e) = 0$ which follows from it; e is a group unity) permits to rewrite (1.2.58) in the form

$$\zeta([X,Y]) = ad^{\otimes 2}_X \zeta(Y) - ad^{\otimes 2}_Y \zeta(X) \; . \tag{1.2.59}$$

Using antisymmetry and Jacobi identity for η one finds that the map ζ satisfies the following conditions

1. co-antisymmetry:

$$\sigma \circ \zeta = -\zeta \; ; \tag{1.2.60}$$

2. the co-Jacobi identity:

$$[1 \otimes 1 \otimes 1 + (1 \otimes \sigma)(\sigma \otimes 1) + (\sigma \otimes 1)(1 \otimes \sigma)](1 \otimes \zeta)\zeta = 0 \; . \tag{1.2.61}$$

Definition 1.10 *A Lie algebra L together with a map $\zeta : L \to L \otimes L$ satisfying the properties (1.2.59) - (1.2.61) are called a Lie bialgebra.*

Thus from the consideration above it follows that a Lie algebra of a Poisson Lie group is a Lie bialgebra.

Drinfel'd has proved[87] that there is one-to-one correspondence between simply connected Poisson-Lie groups and Lie bialgebras. This is an analog of the third Lie theorem on relation between Lie groups and algebras.

Above we have presented the explicit construction of a Lie bialgebra from a Poisson-Lie group (i.e. the construction of the map $\zeta : L \to L \otimes L$, using the map $\eta : G \to L \otimes L$, which defines the Poisson structure on G (see (1.2.42)). In more general terms this relation looks as follows. If one has a Poisson-Lie group G with a Poisson bracket $\{\cdot,\cdot\}_p$, one can define the Lie algebra structure on L^* (dual space, i.e. space of linear functions on the Lie algebra L)

$$[x,y] := d\{\phi,\psi\}_p(e) \; ,$$

where $x = d\phi(e)$, $y = d\psi(e)$; $x,y \in L^*$, $\phi,\psi \in Fun(G)$, e is the group unity. As usual the map $[\cdot,\cdot] : L^* \otimes L^* \to L^*$ defines *by duality* the map $\zeta : L \otimes L \to L$.

Conversely, one can start from a Lie bialgebra and construct Poisson structure on a group G using classical r-matrix derived from the map ζ. The latter can be constructed independently by use of *Manin triple*.

Definition 1.11 *A Manin triple (M, M_1, M_2) consists of a Lie algebra M with a non-degenerate ad-invariant scalar product on it and isotropic Lie subalgebras M_1, M_2 such that M is a direct sum of M_1 and M_2 as a vector space.*

Note that due to the non-degenerate scalar product and isotropicity of the subalgebras, M_2 is naturally isomorphic to M_1^*. If (M, M_1, M_2) is a Manin triple then one can put $L = M_1$ and define ζ as the map dual to the commutator in M_2.

On the other hand, if a Lie bialgebra is given, one can put $M = L \oplus L^*$, $M_1 = L$, $M_2 = L^*$ and define $[X, y]$, where $X \in L$, $y \in L^*$ so that the natural scalar product $\langle \cdot, \cdot \rangle$ on M, i.e.

$$\langle (X_1, y_1), (X_2, y_2) \rangle := \langle \langle X_1, y_2 \rangle \rangle + \langle \langle X_2, y_1 \rangle \rangle ,$$

will be invariant. Thus there is one-to-one correspondence between Lie bialgebras and Manin triples.

This completes the construction of Poisson Lie group on the basis of Lie group, and the dual co-Poisson structure for UEA $\mathcal{U}(L)$ (the corresponding Lie algebra being a Lie bialgebra).

We summarize the essential constructions of this Section and the relations among them in Fig.1.6. The algebraic constructions introduced in this Section are indicated in the boxes; structure maps, which will lead to a new object when added to another object, are indicated in the dashed boxes. Note that the term *"co-Poisson Lie algebra"* is not quite conventional but inspired by the duality (it means Lie bialgebra L with UEA $\mathcal{U}(L)$ being a co-Poisson Hopf algebra).

Now we are ready to quantize classical Lie algebras and groups.

1.3 Deformation of co-Poisson structures

Definition 1.12 *An algebra deformation of a commutative associative algebra A is the linear space $A[[\chi]]$ of formal power series in χ with coefficients in A together with a non-commutative associative algebra multiplication of the form*

$$m_\chi(a \otimes b) := a \star b := ab + \sum_{n=1}^{\infty} \chi^n W_n(a, b) , \qquad \forall a, b \in A \subset A[[\chi]] ,$$

extended in an obvious way to all $A[[\chi]]$. Here W_n is bilinear and

$$\sum_{\substack{l,m \geq 0 \\ l+m=k}} \left(W_l \left(W_m(a, b), c \right) - W_l \left(a, W_m(b, c) \right) \right) = 0 , \qquad \textit{(associativity)}$$

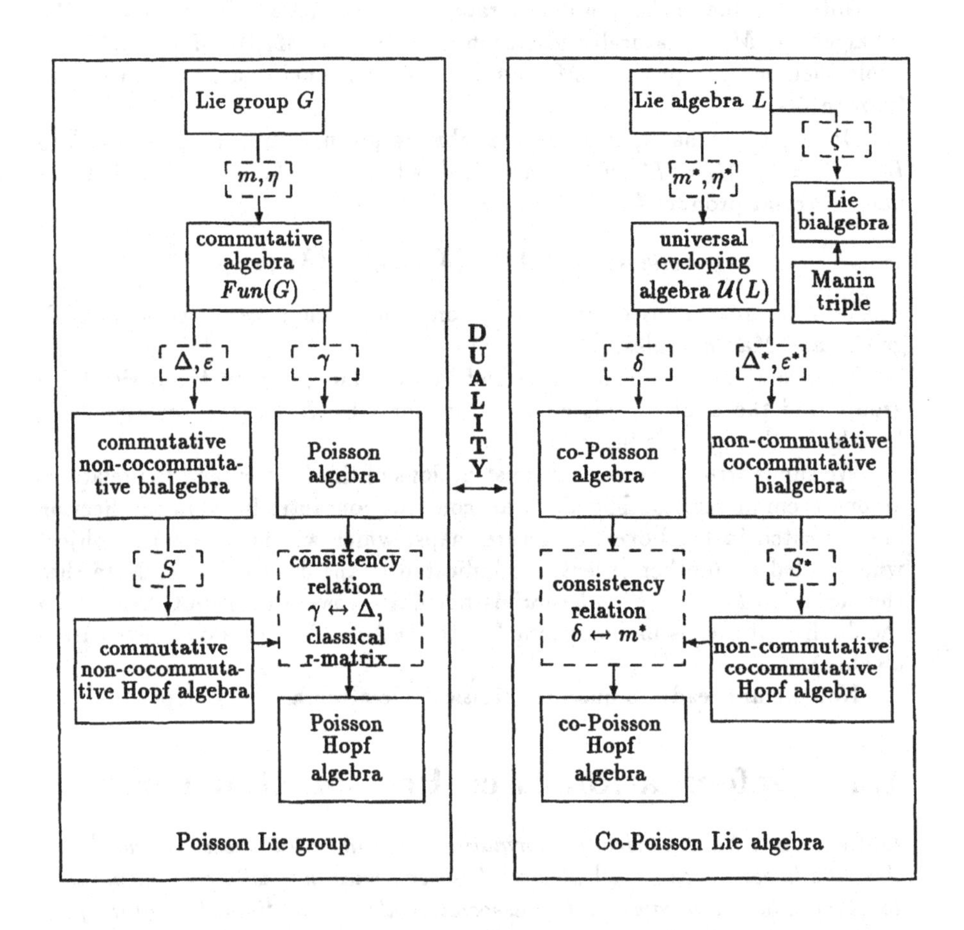

Figure 1.6: Prequantization algebraic structure of Lie groups and algebras

$$W_k(a,1) = W_k(1,a) = 0\ , \qquad k \geq 1 \qquad (\exists \text{ an algebra unity } \mathbf{1})\ .$$

Note that generally $a\,[[u_1,\ldots,u_n]]$ (respectively $a[u_1,\ldots,u_n]$) denotes in mathematical literature a set of all power series (respectively polynomials) in variables $u_1,\ldots,u_n$ with coefficients in a.

If $(A, m, \eta, \{\cdot,\cdot\}_p)$ is a Poisson algebra, then $(A\,[[\chi]]\,, m_\chi, \eta_\chi)$ is called the Poisson algebra deformation if

$$W_1(a,b) = \{a,b\}_p\ , \qquad \forall\, a,b \in A \subset A\,[[\chi]]$$

and η_χ is the corresponding extension of the unity η of A. This means that

$$a \star b - b \star a = \chi\{a,b\} + \mathcal{O}(\chi^2)\ .$$

The most known example of Poisson algebra deformation is the deformation [14] of algebra $Fun(\mathcal{X})$ of functions on classical phase space $\mathcal{X}_{2n}$ which we discussed in Section 1.1. In this case

$$W_k(a,b) = \frac{1}{k!}\left(\frac{i}{2}\right)^k \omega^{\alpha_1\beta_1}\ldots\omega^{\alpha_k\beta_k}(\partial_{\alpha_1}\ldots\partial_{\alpha_k}a)(\partial_{\beta_1}\ldots\partial_{\beta_k}b)\ ,$$

where ω is a symplectic form on $\mathcal{X}$ (cf. (1.1.29)). For this deformation the parameter χ is equal, of course, to the Planck constant $\hbar$.

The next step is the construction of appropriate deformation

$$(A\,[[\chi]]\,, m_\chi, \eta_\chi, \Delta_\chi, \varepsilon_\chi, S_\chi)$$

of a Poisson-Hopf algebra $(A, m, \eta, \Delta, \varepsilon, S, \{\cdot,\cdot\}_p)$. This problem is solved for matrix groups[227, 28]. But experience in ordinary Lie groups and algebras prompts us that it is easier to carry out some construction for Lie algebras at first and then to find appropriate analog for Lie groups using some relations between groups and Lie algebras (in quantum case – duality). So we are going to consider a quantization of co-Poisson structure on UEA $\mathcal{U}(L)$ which is developed in full completeness and generality due to Drinfel'd [86].

In general, UEA $\mathcal{U}(L)$ is non-commutative algebra (for a non-abelian L) but cocommutative (see p.28). So the main object in the process of a Lie algebra quantization is the dual analog of commutator $[\cdot,\cdot]$ in an algebra $Fun(G)$ (which can be written in algebraic terms as $m - m\circ\sigma$), i.e. the map $\Delta - \sigma\circ\Delta$, where Δ is a comultiplication in $\mathcal{U}(L)$. After the quantization, $\mathcal{U}(L)$ becomes a non-commutative algebra (dual analog of non-commutativity in the case of quantization of Poisson algebras). Of course, other structure maps (including multiplication) considered in the preceding Section also must be adjusted appropriately (as we pointed out already, all structure maps of a co-Poisson Hopf algebra are strongly correlated).

Let π denote the natural map $A[[\chi]] \to A$ (which is equivalent to putting $\chi = 0$).

Definition 1.13 *A quantization of a co-Poisson bialgebra* $(A, m, \Delta, \eta, \varepsilon; \delta)$ *is a non-cocommutative bialgebra* $(A[[\chi]], m_\chi, \Delta_\chi, \eta_\chi, \varepsilon_\chi)$ *such that*

$$\begin{aligned}
(\pi \otimes \pi) \circ \Delta_\chi &= \Delta \circ \pi \, , \\
\pi \circ m_\chi &= m \circ (\pi \otimes \pi) \, , \\
\pi \circ \eta_\chi &= \eta \, , \\
\varepsilon_\chi \circ \pi &= \varepsilon \, , \\
\delta(\pi(X)) &= \pi \left[\frac{1}{\chi} (\Delta_\chi(X) - \sigma \circ \Delta_\chi(X)) \right] \, , \quad \forall \, X \in A \, . \qquad (1.3.1)
\end{aligned}$$

Since Δ is cocommutative, it is clear that $(\Delta_\chi(X) - \sigma \circ \Delta_\chi(X)) \sim O(\chi)$ and rhs of the last relation (1.3.1) is nonsingular in the limit $\chi \longrightarrow 0$.

Let us consider now the quantization procedure on the example of $\mathcal{U}(sl(2))$ algebra.

1. Construction of co-Poisson structure.

As we have learned in Section 1.2, on a Lie algebra $L \subset \mathcal{U}(L)$ a co-Poisson structure δ coincides with the map $\zeta : L \longrightarrow L \otimes L$ defined by (1.2.57). To construct the latter, we must know the solution of CYBE or MCYBE. In the case of simple Lie algebras this problem has been solved in [16]. For the algebra $sl(2)$ this solution looks as follows

$$r = \frac{1}{2} J_0 \otimes J_0 + 2 J_+ \otimes J_- \, . \qquad (1.3.2)$$

According to (1.2.57), for arbitrary Lie algebra the map ζ has the form

$$\begin{aligned}
\zeta(Y) &= \frac{d}{dt} (\delta_G r)(e^{tY}) \mid_{t=0} = \frac{d}{dt} (r - e^{tY} \diamond r) \mid_{t=0} \\
&= -r^{\mu\nu} ([Y, X_\mu] \otimes X_\nu + X_\mu \otimes [Y, X_\nu]) \\
&= [r, 1 \otimes Y + Y \otimes 1] \, ,
\end{aligned}$$

where $r = r^{\mu\nu} X_\mu \otimes X_\nu$. So using (1.1.35) one obtains

$$\begin{aligned} \zeta(J_0) &= 0\,, \\ \zeta(J_\pm) &= (J_0 \otimes J_\pm - J_\pm \otimes J_0)\,. \end{aligned} \tag{1.3.3}$$

2. Construction of quantized comultiplication.
The comultiplication Δ_χ has the general form

$$\Delta_\chi = \sum_{n=0}^{\infty} \frac{\chi^n}{n!} \Delta_{(n)}\,. \tag{1.3.4}$$

The coassociative condition (1.2.8) gives the recursive relations

$$\sum_{k=0}^{n} \binom{n}{k} (\Delta_{(k)} \otimes 1 - 1 \otimes \Delta_{(k)}) \Delta_{(n-k)} = 0\,.$$

Two first approximations $\Delta = \Delta_{(0)}$ and $\Delta_{(1)}$ are defined correspondingly by Δ for non-quantized algebra $\mathcal{U}(L)$ (see p.28) and by the condition (1.3.1) (recall that on L the map δ coincides with ζ from (1.3.3)) so that

$$\Delta_{(1)}(H) = 0\,,$$

$$\Delta_{(1)}(X^\pm) = \frac{1}{2}(H \otimes X^\pm - X^\pm \otimes H)\,.$$

Again as in Section 1.1, we use the renotation $J_\pm \longrightarrow X^\pm$, $J_0 \longrightarrow H$ to stress the deformation of the algebra (see remark after (1.1.34)).

One can show by induction that for arbitrary n

$$\Delta_{(n)}(H) = 0\,,$$

$$\Delta_{(n)}(X^\pm) = \left(X^\pm \otimes \left(-\frac{H}{2}\right)^n + \left(\frac{H}{2}\right)^n \otimes X^\pm \right)\,,$$

where H^n is defined by the multiplication m_χ. Introducing these expressions into expansion (1.3.4), one obtains finally (cf. (1.1.34))

$$\begin{aligned} \Delta_\chi(H) &= H \otimes \mathbf{1} + \mathbf{1} \otimes H\,, \\ \Delta_\chi(X^\pm) &= X^\pm \otimes q^{-\frac{H}{2}} + q^{\frac{H}{2}} \otimes X^\pm\,,\quad q := e^\chi\,. \end{aligned} \tag{1.3.5}$$

It is easy to see that this comultiplication is non-cocommutative: $\Delta_\chi \neq \sigma \circ \Delta_\chi$.

3. Adjustment of other Hopf algebra structure maps.

All structure maps of a Hopf algebra are related to each other according to the Hopf algebra axioms (Definitions 1.1–1.3). This was explicitly demonstrated in the preceding section in terms of their structure constants. So a deformation of a comultiplication Δ has to be accompanied by an appropriate deformation of other maps.

In particular, as Δ_χ must be a homomorphism $\mathcal{U}_q \longrightarrow \mathcal{U}_q \otimes \mathcal{U}_q$, a deformation of Δ leads to the corresponding deformation of multiplication, so that

$$\Delta_\chi(m_\chi(X \otimes Y)) = m_\chi(\Delta_\chi(X) \otimes \Delta_\chi(Y)) .$$

This in turn, gives the relation

$$\Delta_\chi([X,Y]) = [\Delta_\chi(X), \Delta_\chi(Y)] , \tag{1.3.6}$$

where the commutators are defined with respect to deformed multiplication m_χ. It is easy to see that in the case of $s\ell(2)$ algebra, the undeformed Lie algebra relations (1.1.35) do not satisfy (1.3.6) with the comultiplication (1.3.5), while appropriate deformation leads just to the algebra (1.1.36).

Let us consider another example. Keeping unity and counity maps as in undeformed case (p.28), one has from(1.2.9) after simple calculations

$$S_\chi(X^\pm) = -q^{-\frac{H}{2}} X^\pm q^{\frac{H}{2}} = -q^{\mp 1} X^\pm ,$$

$$S_\chi(H) = -H .$$

We summarize all the relations for quantum algebra $\mathcal{U}_q(s\ell(2))$

$$\begin{aligned}
[H, X^\pm] &= \pm 2X^\pm , & [X^+, X^-] &= [H]_q \\
\Delta_\chi(H) &= H \otimes \mathbf{1} + \mathbf{1} \otimes H , & \Delta_\chi(X^\pm) &= X^\pm \otimes q^{-\frac{H}{2}} + q^{\frac{H}{2}} \otimes X^\pm , \\
S_\chi(X^\pm) &= -q^{\mp 1} X^\pm , & S_\chi(H) &= -H , \\
\varepsilon_\chi(X^\pm) &= \varepsilon_\chi(H) = 0 , & \varepsilon_\chi(\mathbf{1}) &= 1 , \quad \eta(\alpha) = \alpha \mathbf{1} ,
\end{aligned} \tag{1.3.7}$$

$$[x]_q := \frac{q^x - q^{-x}}{q - q^{-1}} , \qquad q = e^\chi , \qquad \alpha \in \mathbb{C} .$$

The deformation procedure can be applied to any simple Lie algebra, but the corresponding calculations prove to be more complicated. Another possibility is to use an analog of operator quantization of classical mechanics: one can try to find directly a non-commutative Hopf algebra which satisfies all the requirements of the Definition 1.13.

Drinfel'd [88, 87] and Jimbo [129] have independently constructed the quantum universal enveloping algebras (QUEA) for any simple Lie algebra (see also [167, 168, 169, 204, 243]). In their construction, they start from *Chevalley basis* [250, 219, 124] of classical simple Lie algebra L. In this basis UEA $\mathcal{U}(L)$ is generated by $\{X_i^+, X_i^-, H_i\}_{i=1}^r$ (r is rank of the algebra) with the relations

$$\begin{aligned}
[H_i, H_j] &= 0 , \\
[H_i, X_j^\pm] &= \pm a_{ij} X_j^\pm , \\
[X_i^+, X_j^-] &= \delta_{ij} H_i , \\
(\mathrm{ad}_{X_i^\pm})^{1-a_{ij}} X_j^\pm &= 0 , \quad i \neq j ,
\end{aligned} \tag{1.3.8}$$

where $\alpha_1, \ldots, \alpha_r$ are simple roots of the algebra, the (α_i, α_j) are their invariant scalar products. The root scalar products form so-called symmetrized Cartan matrix $a_{ij}^S := (\alpha_i, \alpha_j)$, $i, j = 1, \ldots, r$ [135], which can be obtained from the usual Cartan matrix $a_{ij} = 2(\alpha_i, \alpha_j)/(\alpha_j, \alpha_j)$ by left multiplication with the diagonal matrix. By a rescaling of the generators, the defining relations can be written in terms of the non-symmetrized Cartan matrix.

The last relations in (1.3.8) are called *the Serre relations.* These relations are the price to pay for use of the small (Chevalley) basis and provide that a Lie algebra with CR written in Chevalley basis is semisimple one. Recall, that in the well known *Cartan-Weyl basis,* one uses all the generators corresponding to a root diagram and not only those associated with simple roots. Of course, for $s\ell(2)$ Lie algebra both bases coincide and the Serre relations are trivial. For the simple algebras $so(5) = sp(4)$ of rank two with the Cartan matrix

$$a_{ij} = \begin{pmatrix} 2 & -2 \\ -1 & 2 \end{pmatrix} ,$$

and Chevalley basis $\{H_1, H_2, X_1^\pm, H_2^\pm\}$, the Serre relations are the following

$$\begin{aligned}
(\mathrm{ad}X_1)^{1-a_{12}} X_2 &= (\mathrm{ad}X_1)^3 X_2 \equiv [X_1, [X_1, [X_1, X_2]]] \\
&= X_1^3 X_2 - 3X_1^2 X_2 X_1 + 3X_1 X_2 X_1^2 - X_2 X_1^3 = 0 ,
\end{aligned} \tag{1.3.9}$$

$$\begin{aligned}
(\mathrm{ad}X_2)^{1-a_{21}} X_1 &= (\mathrm{ad}X_2)^2 X_1 \\
&= X_2^2 X_1 - 2X_2 X_1 X_2 + X_1 X_2^2 = 0 ,
\end{aligned} \tag{1.3.10}$$

(here and in what follows the superscripts + and − are omitted if relations are the same for X_i^+ and X_i^-).

Cartan-Weyl basis [250, 124, 13, 31] of $so(5)$ consists, in addition to $H_j, X_i^\pm$ $(i = 1,2)$, of four generators $X_3^\pm, X_4^\pm$. The latter correspond to non-simple roots and the additional CR in the Cartan-Weyl basis

$$X_3 = [X_1, X_2] , \qquad X_4 = [X_1, X_3] , \tag{1.3.11}$$

permit to express them (as elements of the UEA $\mathcal{U}(so(5))$) in terms of the simple root generators X_1, X_2. As $\alpha_1 + \alpha_4$ and $\alpha_2 + \alpha_3$ are *not* roots, one has

$$[X_1, X_4] = 0 , \qquad [X_2, X_3] = 0 . \tag{1.3.12}$$

Taking into account (1.3.11), one can easily check that the relations (1.3.12) are equivalent to the Serre relations (1.3.9),(1.3.10).

In general case of a simple Lie algebra the corresponding QUEA $\mathcal{U}_q$, $q \in \mathbb{C}$ is the $C[[\chi]]$-algebra with generators $\{X_i^+, X_i^-, H_i\}_{i=1}^r$ and the commutation relations

$$\begin{aligned} [H_i, H_j] &= 0 , \\ \left[H_i, X_j^\pm\right] &= \pm a_{ij} X_j^\pm , \\ \left[X_i^+, X_j^-\right] &= \delta_{ij} [H_i]_q , \end{aligned} \tag{1.3.13}$$

which are the straightforward generalization of the $s\ell_q(2)$ relations (this is the advantage of the Chevalley basis). One has to complete them by the appropriately quantized Serre relations. For this aim we need *q-deformed adjoint action* of elements of $\mathcal{U}_q(L)$. To find it we can start from the expression for a classical adjoint action in terms of maps. By using the explicit form of the Hopf algebra maps for the classical UEA it is easy to check that there are two possibilities

$$\mathrm{ad}_X^+ := (m_L \otimes m_R) \circ (\mathbf{1} \otimes S) \circ \Delta(X) , \tag{1.3.14}$$

$$\mathrm{ad}_X^- := (m_L \otimes m_R) \circ (S \otimes \mathbf{1}) \circ \Delta(X) , \tag{1.3.15}$$

where $(m_L \otimes m_R)$ acts in $\mathcal{U}_q(L) \otimes \mathcal{U}_q(L) \otimes \mathcal{U}_q(L)$

$$(m_L \otimes m_R)(X \otimes Y) \otimes Z := XZY , \qquad X, Y, Z \in \mathcal{U}_q(L) .$$

In the case of a classical UEA, ad_X^+ and ad_X^- differ from each other by a sign only. In the quantum case, the difference is more essential because of non-cocommutativity.

Writing, as usual,

$$\Delta_\chi(X) = \sum_\mu X_\mu \otimes X^\mu , \qquad X_\mu, X^\mu \in \mathcal{U}_q(L) ,$$

we have for the action of ad^+_X on arbitrary element Y of $\mathcal{U}_q(L)$

$$\begin{aligned}\mathrm{ad}^+_X Y &= (m_L \otimes m_R)(\sum_\mu X_\mu \otimes S(X^\mu))Y \\ &= \sum_\mu X_\mu Y S(X^\mu)\,, \end{aligned} \tag{1.3.16}$$

($\mathrm{ad}^-_X Y$ has an analogous form). In what follows we will consider mainly co-multiplications in deformed algebras and in order to simplify the notations, we shall omit the subscript χ in Δ_χ.

To find the *explicit* form of this action in quantum case we must know the explicit form of the Hopf algebra maps. Using again an analogy of the triple $\{X_i^\pm, H_i\}$ with $\mathcal{U}_q(s\ell(2))$, one can write immediately

$$\begin{aligned}\Delta(H_i) &= H_i \otimes \mathbf{1} + \mathbf{1} \otimes H_i\,, \\ \Delta(X_i^\pm) &= X_i^\pm \otimes q_i^{-H_i} + q_i^{H_i} \otimes X_i^\pm\,, \\ \varepsilon(\mathbf{1}) &= 1\,, \quad \varepsilon(X_i^\pm) = \varepsilon(H_i) = 0\,, \end{aligned} \tag{1.3.17}$$

where $q_i := q^{(\alpha_i,\alpha_i)/2}$. The antipode axiom (1.2.9) gives

$$\begin{aligned}S(H_i) &= -H_i\,, \\ S(X_i^\pm) &= -q_i^\mp X_i^\pm\,. \end{aligned} \tag{1.3.18}$$

The $\mathrm{ad}^\pm_X$ operations now take the form

$$\begin{aligned}\mathrm{ad}^\pm_{H_i} Y &= [H_i, Y]\,, \\ \mathrm{ad}^+_{X_i^\pm} Y &= X_i^\pm Y q_i^{H_i} - q_i^{H_i \mp 1} Y X_i^\pm\,, \\ \mathrm{ad}^-_{X_i^\pm} Y &= -\left(X_i^\pm Y q_i^{-H_i \mp 1} - q_i^{-H_i} Y X_i^\pm\right)\,. \end{aligned}$$

The exponential factors of the type $q_i^{\pm H_i}$ complicate the form of the adjoint operators and the Serre relations written in terms of $\mathrm{ad}^\pm_X$. One can eliminate these factors by going over to another set of $\mathcal{U}_q(L)$ generators: $\{\widetilde{X}_i^\pm, k_i\}$, where

$$\widetilde{X}_i^+ = q_i^{H_i} X_i^+\,,$$

$$\widetilde{X}_i^- = X_i^- q_i^{-H_i}\,,$$

$$k_i = q_i^{2H_i}\,.$$

The CR for this new set of generators are

$$\begin{aligned} k_i k_j &= k_j k_i \ , \\ k_i \widetilde{X}_j^{\pm} k_i^{-1} &= q^{\pm(\alpha_i,\alpha_j)} \widetilde{X}_j^{\pm} \ , \\ [\widetilde{X}_i^{+}, \widetilde{X}_j^{-}] &= \delta_{ij} \frac{k_i - k_i^{-1}}{q_i - q_i^{-1}} \ , \end{aligned} \tag{1.3.19}$$

and the Hopf algebra maps take the form

$$\begin{aligned} \Delta(k_i) &= k_i \otimes k_j \ , \\ \Delta(\widetilde{X}_i^{+}) &= \widetilde{X}_i^{+} \otimes \mathbf{1} + k_i \otimes \widetilde{X}_i^{+} \ , \\ \Delta(\widetilde{X}_i^{-}) &= \widetilde{X}_i^{-} \otimes k_i^{-1} + \mathbf{1} \otimes \widetilde{X}_i^{-} \ , \\ S(\widetilde{X}_i^{+}) &= -k_i^{-1} \widetilde{X}_i^{+} \ , \quad S(\widetilde{X}_i^{-}) = -\widetilde{X}_i^{-} k_i \ , \\ S(k_i) &= k_i^{-1} \ , \\ \varepsilon(\widetilde{X}_i^{\pm}) &= 0 \ , \quad \varepsilon(k_i) = 1 \ . \end{aligned} \tag{1.3.20}$$

In terms of these generators the adjoint action takes the form of so-called *q-commutator* $[\cdot,\cdot]_q$, for example

$$\mathrm{ad}^{\pm}_{\widetilde{X}_i^{+}} = \widetilde{X}_i^{+} \widetilde{X}_j^{\pm} - k_i \widetilde{X}_j^{\pm} k_i^{-1} \widetilde{X}_i^{+} = [\widetilde{X}_i^{+}, \widetilde{X}_j^{\pm}]_q \ ,$$

where

$$[\widetilde{X}_i^{\pm}, \widetilde{X}_j^{\pm}]_q := \widetilde{X}_i^{+} \widetilde{X}_j^{\pm} - q^{(\alpha_i,\alpha_j)} \widetilde{X}_j^{\pm} \widetilde{X}_i^{+} \ .$$

Using the q-adjoint action (q-commutator), the deformed Serre relations are written as follows

$$\left(\mathrm{ad}^{(q)}_{\widetilde{X}_i^{\pm}}\right)^{1-a_{ij}} \widetilde{X}_j^{\pm} = 0 \ , \quad \text{for } i \neq j \ , \tag{1.3.21}$$

where $\mathrm{ad}^{(q)}_{\widetilde{X}_i^{\pm}} \widetilde{X}_j^{\pm} := \left[\widetilde{X}_i^{\pm}, \widetilde{X}_j^{\pm}\right]_q$. A more explicit form of the deformed Serre relation is

$$\sum_{n=0}^{1-a_{ij}} (-1)^n \begin{pmatrix} 1-a_{ij} \\ n \end{pmatrix}_{q_i} \left(\widetilde{X}_i^{\pm}\right)^{1-a_{ij}-n} \left(\widetilde{X}_j^{\pm}\right) \left(\widetilde{X}_i^{\pm}\right)^{n} = 0 \ , \quad i \neq j \ , \tag{1.3.22}$$

where the *q-binomial coefficients* are

$$\begin{pmatrix} m \\ k \end{pmatrix}_q := \frac{[m]_q!}{[k]_q! [m-k]_q!} \ , \tag{1.3.23}$$

$$[n]_q! := [1]_q[2]_q \cdots [n-1]_q[n]_q \,, \quad \text{with} \quad [0]_q! := 1 \,.$$

The more familiar to physicists Cartan-Weyl basis can be constructed also in the quantum case [229, 30, 141]. Essentially, the new ingredient of the construction compared with the classical case is an ordering of the root system:

- the system Δ_+ of positive roots (not to be confused with a comultiplication Δ !) is in the normal order if each non-simple root $\gamma = \alpha + \beta \in \Delta_+$, where $\alpha \neq \lambda\beta$, $\alpha, \beta \in \Delta_+$, $\lambda \in \mathbb{C}$ is written *between* α and β.

A deformed Cartan-Weyl basis is constructed by the method of induction:

- fix some normal ordering in Δ_+;
- let $\alpha_i, \alpha_j, \alpha_k \in \Delta_+$ be pairwise non-collinear roots, such that $\alpha_k = \alpha_i + \alpha_j$ and between α_i and α_j (according to the ordering) there are no other roots α_i' and α_j', such that $\alpha_i' + \alpha_j' = \alpha_k$;
- if $X_i^\pm$ and $X_j^\pm$ have been constructed already, set

$$X_k^+ = [X_i^+, X_j^+]_q \,, \qquad X_k^- = [X_i^-, X_j^-]_{q^{-1}} \,.$$

The new generators satisfy the CR

$$\begin{aligned} [H_i, X_k^\pm] &= (\alpha_i, \alpha_k) X_k^\pm \,, \\ [X_k^+, X_k^-] &= C_k [H_k]_q \quad \text{(no summation)} \,, \end{aligned}$$

where C_k is some function of q.

In this basis the q-Serre relations are equivalent to the statement that if $\alpha_i + \alpha_j$ is not a root, corresponding q-commutator is zero

$$[X_i^\pm, X_j^\pm]_q = 0 \,, \qquad \alpha < \beta \,,$$

where by definition $\alpha < \beta$ if α is located to the left of β in the normal-ordered Δ_+.

Co-algebra maps for new generators are derived from those for initial (Chevalley) ones using the homomorphism or antihomomorphism properties of the maps.

Now to quantize the algebra of functions on a group $Fun(G)$, one can use the duality. A very useful fact to do this is that an algebra $Fun(G)$ is isomorphic to the algebra of matrix elements of finite dimensional representations of UEA $\mathcal{U}(L)$, where L is the Lie algebra of a group G. In fact, this is the Peter-Weyl type theorem [13]. In a standard way one can use this theorem as a definition in the quantum case [222, 223].

Definition 1.14 *An algebra $Fun_q(G)$ of matrix elements of QUEA $\mathcal{U}_q(L)$, where L is a Lie algebra of a group G, is called an algebra of functions on a quantum group G_q.*

The duality between an algebra A and matrix elements of its finite dimensional representations is easy to understand. Indeed, let $\rho : A \to End\,V$ (endomorphisms, i.e. homomorphic maps into itself, of a finite dimensional linear space) is a finite dimensional representation of a complex algebra A. Then the matrix elements T_{ij} are linear functionals on A defined by the map $\rho : a \to T_{ij}(a)$, $a \in A$. If A is a Hopf algebra then there is the Hopf algebra structure on the set of matrix elements. For example, multiplication is defined as follows (cf. (1.2.10))

$$m(T_{ij}(a), T_{kl}(a)) = (T_{ij} \otimes T_{kl})\Delta(a) \ . \tag{1.3.24}$$

The family of $Fun_q(G)$ gives quantization of the Poisson algebra $Fun(G)$ with the brackets

$$\{T_{ij}(Y), T_{kl}(Z)\} = [(T \otimes T)(-ir_0), T(Y) \otimes T(Z)]_{ij,kl} \ ,$$

where

$$-ir_0 := \frac{1}{2} \sum_{\alpha \in \Delta_+} (X_{-\alpha} \otimes X_\alpha - X_\alpha \otimes X_{-\alpha}) \ ,$$

$X, Y, Z \in L$ and Δ_+ is a set of positive roots. Moreover, the Hopf structure in $Fun_q(G)$ is a quantization of the Hopf structure in $Fun(G)$.

In Section 1.5 we will consider the Hopf structure of matrix quantum groups $Fun_q(G)$ in detail. But at first, in the next Section, we have to discuss a very important notion of quasi-triangular Hopf algebras.

1.4 Quasi-triangular Hopf algebras and quantum double construction

Since a comultiplication Δ for a *quantized* algebra $\mathcal{U}_q(L)$ is not symmetric, one can define a second comultiplication Δ' as its transpose

$$\Delta' := \sigma \circ \Delta \ ,$$

$$\sigma(a \otimes b) := b \otimes a \ , \quad \forall a, b \in \mathcal{U}_q(L) \ .$$

For example, Δ' of the generators of $\mathcal{U}_q(s\ell(2))$ reads (cf. (1.3.5))

$$\Delta'(H) = H\otimes\mathbf{1} + \mathbf{1}\otimes H ,$$

$$\Delta'(X^{\pm}) = X^{\pm}\otimes q^{\frac{H}{2}} + q^{-\frac{H}{2}}\otimes X^{\pm}$$

(we recall again that to simplify the notations we omit here and below the subscript χ indicating a quantization of Δ, Δ' and other structure maps).

In an arbitrary Hopf algebras the maps Δ and Δ' are not in general related. However, in the special and very important class of so-called *quasi-triangular Hopf algebras*, they are related by the invertible element $\mathcal{R} \in A\otimes A$. Quasi-triangular Hopf algebras (QTHA) are closely related to the famous quantum Yang-Baxter equation (QYBE) (1.1.20) which plays very important role in theory of integrable models [100, 101, 105] (see also the review [60] and refs. therein) and conformal field theory [3].

Definition 1.15 *Let $(A, m, \Delta, \eta, \varepsilon, S)$ be a Hopf algebra and $\mathcal{R}$ an invertible element of $A\otimes A$. Then the pair $(A, \mathcal{R})$ is called a QTHA if the following relations are fulfilled*

$$\Delta'(a) := \sigma \circ \Delta = \mathcal{R}\Delta(a)\mathcal{R}^{-1} , \tag{1.4.1}$$

$$(\Delta\otimes 1)\mathcal{R} = \mathcal{R}_{13}\mathcal{R}_{23} , \tag{1.4.2}$$

$$(1\otimes\Delta)\mathcal{R} = \mathcal{R}_{13}\mathcal{R}_{12} . \tag{1.4.3}$$

The meaning of the indices in (1.4.2),(1.4.3) was explained in Section 1.1 (cf.(1.1.21)) and is the same as in the case of classical r-matrix (1.2.53).

The relation (1.4.1) shows that in the case of QTHA the tensor product representations $V_1\hat{\otimes}V_2$ and $V_2\hat{\otimes}V_1$ are isomorphic, the isomorphism being provided by the element $\mathcal{R}$, called the *universal R-matrix.*

There is a simpler but still important for applications subclass of QTHA, namely, the triangular Hopf algebras.

Definition 1.16 *A quasi-triangular Hopf algebra is called triangular Hopf algebra if the element $\mathcal{R}$ satisfies the additional relation*

$$\mathcal{R}_{12}\mathcal{R}_{21} = 1 ,$$

where $\mathcal{R}_{21} := \sigma \circ \mathcal{R}_{12} \circ \sigma$.

This class of Hopf algebras is used in the theory of knots and conformal field theory. In Chapter 3 we will consider an application of triangular Hopf algebras to the problem of space-time quantization and quantum space-time symmetries.

The relevance of QTHA to the QYBE becomes clear from the calculation of $(1\otimes\Delta')\mathcal{R}$ in two different ways. Let $\mathcal{R}=\sum_i \mathcal{R}_i^{(1)}\otimes\mathcal{R}_i^{(2)}$. Hence,

$$\begin{aligned}(1\otimes\Delta')(\mathcal{R}) &= \sum_i \mathcal{R}_i^{(1)}\otimes\Delta'(\mathcal{R}_i^{(2)}) = \sum_i \mathcal{R}_i^{(1)}\otimes\mathcal{R}\Delta(\mathcal{R}_i^{(2)})\mathcal{R}^{-1} \\ &= \sum_i (1\otimes\mathcal{R})(\mathcal{R}_i^{(1)}\otimes\Delta(\mathcal{R}_i^{(2)}))(1\otimes\mathcal{R}^{-1}) = \mathcal{R}_{23}(1\otimes\Delta)\mathcal{R}\mathcal{R}_{23}^{-1} \\ &= \mathcal{R}_{23}\mathcal{R}_{13}\mathcal{R}_{12}\mathcal{R}_{23}^{-1}\ . \end{aligned} \tag{1.4.4}$$

The last equality follows from (1.4.3). On the other hand one has

$$\begin{aligned}(1\otimes\Delta')\mathcal{R} &= (1\otimes\sigma)(1\otimes\Delta)\mathcal{R} \\ &= (1\otimes\sigma)\mathcal{R}_{13}\mathcal{R}_{12} = \mathcal{R}_{12}\mathcal{R}_{13}\ . \end{aligned} \tag{1.4.5}$$

Equating rhs of (1.4.4) and (1.4.5) one obtains QYBE (1.1.20) which we write down here again

$$\mathcal{R}_{12}\mathcal{R}_{13}\mathcal{R}_{23} = \mathcal{R}_{23}\mathcal{R}_{13}\mathcal{R}_{12}\ . \tag{1.4.6}$$

The difference between (1.1.20) and (1.4.6) is that the latter puts the restriction on the element $\mathcal{R}$ of an algebra $A\otimes A$ as *an abstract algebraic element* and the former deals with its *matrix representation* R in some finite dimensional (four-dimensional in the case of (1.1.20)) space.

As we described in the preceding section, an algebra of matrix elements of $\mathcal{U}_q(L)$ is considered as an algebra $Fun_q(G)$ of functions on a quantum group. The QTHA defining property (1.4.1) written in some representation, gives a very elegant way of describing such a $Fun_q(G)$. Indeed, in a representation ρ, (1.4.1) looks as follows

$$R(\rho\otimes\rho)\Delta'(a) = (\rho\otimes\rho)\Delta(a)R\ , \tag{1.4.7}$$

where

$$R = (\rho\otimes\rho)\mathcal{R}\ . \tag{1.4.8}$$

Using (1.3.24), one can write

$$\begin{aligned}(\rho\otimes\rho)\Delta &=: (\mathbf{1}\otimes T)(T\otimes\mathbf{1}) =: T_2T_1\ , \\ (\rho\otimes\rho)\Delta' &=: (T\otimes\mathbf{1})(\mathbf{1}\otimes T) =: T_1T_2\ . \end{aligned} \tag{1.4.9}$$

Here $\mathbf{I}$ is the unit matrix and sign $\otimes$ corresponds to the matrix cross-product. Using the explicit form (1.1.23) for T_1, T_2 and the definition $\Delta' := \sigma \circ \Delta$, it is easy to see that (1.4.9) is indeed equivalent to (1.3.24). Thus, from (1.4.7)-(1.4.9) one has

$$RT_1T_2 = T_2T_1R\,. \tag{1.4.10}$$

In the particular case of $SL_q(2)$, this equation was already obtained in the Section 1 from some postulates (e.g., the possibility of T-matrix entries ordering etc.) and self-consistency relations. Now we have derived it for an arbitrary QTHA and from essentially more general point of view.

To apply the CR (1.4.10) for concrete problems, we must know the explicit form of R-matrix. One way is obvious: to use the anzatz

$$\mathcal{R} = 1\otimes 1 + \sum_{n=1}^{\infty} \mathcal{R}^{(n)} \chi^n \tag{1.4.11}$$

and, inserting it in the axioms (1.4.1)-(1.4.3), to solve a set of recursive relations for the $\mathcal{R}^{(n)}$. Note that substitution of the expansion (1.4.11) into the QYBE, gives for the first approximation $\mathcal{R}^{(1)}$ the classical YBE $(r,r)_S = 0$ (cf. (1.2.55)). Unfortunately, this method of R-matrix construction is extremely cumbersome.

Solutions of QYBE (1.4.6) associated to classical groups of A_n, B_n, C_n, D_n types have been constructed [129],[130],[12],[197] from – already known in frame of quantum integrable systems theory – solutions $R(\lambda)$, which depend on the so-called spectral parameter λ, by the appropriate limit $\lambda \longrightarrow \infty$. Some others have been found by direct computer calculations [121] or starting from quantum linear spaces as we illustrated in Section 1.1 [178, 75, 225].

The consistent method of R-matrices construction at least for simple quantum groups is provided by a quantum double construction [86, 205, 144, 161, 104]. A quantum double construction has its own important and profound meaning (not only for R-matrix computations) both for general theory and for applications.

The idea of the construction is the following. Let A be a Hopf algebra and A^* its dual Hopf algebra. Choose in A a basis $\{e_i\}$ and a dual basis $\{e^i\}$ in A^*.

In $A\otimes A^*$ there is a *canonical* element $K := e_i \otimes e^i$. Simple calculations with the use of structure constants (1.2.11), show that

$$(\Delta\otimes 1)K = K_{13}K_{23}\,.$$

This relation is identical to the axiom (1.4.2) for $\mathcal{R}$, while the analog of (1.4.3) gives (cf. (1.2.22))

$$(1\otimes\Delta)K \;=\; (1\otimes\Delta)(e_i\otimes e^i) = e_i\otimes m^i_{kl} e^k \otimes e^l$$

$$= e_k e_l \otimes e^k \otimes e^l = (e_k \otimes 1 \otimes e^l)(e^l \otimes e^k \otimes 1) \neq K_{13}K_{12} .$$

It is easy to see from these calculations that to obtain the analog of the relation (1.4.3) one must change the order of basis elements in the first factor of the triple tensor product of algebras. This means that instead of A^* one has to use an algebra A^0 dual to A *and* with an additional permutation in the comultiplication

$$\Delta(e^i) := m^i_{lk} e^k \otimes e^l ,$$

which was introduced in Section 1.2 (see (1.2.24),(1.2.25)).

With this comultiplication, K satisfies both the relations (1.4.2) and (1.4.3). But it cannot be considered as an element $\mathcal{R}$ because by definition $\mathcal{R} \subset A \otimes A$ and $K \subset A \otimes A^0$. To overcome this problem, consider a space $D(A)$ which, as a vector space, is isomorphic to $A \otimes A^0$ and contains A and A^0 as Hopf subalgebras. This means that $E^j_i = e_i \hat{\otimes} e^j$ (recall that $\hat{\otimes}$ denotes vector space tensor product) constitute a linear basis in $D(A)$ and

$$\Delta_D(E^j_i) = \Delta^{pt}_i m^j_{rs} E^s_p \otimes E^r_t . \qquad (1.4.12)$$

The obvious generalization of the element K for the space $D(A)$ is

$$\mathcal{R} = (e_i \hat{\otimes} 1) \otimes (1 \hat{\otimes} e^i) . \qquad (1.4.13)$$

The expression (1.4.13) satisfies both the axioms (1.4.2),(1.4.3) for the element $\mathcal{R}$. But $D(A)$ is not an algebra yet, as we have not defined a multiplication between elements of A and A^0. One can use this freedom and define it so that the first axiom (1.4.1) would be also fulfilled. This leads to the relations

$$\Delta^{rp}_s m^q_{pt} (e_r \hat{\otimes} 1)(1 \hat{\otimes} e^t) = m^q_{tp} \Delta^{pr}_s (1 \hat{\otimes} e^t)(e_r \hat{\otimes} 1) . \qquad (1.4.14)$$

Using Hopf algebra axioms, one can rewrite it in the form

$$E^a_i E^j_d = m^a_{qb} (S^{-1})^b_c \Delta^{cs}_d \Delta^{up}_s m^q_{pt} m^k_{iu} \Delta^{jt}_l E^l_k . \qquad (1.4.15)$$

To derive this, one has to multiply the relation (1.4.14) by $(e_i \hat{\otimes} 1)$ from the left and by $(1 \hat{\otimes} e^j)$ from the right and to use the identities (1.2.28), (1.2.13), (1.2.14), (1.2.12) and (1.2.15) to find the relations

$$\left(m^p_{qj} (S^{-1})^j_l \Delta^{ls}_m\right) \left(m^q_{ri} \Delta^{in}_s\right) = \delta^p_r \delta^n_m , \qquad (1.4.16)$$

which gives the inverse of the matrix

$$Q_{(qs)(rt)} := m^q_{tp} \Delta^{pr}_s .$$

Applying (1.4.16), the relations (1.4.14) can be rewritten in the form

$$e^i e_j = \Delta_b^{ls} \Delta_s^{rp} (S^{-1})_l^n m_{qn}^i m_{pt}^q e_r e^t ,$$

(we have used here the simplified notations $e_r \hat{\otimes} 1 \to e_r$, $1 \hat{\otimes} e^t \to e^t$) which can be presented, in turn, as follows

$$e^i e_j = \sum \langle\langle e_j^{(1)}, S'(e^i_{(1)}) \rangle\rangle \langle\langle e_j^{(3)}, e^i_{(3)} \rangle\rangle e_j^{(2)} e^i_{(2)} , \qquad (1.4.17)$$

where

$$\sum e_j^{(1)} \otimes e_j^{(2)} \otimes e_j^{(3)} := \Delta^{(2)}(e_j) := (\mathrm{id} \otimes \Delta)\Delta(e_j) ,$$

$$\sum e^i_{(1)} \otimes e^i_{(2)} \otimes e^i_{(3)} := \Delta'^{(2)}(e^i) := (\mathrm{id} \otimes \Delta')\Delta'(e^i) ,$$

(Δ', S' is the comultiplication and antipode in A^0).

The antipode, unity and counity are defined in the obvious way, e.g.

$$S(E_i^j) = S_i^k (S^{-1})_l^j E_l^k$$

(S_i^k are structure constants for an antipode in A). A Hopf algebra $D(A)$ constructed in this way satisfies all the properties of QTHA.

Thus, with every Hopf algebra A there is associated a quasi-triangular Hopf algebra $D(A)$ which contains A and A^0 as Hopf subalgebras and is isomorphic to $A \hat{\otimes} A^0$ as a vector space. Here is the precise definition:

Definition 1.17 *Let A be a Hopf algebra and A^0 its dual algebra with opposite comultiplication. The canonically associated with A a Hopf algebra $D(A)$ with the properties*

1. *as a coalgebra $D(A) = A \otimes A^0$ (cf. (1.4.12));*
2. *A and A^0 are embedded in $D(A)$ as Hopf subalgebras;*
3. *$\mathcal{R}\Delta_D(a) = (\sigma \circ \Delta_D)(a)\mathcal{R}$, $\forall a \in D(A)$, where Δ_D is comultiplication in $D(A)$ and $\mathcal{R}$ is image of the canonical element in $A \otimes A^0$ under the embedding $A \otimes A^0 \hookrightarrow D(A) \otimes D(A)$, cf.(1.4.13);*

is called the quantum double of the algebra A.

So if a Hopf algebra can be represented as a quantum double of some of its Hopf subalgebra, the construction gives explicit form of the element $\mathcal{R}$ which defines its quasi-triangular structure. Choosing some representation ρ of the algebra, one obtains the corresponding R-matrix of the type considered for $SL_q(2)$ case in Section 1.1 (see (1.1.25), where the fundamental representation was used).

This method can be applied for the R-matrix construction of QUEA $\mathcal{U}_q(L)$ of a simple Lie algebra L, the subalgebra A being a QUEA $\mathcal{U}_q(\mathcal{B}^+)$ (or $\mathcal{U}_q(\mathcal{B}^-)$) of Lie subalgebra $\mathcal{B}^+ \subset L$ ($\mathcal{B}^- \subset L$), generated in Chevalley basis (1.3.8) by H_i and X_j^+ (by H_i and X_j^-), $1 \le i,j \le r = rank\ L$. Such subalgebras $\mathcal{B}^\pm$ are called the Borel subalgebras.

The double $D(\mathcal{B}^+)$ or $D(\mathcal{B}^-)$ coincides with $\mathcal{U}_q(L)$ itself (after the natural elimination of the second Cartan subalgebra).

We consider now this construction on the example of $\mathcal{U}_q(s\ell(2))$.

The bases of $\mathcal{U}_q(\mathcal{B}^+)$ and $\mathcal{U}_q(\mathcal{B}^-)$ can be chosen in the form of monomials:

$$\mathcal{U}_q(\mathcal{B}^\pm) = span\{H^n(X^\pm)^m\}_{n,m=0}^\infty .$$

Let $\{(x^+)^m h^n\}_{n,m=0}^\infty$ be a basis of the algebra $(\mathcal{U}_q(\mathcal{B}^+))^*$ dual to the basis $\{H^n, (X^+)^m\}_{n,m=0}^\infty$ of the positive Borel subalgebra $\mathcal{U}_q(\mathcal{B}^+)$. Using the rules

$$\begin{aligned} \langle\langle fg, Y\rangle\rangle &= \langle\langle f\otimes g, \Delta(Y)\rangle\rangle , \\ \langle\langle \Delta(f), Y\otimes Z\rangle\rangle &= \langle\langle f, YZ\rangle\rangle , \end{aligned} \tag{1.4.18}$$

one can calculate CR and comultiplication in $(\mathcal{U}_q(\mathcal{B}^+))^*$

$$[h, x^+] = \chi x^+ ,$$

$$\Delta(h) = h\otimes\mathbf{I} + \mathbf{I}\otimes h ,$$
$$\Delta(x^+) = x^+\otimes\mathbf{I} + e^{2h}\otimes x^+ .$$

After transposition of the comultiplication and rescaling $h \longrightarrow \hat{H} = -\frac{2}{\chi}h$, these relations coincide with those for $\mathcal{U}_q(\mathcal{B}^-)$ in the basis $\{(\widetilde{X})^m, \hat{H}^n\}_{n,m=0}^\infty$ of generators

$$\widetilde{X}^- = X^- q^{-H/2} = x^+ , \qquad \hat{H} = H = -\frac{2}{\chi}h , \tag{1.4.19}$$

(cf. (1.3.19),(1.3.20)). Hence $(\mathcal{U}_q(\mathcal{B}^+))^0$ is isomorphic to $\mathcal{U}_q(\mathcal{B}^-)$ and the quantum double of the positive Borel subalgebra is therefore the Hopf algebra $\mathcal{U}_q(\mathcal{B}^+)\otimes\mathcal{U}_q(\mathcal{B}^-)$, which is generated by the elements $\{H, X^+, \hat{H}, \widetilde{X}^-\}$. To find the CR between elements of the positive Borel subalgebra and its opposite dual it is convenient to use (1.4.17) and the basis of $\mathcal{U}_q(\mathcal{B}^+)$ generated by H and

$$\widetilde{X}^+ = q^{H/2}X^+ .$$

The result is

$$\begin{aligned} \left[H, \tilde{X}^\pm\right] &= 2\tilde{X}^\pm , \qquad & \left[\tilde{X}^+, \tilde{X}^-\right] &= [H]_q , \\ \left[\hat{H}, \tilde{X}^\pm\right] &= 2\tilde{X}^\pm , \qquad & \left[H, \hat{H}\right] &= 0 . \end{aligned}$$

We leave the calculations to the reader as a useful exercise. To derive these CR, one has to put

$$\langle\langle x^+, X^+ \rangle\rangle = -(q - q^{-1})^{-1} ,$$

and to use (1.4.18) for the calculation of pairing of monomials in the generators. Note that after the transition to a transposed comultiplication $\Delta \longrightarrow \Delta'$, the dual pairing is changed as it must be correlated with the new comultiplication according to (1.4.18).

Thus canonical projection of $D((\mathcal{U}_q(\mathcal{B}^+))$ given by

$$H, \hat{H} \longrightarrow H , \qquad \tilde{X}^{\pm} \longrightarrow \tilde{X}^{\pm} ,$$

coincides with $\mathcal{U}_q(s\ell(2))$.

The canonical element (universal R-matrix) takes the form

$$\mathcal{R} = (C^{-1})_{kl,nm} h^k (x^+)^l \otimes H^n (\tilde{X}^+)^m ,$$

where the matrix C^{-1} transforms the basis $\{(x^+)^l h^k\}_{k,l=0}^{\infty}$ to the one dual to $\{(\widetilde{X}^+)^m H^n\}_{m,n=0}^{\infty}$

$$C^{kl,nm} = \langle\langle (x^+)^l h^k, (\widetilde{X}^+)^m H^n \rangle\rangle .$$

Explicit calculations with the use of (1.4.18), show that

$$C^{kl,nm} = \delta^{kn} \delta^{lm} \frac{k! [l; q^{-2}]!}{(q^{-1} - q)^l} ,$$

$$[l; q^{-2}]! := [1; q^{-2}][2; q^{-2}] \dots [l; q^{-2}] .$$

Here we have used another q-number $[x; q]$

$$[x; q] := \frac{q^x - 1}{q - 1} , \tag{1.4.20}$$

which is related to the former one (1.1.37) by the equality

$$[x; q] = q^{(x-1)/2} [x]_{\sqrt{q}} . \tag{1.4.21}$$

Thus the explicit expression for $\mathcal{R}$ is

$$\begin{aligned} \mathcal{R} = \; & = \sum_{kl} \frac{(q^{-1} - q)^l}{k! [l; q^{-2}]!} H^k (\tilde{X}^+)^l \otimes h^k (x^+)^l \\ & = e^{H \otimes h} \sum_l \frac{(q^{-1} - q)^l}{[l; q^{-2}]!} (X^+)^l \otimes (x^+)^l . \end{aligned}$$

The image of this element under the canonical homomorphism $D(\mathcal{U}_q(\mathcal{B}^+)) \longrightarrow \mathcal{U}_q(s\ell(2))$ is

$$\mathcal{R} = q^{-\frac{1}{2}H\otimes H} \exp_{q^{-2}}\left\{-(q-q^{-1})\widetilde{X}^+\otimes\widetilde{X}^-\right\} , \tag{1.4.22}$$

where

$$\exp_q x \equiv e_q^x := \sum_{n=0}^{\infty} \frac{x^n}{[n;q]!} \tag{1.4.23}$$

is the so-called q-deformed exponent.

Using the well known properties of Pauli matrices, one can check that they form the representation of $\mathcal{U}_q(s\ell(2))$ (as well as $s\ell(2)$) and the element $\mathcal{R}$ (1.4.22) in this representation is equivalent (equals up to rescaling) to the R-matrix (1.1.25) derived from quite a different point of view. In this representation the generators $\widetilde{X}^+, \widetilde{X}^-$ become nilpotent: $(\widetilde{X}^+)^2 = (\widetilde{X}^-)^2 = 0$ and the derivation of the R-matrix is very simple. We leave it to the reader.

1.5 Quantum matrix groups

Now we return to the relation(1.4.10) and in this Section shall consider quantum matrix groups in more details[104, 226]: present expressions for R-matrices of quantum analogs of classical simple groups of A_n, B_n, C_n and D_n series (according to Dynkin classification, see, e.g., [13, 250]) and give the explicit construction of all the Hopf algebra structure maps for them.

Let $\mathbb{C}[[T^i_j]]$ be an algebra freely generated by $T^i{}_j$ $(i, j = 1, ..., n)$. Consider a two-sided ideal I_R in $\mathbb{C}[[T^i_j]]$ generated by the relation (1.4.10), where $T = (T^i_j)^n_{i,j=1} \in M_n(\mathbb{C}[[T^i_j]])$ is an $n \times n$ matrix with elements belonging to $\mathbb{C}[[T^i_j]]$.

Definition 1.18 *A quotient algebra*

$$Fun_R = \mathbb{C}[[T^i_j]]/I_R \ ,$$

is called an algebra of functions on a quantum matrix group of rank n associated with a matrix R.

This means that two monomials in T^i_j are the same element of the algebra Fun_R if they can be transformed to each other by the use of relations (1.4.10). If $R = \mathbf{1}\otimes\mathbf{1}$ the algebra Fun_R coincides with commutative algebra of polynomial functions on $M_n(\mathbb{C})$.

The bialgebra structure maps are defined by the comultiplication

$$\Delta(T^i{}_j) = T^i{}_k\otimes T^k{}_j \ , \tag{1.5.1}$$

and counity

$$\varepsilon(T) = \mathbf{1} . \tag{1.5.2}$$

Corresponding quantum spaces, the example of which was considered in Section 1.1, are defined as follows. Let $\mathbb{C}[[x^i]]$ $(i = 1, ..., n)$ is an algebra freely generated by $x^1, \ldots, x^n$. Set $\hat{R} = PR$, where P is the permutation matrix in $\mathbb{C}^n \otimes \mathbb{C}^n$: $Pu \otimes v = v \otimes u$ $\forall u, v \in \mathbb{C}^n$ (so it has the matrix entries $P^{ij}_{kl} = \delta^i{}_l \delta^j{}_k$). Denote by $I_{f,R}$ the two-sided ideal in $\mathbb{C}[[x^i]]$, generated by

$$f(\hat{R})^{ij}{}_{kl} x^k \otimes x^l = 0 , \tag{1.5.3}$$

where $f(\hat{R})$ is some polynomial.

Definition 1.19 *A quotient algebra*

$$\mathbb{C}^n_{f,R} = \mathbb{C}[[x^1, \ldots, x^n]]/I_{f,R} ,$$

is called an algebra of functions on a quantum n-dimensional vector space , associated with a matrix R and a polynomial $f(\hat{R})$.

The analog of a group action in the classical case, after the quantization looks as follows

$$\delta : \mathbb{C}^n_{f,R} \to Fun_R(G) \otimes \mathbb{C}^n_{f,R} ,$$

and is specified by the formula

$$(x')^i := \delta(x^i) := T^i{}_j \otimes x^j , \qquad i = 1, \ldots, n , \tag{1.5.4}$$

or, in matrix notations,

$$x' := \delta(x) := T \otimes x .$$

The elements ("coordinates" of quantum space) $(x')^i$ obey the same CR (1.5.3). This means that the map δ is an algebra homomorphism and is called *the quantum group coaction.* The proof of this statement demonstrates the power and beauty of the R-matrix approach. Indeed, the CR (1.4.10) written in terms of $\hat{R}$-matrix is the following

$$\hat{R}(T \otimes T) = (T \otimes T)\hat{R} .$$

Hence,

$$f(\hat{R})(\delta(x) \otimes \delta(x)) = f(\hat{R})((T \otimes x) \otimes (T \otimes x))$$
$$= f(\hat{R})((T \otimes T) \otimes (x \otimes x)) = (T \otimes T) \otimes f(\hat{R})(x \otimes x) = 0 .$$

Notice that QYBE written in terms of $\hat{R}$ has the form

$$(\hat{R}\otimes\mathbf{1})(\mathbf{1}\otimes\hat{R})(\hat{R}\otimes\mathbf{1}) = (\mathbf{1}\otimes\hat{R})(\hat{R}\otimes\mathbf{1})(\mathbf{1}\otimes\hat{R}) \tag{1.5.5}$$

Moreover the coaction δ provides $\mathbb{C}^n_{f,R}$ with the left Fun_R-comodule structure. This means that the following properties are fulfilled

- $(id\otimes\delta)\delta = (\Delta\otimes id)\delta$,
- $(\varepsilon\otimes id)\delta = id$.

If $\hat{R} = P$ and $f(P) = P - 1$, the algebra $\mathbb{C}^n_{f,R}$ turns into a commutative algebra $\mathbb{C}[[x^i]]$ and the coaction becomes a usual matrix action.

In the preceding section we explained how the R-matrix can be derived from the quantum double construction. Actually, solution of QYBE associated with simple Lie algebras of all the classical types were constructed in [129, 130, 12] by a method motivated by the quantum inverse scattering method (QISM).

A detailed description of QISM and, in particular, the construction of R-matrices in its frame, lie outside the scope of the present book. Instead, we present the results of the construction following [104, 226].

1.5.1 Quantum groups $GL_q(n)$ and $SL_q(n)$

The R-matrices associated with A_{n-1} type Lie algebras ($n \geq 2$) have the form

$$\begin{aligned} R_q &= R^{ij}{}_{kl} e_i{}^k \otimes e_j{}^l \\ &= \sum_{i=1}^{n} e_i{}^i \otimes e_i{}^i + + \frac{1}{q} \sum_{\substack{i,j=1 \\ i\neq j}}^{n} e_i{}^i \otimes e_j^j + (1-q^{-2}) \sum_{\substack{i,j=1 \\ i>j}}^{n} e_i{}^j \otimes e_j{}^i \,, \end{aligned} \tag{1.5.6}$$

where the matrices $e_i{}^j$ have only one nonzero entry:

$$(e_i{}^j)^a{}_b = \delta_i{}^a \delta_b{}^j \,. \tag{1.5.7}$$

The R-matrix entries $R^{ij}{}_{kl}$ can be written in the form

$$R^{ij}{}_{kl} = \delta^i{}_k \delta^j{}_l (\delta^{ij} + (1-\delta^{ij})q^{-1}) + \delta^i{}_l \delta^j{}_k \Theta^{ij} (1-q^{-2}), \tag{1.5.8}$$

here

$$\Theta^{ij} = \begin{cases} 1 & i > j \,, \\ 0 & i \leq j \,. \end{cases} \tag{1.5.9}$$

Define the map

$$S(T_{ij}) = (-q)^{i-j} \widetilde{T}_{ij} \,, \tag{1.5.10}$$

where

$$\tilde{T}_{ij} = \sum_{\sigma \in S_{n-1}} (-q)^{l(\sigma)} T_{1\sigma_1} \dots T_{i-1\sigma_{i-1}} T_{i+1\sigma_{i+1}} \dots T_{n\sigma_n} , \quad i,j = 1, \dots n .$$

Here S_{n-1} is the symmetric group and $l(\sigma)$ is a "length" (minimal number of transpositions) of the substitution $\sigma = \sigma(1, \dots, j-1, j+1, \dots, n) = (\sigma_1, \dots, \sigma_{i-1}, \sigma_{n+1}, \dots, \sigma_n)$. One can recognize in $\tilde{T}_{ij}$ the deformation of classical matrix minors and the definition (1.5.10) is motivated by the relation

$$TS(T) = S(T)T = \det_q T \cdot \mathbf{1} , \tag{1.5.11}$$

where

$$\det_q T := \sum_{\sigma \in S_n} (-q)^{l(\sigma)} T_{1\sigma_1} \dots T_{n\sigma_n}$$

is called the *quantum determinant.* This is a central element, i.e., commutes with all the other algebra elements. The action of comultiplication has a form

$$\Delta(\det_q T) = \det_q T \otimes \det_q T .$$

Definition 1.20 *The algebra of functions $Fun_q(SL(n))$ on the quantum group $SL_q(n)$ is defined as the quotient algebra Fun_R modulo the relation* $\det_q T = 1$ *with comultiplication Δ (1.5.1), counity ε (1.5.2) and the antipode $S(T)$ (1.5.10).*

Note that though the antipode looks analogous to the matrix inverse, its square is not equal to identity operation

$$S^2(T) = \mathcal{D} T \mathcal{D}^{-1} , \tag{1.5.12}$$

where $\mathcal{D} := diag(1, q^2, \dots, q^{2(n-1)})$.

To define $GL_q(n)$, one must add the inverse element t for the quantum determinant

$$t \det_q T = (\det_q T) t = 1 .$$

Definition 1.21 *The algebra of functions $Fun_q(GL(n))$ on the quantum group $GL_q(n)$ is defined as the quotient algebra Fun_R modulo the relation $tT^i{}_j = T^i{}_j t$, with comultiplication Δ, counity ε and the antipode*

$$\tilde{S}(T_{ij}) := t(-q)^{i-j} \tilde{T}_{ij}, \; \tilde{S}(t) = \det_q T .$$

One can prove that the algebras $Fun_q(SL(n)$ and $Fun_q(GL(n))$ are Hopf algebras [104].

Before going further, we remark that for brevity we often use the term "quantum group" instead of mathematically more correct "an algebra of functions on a quantum group" and correspondingly the notation G_q instead of $Fun_q(G)$. More precisely, a quantum group G_q itself should be interpreted as a spectrum (set of all the representations) of the non-commutative algebra $Fun_q(G)$.

The matrix $\hat{R}_q := PR_q$, for the R_q (1.5.6), satisfies the *Hecke condition*

$$q^2\hat{R}_q^2 = q(q-q^{-1})\hat{R}_q + \mathbf{1} \,. \tag{1.5.13}$$

This permits to present $\hat{R}_q$ as a linear combination of the two projectors

$$\hat{R}_q = P_q^{(+)} - q^{-2}P_q^{(-)} \,, \tag{1.5.14}$$

where

$$P_q^{(\pm)} = \frac{\pm q\hat{R}_q + q^{\mp 1}\mathbf{1}}{q+q^{-1}} \,.$$

Using (1.5.13), one can easily check that $P_q^{(\pm)}$ are indeed projectors: $\left(P_q^{(\pm)}\right)^2 = P_q^{(\pm)}$. They are the quantum analogs of symmetrizer and antisymmetrizer. Now putting in the Definition 1.19 $f(\hat{R}) = \hat{R} - \mathbf{1} = -(1+q^{-2})P_q^{(-)}$, one obtains

Definition 1.22 *An algebra $\mathbb{C}_q^n$ with generators $x^1,\ldots,x^n$ and the relations*

$$x^ix^j = qx^jx^i \,, \qquad 1 \le i < j \le n \,, \tag{1.5.15}$$

is called an algebra of functions on a quantum n-dimensional vector space associated with the q-group $GL_q(N)$.

Algebras $SL_q(n)$ and $GL_q(n)$ act on $\mathbb{C}_q^N$ according to (1.5.4).

Putting $f(\hat{R}) = q\hat{R} + q^{-1} = (q+q^{-1})P_q^{(+)}$, one obtains the following definition.

Definition 1.23 *A finite dimensional algebra $\wedge\mathbb{C}_q^n$ with generators $\xi^1,\ldots,\xi^n$ and the relations*

$$(\xi^i)^2 = 0 \,, \qquad \xi^i\xi^j = -q^{-1}\xi^j\xi^i \,, \qquad 1 \le i < j \le n \,, \tag{1.5.16}$$

is called q-exterior algebra of the quantum vector space $\mathbb{C}_q^n$.

Obviously, the quantized ξ^i must be interpreted as the deformation of an algebra of usual differentials.

The use of quantum differentials gives a new interpretation and derivation for quantum matrix determinant, which are quite similar to that in the classical case. Indeed, the coaction of $GL_q(n)$ on the elements $\xi^1, \ldots, \xi^n \in \mathbb{C}^n_q$ leads to the formula [177, 178]

$$\delta(\xi^1 \ldots \xi^n) = \det{}_q T \otimes \xi^1 \ldots \xi^n \,. \tag{1.5.17}$$

As is well known, real forms of classical groups play an important role in physical applications. Let us consider real forms of the quantum group $SL_q(n)$. For this aim one needs the generalization of complex conjugation operation for non-commutative algebras. It is called a *$*$-involution (star-involution)* and in the case of Hopf algebras is defined as follows.

Definition 1.24 *A map $* : A \longrightarrow A$ which is an algebra anti-automorphism, coalgebra automorphism and satisfies the involution conditions*

$$\begin{aligned} (a^*)^* &= a \,, \\ S(S(a^*)^*) &= a \,, \ \forall a \in A \,, \end{aligned} \tag{1.5.18}$$

is called a $*$-involution of a Hopf algebra A.

The involution of defining CR (1.4.10) gives

$$\bar{R} T_2^* T_1^* = T_1^* T_2^* \bar{R} \,, \tag{1.5.19}$$

where $T^* := (T^*{}^i_j)^n_{i,j=1}$ and the bar denotes the usual complex conjugation. Thus CR of conjugate elements $T^*{}^i_j$ depend on the properties of R-matrix under the complex conjugation. For $SL_q(2)$ group there are two possibilities

1) $q \in \mathbb{C}$ *and* $|q| = 1$.

In this case $\bar{R}_q = R_{q^{-1}} = R_q^{-1}$. The relation (1.5.19) becomes

$$R_q T_1^* T_2^* = T_2^* T_q^* R_q \,,$$

and, hence, $T^* = UTU^{-1}$, where $U \in M_n(\mathbb{C})$ is a diagonal matrix that satisfies the condition $U^\dagger U = \mathbf{I}$. The properties (1.5.18) lead to the same condition on the matrix U. The map $T^i{}_j \longrightarrow \alpha_i \alpha_j T^i{}_j$, $\alpha_i \in \mathbb{C}$ is obviously an automorphism of the algebra $Fun_q(SL(n))$ and one can put $U = \mathbf{I}$, so that $T^* = T$.

Definition 1.25 *The algebra $Fun_q(SL(n))$ together with the $*$-involution*

$$T^i{}_j = T^*{}^i_j \,, \qquad i, j = 1, \ldots, n \,,$$

is called $Fun_q(SL(n, \mathbb{R}))$.

The latter is the deformation of classical real group $SL(n, \mathbb{R})$.

Definition 1.26 *The algebra $\mathbb{C}_q^n$ with $(x^*)^i = x^i$, is the algebra of functions on a quantum n-dimensional real Euclidean space and is denoted by $\mathbb{R}_q^n$.*

The map δ is compatible with the involution: $\delta^*(x) = \delta(x^*)$ and defines the coaction of $SL_q(n, \mathbb{R})$ on $\mathbb{R}_q^n$.

Now consider the second possibility.

2) $q \in \mathbb{R}$.

In this case $\bar{R}_q = R_q$ and corresponding $*$-involution has the form $T^* = US(T)^{\mathsf{T}} U^{-1}$, where diagonal matrix $U \in M_n(\mathbb{R}) \ : \ U^2 = 1$. To derive this one has to use the property $R^{\mathsf{T}} = PRP$. Using the same automorphism as in the first case, one can achieve that $U = diag(\varepsilon_1, \ldots, \varepsilon_n)$, $\varepsilon_i = \pm 1 \ \forall i$.

Definition 1.27 *The algebra $Fun_q(SL(n))$ together with $*$-involution*

$$T^*{}_j{}^i = \varepsilon_i \varepsilon_j S(t^i{}_j) \ ,$$

is called $Fun_q(SU(\varepsilon_1, \ldots, \varepsilon_n))$, the algebra of functions on quantum unitary group, where $\varepsilon_i = \pm 1 \ \forall i$. In particular, for $\varepsilon_1 = \ldots = \varepsilon_n = 1$ one obtains the deformation of compact form $Fun_q(SU(n))$ given by the relation $TT^{*\mathsf{T}} = T^{*\mathsf{T}} T = \mathbf{1}$.

1.5.2 Quantum groups $SO_q(N)$ and $Sp_q(n)$

The R-matrix for these quantum groups can be written in the unified form

$$\begin{aligned} R_q \ = \ & q \sum_{\substack{i=1 \\ i \neq i'}}^{N} e_i{}^i \otimes e_i{}^i \overset{odd}{+} e_{(N+1)/2}{}^{(N+1)/2} \otimes e_{(N+1)/2}{}^{(N+1)/2} + \sum_{\substack{i,j=1 \\ i \neq j,j'}}^{N} e_i{}^i \otimes e_j{}^j + \frac{1}{q} \sum_{\substack{i=1 \\ i \neq i'}}^{N} e_{i'}{}^{i'} \otimes e_i{}^i \\ & + \ (q - q^{-1}) \sum_{\substack{i,j=1 \\ i > j}}^{N} e_i{}^j \otimes e_j{}^i - (q - q^{-1}) \sum_{\substack{i,j=1 \\ i > j}}^{N} q^{\rho_i - \rho_j} \varepsilon_i \varepsilon_j e_i{}^j \otimes e_{i'}{}^{j'} \ , \end{aligned} \qquad (1.5.20)$$

where

- $N = 2n + 1$ for the series B_n (i.e., the groups $SO_q(2n+1)$);
- $N = 2n$ for groups of the series C_n (i.e., $Sp_q(n)$) and D_n ($SO_q(2n)$);
- $\varepsilon_i = 1 \ \forall i$ for the series B_n and D_n, $\varepsilon_i = 1$, $i = 1, \ldots, N/2$;
- $\varepsilon_i = -1$, $i = N/2 + 1, \ldots, N$ for the series $\mathbb{C}_n$;

$$(\rho_1,\ldots,\rho_N)=\begin{cases}(n-\frac{1}{2},n-\frac{3}{2},\ldots,\frac{1}{2},0,-\frac{1}{2},\ldots,-n+\frac{1}{2}) & \text{for } B_n \text{ series}\\ (n,n-1,\ldots,1,-1,\ldots,-n) & \text{for } C_n \text{ series}\\ (n-1,n-2,\ldots,1,0,0,-1,\ldots,-n+1) & \text{for } D_n \text{ series.}\end{cases}$$

Primed indices are defined as follows

$$i' := N+1-i\ ,$$

and the matrices $e_i{}^j$ are defined as in the linear group case (1.5.7).

This R-matrix satisfies the relations

$$R_q = C_1(R_q^{\mathsf{T}_1})^{-1}C_1^{-1} = C_2(R_q^{-1})^{\mathsf{T}_2}C_2^{-1}\ ,$$

where for any matrix K acting in $\mathbb{C}^N\otimes\mathbb{C}^N$, one defines partial transpositions

$$(K^{mn}{}_{kl})^{\mathsf{T}_1} := K^{kn}{}_{ml}\ ,$$

$$(K^{mn}{}_{kl})^{\mathsf{T}_2} := K^{ml}{}_{kn}\ ,$$

$C_1 := C\otimes\mathbf{I}$, $C_2 := \mathbf{1}\otimes C$ and $C := q^\rho C_0$ with $\rho = diag(\rho_1,\ldots,\rho_N)$, $(C_0)^i{}_j = \varepsilon_i\delta_j^{i'}$. Thus, $C^2=\varepsilon\mathbf{I}$, $\varepsilon=1$ for types B_n, D_n and $\varepsilon=-1$ for type C_n.

Now the basic CR (1.4.10) can be rewritten in the form

$$C_1T_1^{\mathsf{T}}C_1^{-1}R_q^{-1}T_2 = T_2R_q^{-1}C_1T_1^{\mathsf{T}}C_1^{-1}\ .$$

Multiplying this equality by T_1 from the left and from the right and comparing with (1.4.10) in the form

$$R_q^{-1}T_2T_1 = T_1T_2R_q^{-1}\ ,$$

one arrives at the conclusion that it is natural to introduce the additional relations

$$TCT^{\mathsf{T}}C^{-1} = CT^{\mathsf{T}}C^{-1}T = 1\ , \qquad (1.5.21)$$

which is the analog (deformation) of the condition for invariance of the metric C.

Definition 1.28 *The quotient algebra of the algebra Fun_R by the relations (1.5.21) is called the algebra $Fun_q(G)$ of functions either on the quantum group $G_q = SO_q(N)$ if R_q corresponds to the types B_n, D_n, or on a group $G_q = Sp_q(n)$ if R_q corresponds to the type C_n.*

These algebras are Hopf algebras with the standard comultiplication Δ, counit ε and the antipode which follows from (1.5.21)

$$S(T) = CT^{\top}C^{-1} . \tag{1.5.22}$$

It has the property

$$S^2(T) = \mathcal{D}T\mathcal{D}^{-1} , \quad \mathcal{D} = (CC^{\top}) . \tag{1.5.23}$$

The matrix $\hat{R}_q$ in this case satisfies the equality

$$(\hat{R}_q - q\mathbf{1})(\hat{R}_q + q^{-1}\mathbf{1})(\hat{R}_q - \varepsilon q^{\varepsilon - N}\mathbf{1}) = 0 ,$$

which permits to decompose $\hat{R}_q$ in the sum of three projectors

$$\hat{R}_q = qP_q^{(+)} - q^{-1}P_q^{(-)} + \varepsilon q^{\varepsilon-N}P_q^{(0)} , \tag{1.5.24}$$

$$P_q^{(+)} = \frac{\hat{R}_q^2 - (\varepsilon q^{\varepsilon-N} - q^{-1})\hat{R}_q - \varepsilon q^{\varepsilon-N-1}\mathbf{1}}{(q+q^{-1})(q-\varepsilon q^{\varepsilon-N})} ,$$

$$P_q^{(-)} = \frac{\hat{R}_q^2 - (q + \varepsilon q^{\varepsilon-N})\hat{R}_q - \varepsilon q^{\varepsilon-N+1}\mathbf{1}}{(q+q^{-1})(q^{-1}+\varepsilon q^{\varepsilon-N})} ,$$

$$P_q^{(0)} = \frac{\hat{R}_q^2 - (q - q^{-1})\hat{R}_q - \mathbf{1}}{(\varepsilon q^{\varepsilon-N} - q)(q^{-1} - \varepsilon q^{\varepsilon-N})} .$$

These are projectors of the ranks

$$\frac{N(N+1)}{2} - 1 , \quad \frac{N(N-1)}{2} , \quad 1 \text{ for } B_n \text{ and } D_n \text{ series}$$

and of the ranks

$$\frac{N(N+1)}{2} , \quad \frac{N(N-1)}{2} - 1 , \quad 1 \text{ for } C_n \text{ series}$$

The choice of the function $f(\hat{R})$ in the CR (1.5.3) in the form $f(\hat{R}_q) = (q+q^{-1})P_q^{(-)}$ leads to the notion of a quantum Euclidean space.

Definition 1.29 *An algebra $O_q^N(\mathbb{C})$ with generators $\{x^i\}_{i=1}^N$ and the CR*

$$\begin{aligned} x^i x^j &= q x^j x^i , \qquad 1 \le i < j \le N , \ i \neq j' , \\ x^{i'} x^i &= x^i x^{i'} + (q^2 - 1) \sum_{j=1}^{i'-1} q^{\rho_{i'} - \rho_j} x^j x^{j'} \\ &\quad - \frac{q^2 - 1}{1 - q^{N-2}} q^{\rho_{i'}} \sum_{kl} x^k C_{kl} x^l , \qquad 1 \le i < i' \le N , \end{aligned} \tag{1.5.25}$$

is called the algebra of functions on a quantum N-dimensional complex Euclidean space.

The number of independent monomials of given order for the algebra $O_q^N(\mathbb{C})$ is the same as for the commutative algebra $\mathbb{C}[[x^1,\ldots,x^N]]$.

The quantum analog of the *Euclidean length*, the element

$$l_q := x^i C_{ij} x^j = \sum_{j=1}^{N} q^{-\rho_j} x^j x^{j'} \tag{1.5.26}$$

is a central element and invariant with respect to the quantum group coaction

$$\delta(l_q) = 1 \otimes l_q \,.$$

Now it is clear why the metric C_{ij} was chosen in the antidiagonal form even for orthogonal groups: it is obvious that in the case of diagonal metric the q-length l_q could not be central. Of course, in the classical limit the metric can be diagonalized and $O_q^N(\mathbb{C})$ are deformations of usual Euclidean spaces but in bases with antidiagonal metrics.

The CR for $x^i \in O_q^N(\mathbb{C})$ are more complicated than for the quantum spaces $\mathbb{C}_q^N$. In the simplest non-trivial case $N = 3$, they take the form

$$xy = qyx, \quad yz = qzy, \quad xz - zx = (q^{-1/2} - q^{1/2})y^2, \tag{1.5.27}$$

where we denote as usual $x := x^1$, $y := x^2$, $z := x^3$. The metric C of this q-space is

$$C = \begin{pmatrix} 0 & 0 & q^{1/2} \\ 0 & 1 & 0 \\ q^{-1/2} & 0 & 0 \end{pmatrix}$$

and the Euclidean length takes the form

$$l_q^{(3)} = x^\top C c = q^{1/2} xz + y^2 + q^{-1/2} zx \,.$$

The function $f(\hat{R}_q) = (q + q^{-1}) P_q^{(+)}$ gives the

Definition 1.30 *The algebra $\wedge O_q^N$, $q^2 \neq -1$ with generators $\xi^1,\ldots,\xi^N$ and the relations*

$$\begin{aligned}
(\xi^i)^2 &= 0\,, \quad i = 1,\ldots,N; \quad i \neq i'; \\
\xi^i \xi^j &= -q^{-1} \xi^j \xi^i\,, \quad 1 \leq i < j \leq N\,, \quad i \neq j'; \\
\xi^{i'} \xi^i &= -q^2 \xi^i \xi^{i'} + (q^2 - 1) \sum_{j=1}^{i'-1} q^{\rho_{i'} - \rho_j} \xi^j \xi^{j'} \\
&\quad - \frac{q^2 - 1}{1 - q^N} q^{\rho_{i'}} \xi^k C_{kl} \xi^l\,, \quad 1 \leq i < i' \leq N\,,
\end{aligned}$$

is called the q-exterior algebra of the quantum space $O_q^N(\mathbb{C})$.

In the case $N = 3$ this CR becomes

$$(\xi_x)^2 = (\xi_z)^2 = 0\ , \quad \xi_x\xi_y = -q^{-1}\xi_y\xi_x\ , \quad \xi_y\xi_z = -q^{-1}\xi_z\xi_y\ ,$$

$$\xi_x\xi_z = -q^2\xi_z\xi_x + q^{-1}(q^{1/2} + q^{-1/2})\xi_y^2\ .$$

One can also define the quantum analog of symplectic spaces and the q-exterior algebras [104].

Let us describe the real forms of $SO_q(N)$ and $Sp_q(n)$.

1) Case $\mid q \mid = 1$.

The $*$-involution of the form $T^* = T$ leads to q-groups

$$SO_q(n,n)\ , \ SO_q(n,n+1) \ \text{and} \ Sp_q(n,\mathbb{R})\ .$$

Note that a quantum Lorentz group (the deformation of $SO(1,3)$) does not appear in this way. Fortunately, relativistic quantum mechanics deals not with the group $SO(1,3)$ itself but with its universal covering group $SL(2,\mathbb{C})$ and the deformation of the latter has been described already. q-Lorentz (and other quantum space-time groups) will be considered in the Chapter 3.

2) Case $q \in \mathbb{R}$.

The $*$-involution has the form $T^* = US(T)^{\top}U^{-1}$, where $U = diag(\varepsilon_1, \ldots, \varepsilon_N), \varepsilon_i^2 = 1,\ \varepsilon_i = \varepsilon_{i'},\ i = 1, \ldots, N$ and $\varepsilon_i = 1$ if $i = i'$. For $U = \mathbf{I}$ one obtains the compact form $SO_q(N,\mathbb{R})$. The involution $x_i^* := (x^i)^* = (C^{\top})_{ij}x^j$ turns $O_q^N(\mathbb{C})$ into $O_q^N(\mathbb{R})$ and the coaction of $SO_q(N,\mathbb{R})$ preserves the q-length $l_q = x^iC_{ij}x^j = x_i^*x^i\ :\ \delta(l_q) = 1\otimes l_q$.

The quotient algebra of $O_q^N(\mathbb{R})$ by the relation $x_i^*x^i = 1$ is called a quantum $(N-1)$dimensional sphere S_q^{N-1} [190],[223].

1.5.3 Twists of quantum groups and multiparametric deformations

Up to now we have considered one-parametric deformations only. It has been shown [198, 90] that multiparametric quantum groups can be obtained from the one-parametric q-groups described above via the so-called twists of the Hopf algebras.

Let A be a quasi-triangular Hopf algebra and assume that there exists an element $F = \sum_i f^i \otimes f_i \in A\otimes A$, such that

$$\begin{aligned} (\Delta\otimes\mathrm{id})F &= F_{13}F_{23}\ , & (\mathrm{id}\otimes\Delta)F &= F_{13}F_{12}\ , \\ F_{12}F_{13}F_{23} &= F_{23}F_{13}F_{12}\ , & F_{12}F_{21} &= 1\ , \end{aligned} \tag{1.5.28}$$

where $F_{21} := \sum_i f_i \otimes f^i$.

Define

$$\Delta^{(F)}(a) := F\Delta(a)F^{-1}\,, \quad \mathcal{R}^{(F)} := F^{-1}\mathcal{R}F^{-1}\,, \quad u := f^i S(f_i)\,. \qquad (1.5.29)$$

Then $(A, \Delta^{(F)}, R^{(F)})$ is a quasi-triangular Hopf algebra with the antipode

$$S^{(F)} = uS(a)u^{-1}$$

and counity

$$\varepsilon^{(F)}(a) = \varepsilon(a)\,.$$

One can check that in the case of QUEA $\mathcal{U}_q(L)$ the element

$$F = \exp\left\{\sum_{i<j}(H_i \otimes H_j - H_j \otimes H_i)\,\phi_{ij}\right\}\,, \qquad (1.5.30)$$

where H_i are the Cartan generators and $\phi_{ij} \in \mathbb{C}$ are new parameters, satisfies the relation (1.5.28).

The comultiplication in twisted QUEA $(\mathcal{U}_q(L), \Delta^{(F)}, R^{(F)})$ has a form (cf. (1.3.5))

$$\Delta^{(F)}(H_i) = H_i \otimes 1 + 1 \otimes H_i\,,$$

$$\begin{aligned}\Delta^{(F)}(X_i^{\pm}) &= X_i^{\pm} \otimes \exp\left(-\chi\frac{(\alpha_i, \alpha_i)H_i}{2} + \sum_{k,j} a_{ik}\phi_{kj}H_j\right) \\ &\quad + \exp\left(\chi\frac{(\alpha_i, \alpha_i)H_i}{2} + \sum_{k,j} a_{ik}\phi_{kj}H_j\right) \otimes X_i^{\pm}\,,\end{aligned}$$

(a_{ij} is the Cartan matrix).

Note that the multiplication map in $\mathcal{U}_q^{(F)}(L)$ remains the same after twists. This means that the dual Hopf algebra $Fun_q^{(F)}(G)$ has the same comultiplication as its untwisted counterpart but it has twisted multiplication. This corresponds to the fact that a multiplication in an algebra of functions on a quantum group is defined by R-matrix, and the latter is modified under the twist procedure according to (1.5.29).

In the case of $GL_q(n)$, a twist of R-matrix in the fundamental representation R is described by a diagonal matrix $F = diag(f_{11}, f_{12}, ..., f_{nn})$ with $f_{ij}f_{ji} = 1$, so that R-matrix $R^{(F)}$ of the twisted group $GL_{r,q_{ij}}(N)$ has the form

$$R^{(F)} = F^{-1}RF^{-1}\,, \qquad (1.5.31)$$

and starting from (1.5.8) one obtains the multiparametric R-matrix

$$R^{(F)ij}{}_{kl} := (R_{r,q_{mn}})^{ij}{}_{kl} = \delta^i{}_k\delta^j{}_l(\delta^{ij} + \Theta^{ji}\frac{1}{q_{ij}} + \Theta^{ij}\frac{q_{ji}}{r}) + \delta^i{}_l\delta^j{}_k\Theta^{ij}(1 - r^{-1}), \tag{1.5.32}$$

where

$$q_{ij} := \sqrt{r}\exp(\phi_{ij} - \phi_{i,j-1} - \phi_{i-1,j} + \phi_{i-1,j-1}) \ , \qquad i < j$$

and r is the initial (former q^2) parameter of deformation. The one-parametric R-matrix R_q is recovered by putting $\phi_{ij} = 0$, so that $q_{ij} = \sqrt{r} := q$. The CR for $\mathbb{C}^N_q[[x^i]]$ and $\wedge\mathbb{C}^N_q[[\xi^i]]$ become now [209]

$$x^i x^j = q_{ij} x^j x^i \ , \qquad i < j \ , \tag{1.5.33}$$

$$\xi^i\xi^j = -\frac{q_{ij}}{r}\xi^j\xi^i \ , \qquad i < j \ . \tag{1.5.34}$$

Thus a group $GL_{r,q_{ij}}(n)$ has $\frac{n(n-1)}{2} + 1$ free parameters of deformation. In general, as is seen from (1.5.30), there are $\frac{k(k-1)}{2}$ twist parameters, k being the rank of the group, plus one initial parameter of deformation. So a multiparametric q-group has totally $\frac{k(k-1)}{2} + 1$ parameters of deformation. Notice, that to construct quantized $SL_{r,q_{ij}}(n)$ starting from $GL_{r,q_{ij}}(n)$ one must require the centrality of q-determinant, defined by (1.5.17),(1.5.34)

$$[T^i{}_k, \det_q T] = 0, \qquad \forall i, k \ .$$

This leads to the conditions [209]

$$\frac{\prod_{\alpha=1}^{k-1} q_{\alpha k} \prod_{\gamma=k+1}^{n} (r^2/q_{k\gamma})}{\prod_{\beta=1}^{i-1} q_{\beta i} \prod_{\delta=i+1}^{n} (r^2/q_{i\delta})} = 1 \ , \qquad \forall k, i = 1, \ldots, n \ , \tag{1.5.35}$$

which results in $(n-1)$ independent conditions for q_{ij} and

$$\frac{(n-1)(n-2)}{2} + 1$$

independent parameters for $SL_{r,q_{ij}}(n)$ (this corresponds to the general formula: rank of $SL(n)$ is $(n-1)$).

The multiparametric R-matrices for B_n, C_n and D_n series of q-groups have the form (cf. (1.5.20) [210]

$$R_{r,q} \;=\; r\sum_{\substack{i=1 \\ i\neq i'}}^{N} e_i{}^i \otimes e_i{}^i + e_{(N+1)/2}{}^{(N+1)/2} \otimes e_{(N+1)/2}{}^{(N+1)/2}$$

$$+ \sum_{\substack{i<j=1 \\ i \neq j'}}^{N} \frac{r}{q_{ij}} e_i{}^i \otimes e_j{}^j + \sum_{\substack{i>j=1 \\ i \neq j'}}^{N} \frac{q_{ji}}{r} e_i{}^i \otimes e_j{}^j$$

$$+ \frac{1}{r} \sum_{\substack{i=1 \\ i \neq i'}}^{N} e_{i'}{}^{i'} \otimes e_i{}^i + (r - r^{-1}) \sum_{\substack{i,j=1 \\ i>j}}^{N} e_i{}^j \otimes e_j{}^i$$

$$- (r - r^{-1}) \sum_{\substack{i,j=1 \\ i>j}}^{N} r^{\rho_i - \rho_j} \varepsilon_i \varepsilon_j e_i{}^j \otimes e_{j'}{}^{i'} . \qquad (1.5.36)$$

Note that the non-diagonal part of the R-matrices is not changed under the twists.

The CR of an algebra of functions on multiparametric Euclidian space and the corresponding exterior products algebra are defined as in Definition 1.29 and Definition 1.30 but with multiparametric $\hat{R}$-matrices. We display the simplest non-trivial example of the group $SO_{r,q}(4)$ only [210]

$$\mathbf{O}^4_{\mathbf{r},\mathbf{q}} \; : \; x_1x_2 = qx_2x_1 \; , \quad x_1x_3 = \frac{r^2}{q} x_3x_1 \; , \quad x_2x_3 = x_3x_2 \; ,$$

$$x_2x_4 = qx_4x_2 \; , \quad x_3x_4 = \frac{r^2}{q} x_4x_3 \; , \quad x_1x_4 = x_4x_1 - (r - \frac{1}{r}) x_2x_3 \; .$$

$$\wedge \mathbf{O}^4_{\mathbf{r},\mathbf{q}} \; : \; \xi_1\xi_2 + \frac{q}{r^2}\xi_2\xi_1 = 0 \; , \quad \xi_1\xi_3 + \frac{1}{q}\xi_3\xi_1 = 0 \; , \quad \xi_2\xi_3 + \xi_3\xi_2 = (r - \frac{1}{r})\xi_1\xi_4 \; ,$$

$$\xi_2\xi_4 + \frac{q}{r^2}\xi_4\xi_2 = 0 \; , \quad \xi_3\xi_4 + \frac{1}{q}\xi_4\xi_3 = 0 \; , \quad \xi_1\xi_4 + \xi_4\xi_1 = 0 \; .$$

The metric has the form

$$C = \begin{pmatrix} 0 & 0 & 0 & r^{-1} \\ 0 & 0 & 1 & 0 \\ 0 & 1 & 0 & 0 \\ r & 0 & 0 & 0 \end{pmatrix}$$

and the central length is

$$l_r = r^{-1}x_1x_4 + x_2x_3 + x_3x_2 + rx_4x_1 = (r + \frac{1}{r})(x_1x_4 + rx_2x_3).$$

Now we will show that R-matrices of all simple quantum groups have properties which permit to present the quantum group twists as transitions to

other coordinate frames on quantum spaces. This implies physical equivalence of field theories invariant with respect to q-groups (considered as q-deformed space-time groups of transformations) connected with each other by the twists. In other words, we will show that the known R-matrices for all simple q-groups have the property which permits to describe the twist procedure as a transformation of q-space coordinates [74].

Consider at first the case of q-deformations of $GL(N)$ groups. In general, R in the rhs of (1.5.31) is also a multiparametric R-matrix of the initial group $GL_{r,q_{ij}}(N)$ and parameters of twisted groups $GL_{r,\tilde{q}_{ij}}(N)$ are expressed through the initial ones as follows

$$\tilde{q}_{ij} = q_{ij} f_{ij}^2 \ . \tag{1.5.37}$$

Coordinates of the initial quantum space $\mathbb{C}_q^n$ satisfy the CR

$$x^i x^j = q_{ij} x^j x^i \tag{1.5.38}$$

and coordinates of the twisted space $\mathbb{C}_q^{(F)n}$ have the CR

$$\tilde{x}^i \tilde{x}^j = \tilde{q}_{ij} \tilde{x}^j \tilde{x}^i \ . \tag{1.5.39}$$

Now we introduce the algebra $E_q^n[[e^i, g_j]]$ with the generators $\{e^i, g_i\}_{i=1}^n$ which commute with coordinates and put

$$\tilde{x}^i = e^i x^i \qquad \text{(no summation)} \ . \tag{1.5.40}$$

The elements e^i play the role of components of a *q-deformed (diagonal) n-bein.* CR for them follows from (1.5.37)-(1.5.40)

$$e^i e^j = f_{ij}^2 e^j e^i \ , \tag{1.5.41}$$

and g_i are the inverse elements

$$g_i e^i = 1 \ . \tag{1.5.42}$$

The coordinates $\tilde{x}^i$ are transformed by a q-matrix $\tilde{T}$

$$\tilde{x}'^{\,i} = \sum_{j=1}^{N} \tilde{T}^i{}_j \tilde{x}^j \ . \tag{1.5.43}$$

Then using (1.5.40),(1.5.42), one obtains from (1.5.43) the transformations of the coordinates x^i

$$x'^{\,i} = \sum_{j=1}^{N} g_i \otimes \tilde{T}^i{}_j \otimes e^j \otimes x^j \ . \tag{1.5.44}$$

We used in (1.5.44) a cross-product sign to stress that the elements from the different sets commute with each other (the elements g_i in (1.5.44) must be considered as inverse elements to generators e^i of another copy of an algebra E_q^n with respect to the elements e^i entering the same formula). The relation (1.5.44) implies that the coordinates x^i are transformed by the matrix T with the entries

$$T^i{}_j = g_i \otimes \tilde{T}^i{}_j \otimes e^j \qquad \text{(no summation)} . \tag{1.5.45}$$

Using (1.5.42), one can express the matrix $\tilde{T}$ through T

$$\tilde{T}^i{}_j = e^i \otimes T^i{}_j \otimes g_j \qquad \text{(no summation)} . \tag{1.5.46}$$

One can check straightforwardly that x'^i defined by (1.5.44) satisfy the correct CR

$$x'^i x'^j = q_{ij} x'^j x'^i .$$

The general reason for this is the following property of the R-matrices: if a q-matrix T satisfies the TT-relation defined by the corresponding R-matrix, then the $\tilde{T}$-matrix defined by (1.5.45) or (1.5.46) satisfies the relation with twisted R-matrix $R^{(F)}$.

To prove this statement, let us use the TT-relation (CR for entries of a matrix T) in explicit form (1.1.24) and substitute $T^i{}_j$ by their expressions (1.5.45) in terms of $\tilde{T}^i{}_j$. This gives the relation for the latter

$$\sum_{p,s} R^{mn}_{ps} g_p g_s e^u e^v \tilde{T}^p{}_u \tilde{T}^s{}_v = \sum_{s,r} \tilde{T}^n{}_s \tilde{T}^m{}_r g_n g_m e^s e^r R^{rs}_{uv} . \tag{1.5.47}$$

Recall that in this relation the elements g_i must be considered as inverse elements for the generators e^i of another copy of the algebra E_q^N and thus they commute with the elements e^i entering the same relation.

The multiparametric R-matrix for $GL_{r,q_{ij}}(N)$ group has the form

$$R^{mn}_{ps} = B^{mn}_{ps} + N^{mn}_{ps} , \tag{1.5.48}$$

where B is the diagonal matrix

$$B^{mn}_{ps} = \delta^m{}_p \delta^n{}_s (\delta^{mn} + \Theta^{nm} q^{-1}_{mn} + \Theta^{mn} q_{nm} r^{-1}) , \tag{1.5.49}$$

with $\Theta^{mn} = 1$ if $m > n$, and $\Theta^{mn} = 0$ if $m \leq n$, and the matrix N is off-diagonal part of the R-matrix

$$N^{mn}_{ps} = \delta^m{}_s \delta^n{}_p \Theta^{mn} (1 - r^{-1}) . \tag{1.5.50}$$

Using these expressions, one easily obtains

$$B^{mn}_{ps} g_p g_s = g_m g_n B^{mn}_{ps} = f^{-2}_{mn} g_n g_m B^{mn}_{ps} = g_n g_m B^{(F)mn}_{ps} \; , \tag{1.5.51}$$

$$N^{mn}_{ps} g_p g_s = g_n g_m N^{mn}_{ps} \; , \tag{1.5.52}$$

so that

$$R^{mn}_{ps} g_p g_s e^u e^v = g_n g_m e^u e^v R^{(F)mn}_{ps} \; ,$$

where $R^{(F)}, B^{(F)}$ are the twisted matrices of the same form (1.5.48)-(1.5.50) but for the twisted parameters $\tilde{q}_{ij} = q_{ij} f^2_{ij}$. An analogous consideration for the rhs of (1.5.47) shows that this relation can be rewritten in the form

$$\sum_{p,s} R^{(F)mn}_{ps} \tilde{T}^p_u \tilde{T}^s_v = \sum_{s,r} \tilde{T}^n_s \tilde{T}^m_r R^{(F)rs}_{uv} \; . \tag{1.5.53}$$

- Thus the twisted q-matrices can be constructed with the help of q-deformed n-beins (1.5.41),(1.5.42) and the formula (1.5.45) which is the direct generalization (q-deformation) of the relation between matrices of transformations in different coordinate frames.

In the case of q-deformation of simple groups of the series B_N, C_N, D_N there is one more structure, namely an invariant length (1.5.26). To preserve l_q, the components of the q-bein must satisfy the additional constraints

$$e^i e^{i'} = e^{i'} e^i = 1 \; , \tag{1.5.54}$$

$i = 1, ..., N/2$ for C_N, D_N series; $i = 1, ..., (N+1)/2$ for B_N series. In particular, for the series B_N

$$e^{(N+1)/2} = 1 \; . \tag{1.5.55}$$

These constraints reduce the number of twist parameters, which from the geometrical point of view define CR for the components of the q-beins, so that the number is equal to $k(k-1)/2$, where k is rank of the group. Multiparametric R-matrices for the B_N, C_N, D_N series have the form (cf. (1.5.20),(1.5.31),(1.5.36))

$$\begin{aligned} R^{ij}_{kl} &= \Big[\delta^i_k \delta^j_l \Big(r\delta^{ij}(1-\delta^{ii'}) + (\Theta^{ji} r q^{-1}_{ij} + \Theta^{ij} q_{ji} r^{-1})(1-\delta^{ii'})\Big) \\ &+ (r - r^{-1})\delta^i_l \delta^j_k \Theta^{ij}\Big] \\ &+ \Big[\frac{1}{r}\delta^i_k \delta^j_l \delta^{ji'}(1-\delta^{ii'}) - \Theta^{ij}(r - r^{-1}) r^{(\rho_i - \rho_j)} \epsilon_i \epsilon_j \delta^{ij'} \delta_{kl'} \\ &\overset{\text{odd}}{+} \delta^i_{(N+1)/2} \delta^j_{(N+1)/2} \delta^{(N+1)/2}_k \delta^{(N+1)/2}_l\Big] \; , \end{aligned}$$

(the last term exists for the B_N series only). Using this explicit form, one can easily show that the R-matrices have the property analogous to that of the A_N groups. Indeed, the terms in the first square brackets have the structure similar to those of the R-matrix for the $GL_{r,q_{ij}}(N)$ groups. So they are transformed properly when the elements e_i, g_j move through them (cf., (1.5.51),(1.5.52)). The terms in the second square brackets are not changed because of Kronecker symbols $\delta^{ji'}$, $\delta^{ij'}$ or $\delta^{(N+1)/2}{}_k$, $\delta^i{}_{(N+1)/2}$ and the relations (1.5.54),(1.5.55).

Thus again the matrices $\tilde{T}$ defined by (1.5.45),(1.5.46) satisfy the CR (1.5.53) for twisted quantum groups.

For further details and description of real forms of multiparametric q-groups, see e.g. [198, 90, 211, 209, 210, 97, 116] and Chapter 3 of this book, where we apply multiparametric q-groups and q-spaces for the description of space-time and its symmetries.

In conclusion of this section we mention that while a certain group G' is a subgroup of a group G, this is not the case, in general, for their q-deformations: the conditions which pick up the subgroup can contradict the algebra or coalgebra relations in $Fun_q(G)$. The reader will find concrete examples of this phenomenon in Sections 2.2 and 3.4.1.

1.6 Quantum deformation of differential and integral calculi

In non-deformed classical case, the Lie algebra generators, being left or right invariant vector fields on a group manifold, act on functions on the group as differential operators. We used this action in Section 1.2 to establish the duality between UEA $\mathcal{U}(L)$ and the set $Fun(G)$. It is this relation that survives after the quantum deformation (contrary to the exponential map). The R-matrix formalism for quasi-triangular quantum algebras permits to clarify further this duality and to give explicit formulas for the action of elements of QUEA on $Fun_q(G)$ and thus to develop differential geometry on q-groups. Bicovariant differential calculus on quantum groups was introduced by Woronowicz [247]. The formalism which provides a connection between the calculus of Woronowicz and R-matrix approach of Faddeev,Reshetikhin and Takhtajan [104] was developed in [131, 255].

Another important ingredient of the group theory applications is a group invariant calculus on homogeneous spaces (on which the group acts as a group of transformations). As we described in the preceding section, the projector decomposition of $\hat{R}$-matrices permits to introduce, besides the algebras of functions on quantum spaces, the corresponding quantum exterior product algebras. Generators of the latter play the role of quantum deformed differentials. Of course, such an interpretation is meaningful because the complete q-group invariant differential calculus can be defined on q-spaces [239].

An integral calculus on quantum groups and q-spaces is not yet developed in full completeness. Nevertheless we shall describe some approaches to its construction following [245, 255, 140].

1.6.1 Differential calculus on q-groups

A basis for QUEA $\mathcal{U}_q(L)$ can be chosen so that being packed in two matrices L^+ and L^-, it allows to write $\mathcal{U}_q(L) \leftrightarrow Fun_q(G)$ pairing in a compact and explicit form [104]

$$\begin{aligned} \langle\langle L_0^+, T_1T_2\dots T_n\rangle\rangle &= \tilde{R}_{10}\tilde{R}_{20}\dots\tilde{R}_{n0}\ , \\ \langle\langle L_0^-, T_1T_2\dots T_n\rangle\rangle &= (\tilde{R}_{01})^{-1}(\tilde{R}_{02})^{-1}\dots(\tilde{R}_{0n})^{-1}\ , \end{aligned} \tag{1.6.1}$$

where

$$T_i = \mathbf{1}\otimes\dots\otimes\underbrace{T}_{i}\otimes\dots\otimes\mathbf{1}\ ,$$

$\tilde{R}_{ij} := P\tilde{R}_{ji}P$; the matrices $\tilde{R}_{ij}$ $(i,j = 0,\dots,n)$ act non-trivially only on the factors with the indices i and j in the tensor product $\mathbb{C}^n\otimes\dots\otimes\mathbb{C}^n$ ($n+1$ times) and coincide there with the matrix $\tilde{R}$, which is a suitably normalized R-matrix; for $SL(n)_q$ it takes the form $\tilde{R} := q^{(n-1)/n}R$ (normalization is chosen so that $L^\pm$ have the q-determinants equal to one).

For example, in the case of $SL_q(2)$ the matrices $L^\pm$ take the form

$$L^+ = \begin{pmatrix} q^{-H/2} & \lambda X^+ \\ 0 & q^{H/2} \end{pmatrix}\ , \tag{1.6.2}$$

$$L^- = \begin{pmatrix} q^{H/2} & 0 \\ -\lambda X^- & -q^{-H/2} \end{pmatrix}\ , \tag{1.6.3}$$

where

$$\lambda := q - q^{-1}\ .$$

A nice property of these matrices is that the CR in $\mathcal{U}_q(L)$ take the form very close to (1.4.10)

$$\begin{aligned} R_{12}L_2^\pm L_1^\pm &= L_1^\pm L_2^\pm R_{12}\ , \\ R_{12}L_2^+ L_1^- &= L_1^- L_2^+ R_{12}\ , \end{aligned} \tag{1.6.4}$$

and comultiplication close to the form (1.5.1)

$$\Delta((L^{\pm})^i{}_j) = (L^{\pm})^i{}_k \otimes (L^{\pm})^k{}_j \,. \tag{1.6.5}$$

Thus this basis makes the duality between $Fun_q(G)$ and $\mathcal{U}_q(L)$ explicit. If S_q is an antipode for $Fun_q(G)$, where G is a group of some classical type, then the antipode for $\mathcal{U}_q(L)$ in the basis $L^{\pm}$ has the form

$$S(L^{\pm}) = S_{q^{-1}}(L^{\pm}) \,.$$

Stating rigorously, the algebra generated by the entries $(L^{\pm})^i{}_j$ becomes isomorphic to $\mathcal{U}_q(L)$ after a certain completion. We shall skip these mathematical details, and istead refer the reader to [104].

The action of $\mathcal{U}_q(L)$ on $Fun_q(G)$ (the analog of vector fields action on functions on a group) is defined by the formula

$$l(t) := \langle\langle l \otimes \mathrm{id}, \Delta(t) \rangle\rangle \ , \quad l \in \mathcal{U}_q(L), \ t \in Fun_q(G) \,,$$

from which the following CR can be derived [104],[255]

$$\begin{aligned} L_1^+ T_2 &= T_2 \tilde{R}_{21} L_1^+ \,, \\ L_1^- T_2 &= T_2 (\tilde{R}_{12})^{-1} L_1^- . \end{aligned} \tag{1.6.6}$$

Now define the matrix

$$Y := L^+ (L^-)^{-1} \,. \tag{1.6.7}$$

In terms of this matrix (1.6.6) takes the form

$$Y_1 T_2 = T_2 \tilde{R}_{21} Y_1 \tilde{R}_{12} \,, \tag{1.6.8}$$

and CR (1.6.4) between $L^{\pm}$ themselves are transformed to

$$R_{21} Y_1 R_{12} Y_2 = Y_2 R_{21} Y_1 R_{12} \,, \tag{1.6.9}$$

(in this relation, as well as in (1.6.4), the R-matrix normalization is not essential so we omit the tilde sign over the R-matrix).

One can check that the CR (1.6.8) and (1.6.9) are *invariant* with respect to "left action"

$$\begin{aligned} T^i{}_j &\to \delta_L(T^i{}_j) := T'^i{}_k \otimes T^k{}_j \,, \\ Y^i{}_j &\to \delta_L(Y^i{}_j) := Y^i{}_j \,, \end{aligned}$$

and *covariant* with respect to "right action"

$$\begin{aligned} T^i{}_j &\to \delta_R(T^i{}_j) := T^i{}_k \otimes T'^k{}_j \ , \\ Y^i{}_j &\to \delta_R(Y^i{}_j) := S(T'^i{}_k)T'^l{}_j \otimes Y^k{}_l \ . \end{aligned}$$

This means that the matrix Y provides with a *bicovariant* calculus on the quantum group. Of course, one can use the calculus based on the matrix $Y^{-1} = L^-(L^+)^{-1}$, which is right invariant and left covariant.

To establish a connection with the classical (non-deformed) Lie algebra, define the matrix X by

$$Y =: \mathbf{I} - \lambda X \ , \tag{1.6.10}$$

(matrices $L^\pm$ and Y tend to a unit matrix as $q \to 1$).

The CR (1.6.9) in terms of X becomes

$$R_{21}X_1R_{12}X_1 - X_2R_{21}X_1R_{12} = \frac{1}{\lambda}(R_{21}R_{12}X_2 - X_2R_{21}R_{12}) \ .$$

One can check that in the limit $q \longrightarrow 1$ the latter CR becomes a usual Lie algebra commutator relations. Thus here the R-matrix plays the role of deformed structure constants.

Having in disposal the covariant vector fields (derivatives), one can define an exterior differential. In the present case the consistent definition is the following

$$d := \mathrm{Tr}(\mathcal{D}^{-1}\Omega X) \ , \tag{1.6.11}$$

where Ω is the matrix of differential one-forms with properties to be defined, $\mathcal{D}$ is a suitable $\mathbb{C}$-matrix, which provides the invariance properties of the trace operation

$$\mathrm{Tr}(\mathcal{D}^{-1}M') = \mathrm{Tr}(\mathcal{D}^{-1}M) \ ,$$

where

$$(M')^i{}_j = S(T^i{}_k)T^l{}_j \otimes M^k{}_l \ , \tag{1.6.12}$$

M is any matrix with (in general) non-commuting entries and Tr is the usual matrix trace. The appearance of this additional matrix $\mathcal{D}$ is related to the fact that square of an antipode is not equal to identity map $S^2(T) \neq T$ (in contrast to the classical operation of a matrix inversion: $(T^{-1})^{-1} = T$). Indeed, the usual trace would give

$$\mathrm{Tr}M' = S(T^i{}_l)T^k{}_i \otimes M^l{}_k \ . \tag{1.6.13}$$

Using the defining property (1.2.9) of an antipode, which for matrix q-groups takes the form $TS(T) = S(T)T = \mathbf{I}$ (cf. (1.5.11),(1.5.22)) and antihomomorphism of S one gets

$$S(T^i{}_j)S^2(T^k{}_i) = S(T^i{}_l)\mathcal{D}^k{}_j T^j{}_n(\mathcal{D}^{-1})^n{}_i = \delta_l{}^k ,$$

where the matrix $\mathcal{D}$ is given by (1.5.12) and (1.5.23). Thus the usual trace (1.6.13) is not invariant with respect to the transformations and the matrix $\mathcal{D}$ inserted in the trace operation just correct this shortcoming. This operation is called q-trace (for general discussion of quantum trace see [154])

$$\mathrm{Tr}_q M := \mathrm{Tr}(\mathcal{D}^{-1}M) .$$

Properties of the entries of matrix Ω in (1.6.11) are defined from the standard postulates (1.1.2) of the differential calculus. For example, from the requirement $d^2 = 0$ in $SU_q(2)$ case (recall that for brevity we use the notation G_q instead of $Fun_q(G)$; see remark in Section 1.5.1) one finds the relations

$$\begin{aligned} &\omega^+\omega^- + \omega^-\omega^+ = 0 , && \omega^2\omega^- + q^{-2}\omega^-\omega^2 = -q^{-1}\lambda\omega^-\omega^1 , \\ &\omega^1\omega^+ + \omega^+\omega^1 = 0 , && \omega^1\omega^2 + \omega^2\omega^1 = -q^{-1}\lambda\omega^+\omega^- , \\ &\omega^1\omega^- + \omega^-\omega^1 = 0 , && (\omega^1)^2 = (\omega^+)^2 = (\omega^-)^2 = 0 , \\ &\omega^2\omega^+ + q^2\omega^+\omega^2 = q\lambda\omega^+\omega^1 , && (\omega^2)^2 = q\lambda\omega^+\omega^- , \end{aligned}$$

for the entries of the matrix

$$\Omega =: \begin{pmatrix} \omega^1 & \omega^- \\ \omega^+ & \omega^2 \end{pmatrix} ,$$

and the quantum version of Maurer-Cartan equations (for the classical case see, e.g., [145])

$$\begin{aligned} d\omega^1 &= -q^{-3}\omega^+\omega^- , & d\omega^+ &= q^{-1}\omega^+(\omega^1 - \omega^2) , \\ d\omega^2 &= q^{-1}\omega^+\omega^- , & d\omega^- &= q^{-1}(\omega^1 - \omega^2)\omega^- . \end{aligned}$$

Further details about differential calculus on q-groups can be found in [247, 131, 255, 211, 68, 33, 83, 84].

1.6.2 Differential calculus on quantum spaces

In order to set up a differential calculus on a quantum space $\mathbb{C}_q^n[[x^i]]$ invariant with respect to the action of $GL_q(n)$ one introduces [239],[209] a linear operator d with the properties (1.1.2) and identifies anticommuting in classical limit generators of the algebra $\wedge\mathbb{C}_q^n[[\xi^i]]$ with quantum differentials $dx^i := \xi$. Rewriting (1.5.16) or (1.5.34) one has

$$dx^i dx^k = -r\hat{R}^{ik}_{lm} dx^l dx^m \ , \tag{1.6.14}$$

(we use multiparametric $\hat{R}$-matrix at once; recall that the one-parametric standard deformation is recovered by putting $q_{ij} = \sqrt{r} =: q$). It remains to determine how the quantum space coordinates and its differentials commute. To do this, we follow the steps described in Section 1.1. Postulating

$$x^i dx^k = C^{ik}_{lm} dx^l x^m \ ,$$

and applying the operator d one finds, using its nilpotency

$$dx^i dx^k = -C^{ik}_{lm} dx^l dx^m \ .$$

Comparison with (1.6.14) gives two possibilities

$$C = r\hat{R} \tag{1.6.15}$$

or

$$C = r^{-1}\hat{R}^{-1} \ . \tag{1.6.16}$$

The second solution follows from the R-matrix property

$$R^{-1}_{r,q_{ij}} = R_{1/r,1/q_{ij}} \ . \tag{1.6.17}$$

Partial derivatives are introduced by

$$d = dx^i \partial_i^{(q)} \ . \tag{1.6.18}$$

In a sense, we are using here the inverse way in comparison with the calculus on a group: there we started from derivatives (generators of $\mathcal{U}_q(L)$) and then defined differentials, here differentials (generators of $\wedge\mathbb{C}_q^n$) are initial objects and properties of derivatives to be derived from those of the differentials.

The superscript "(q)" in (1.6.18) indicates that the relation defines *deformed* or *q-derivatives*. However, since almost all derivatives in this book are deformed ones, we will omit in what follows the explicit indication through the use of the superscript in order to simplify the notations. Instead, in cases of use of non-deformed usual derivatives we will point out this specially.

Application of Leibniz rule to $xf(x)$, where $f(x)$ is a function of the coordinates x^i and comparison of the coefficients in dx^i yield (for the case (1.6.15))

$$\partial_i x^k = \delta_i^k + r\hat{R}^{kl}_{im} x^m \partial_l \ .$$

For CR of ∂_i and dx^k one can try the form

$$\partial_i dx^k = D^{kl}_{ij} dx^j \partial_l \ .$$

Multiplying these relations by x^r from the right and commuting it through to the left one finds terms linear in dx^l which must cancel separately. This gives

$$\partial_i dx^k = \frac{1}{r} (\hat{R}^{-1})^{kl}_{im} dx^m \partial_l \ .$$

Finally, one obtains by a similar way the CR between derivatives

$$\partial_i \partial_k = \hat{R}^{ml}_{ki} \partial_l \partial_m \ .$$

The second possibility (1.6.16) leads to partial derivatives $\tilde{\partial}_i$ with the CR

$$\begin{aligned} \tilde{\partial}_i x^k &= \delta^k_i + \frac{1}{r} (\hat{R}^{-1})^{kl}_{im} x^m \tilde{\partial}_l \ , \\ \tilde{\partial}_i dx^k &= r \hat{R}^{kl}_{im} dx^m \tilde{\partial}_l \ , \\ \tilde{\partial}_i \tilde{\partial}_k &= \hat{R}^{ml}_{ki} \tilde{\partial}_l \tilde{\partial}_m \ , \end{aligned}$$

(the latter relations are the same for both types of derivatives).

For the reader's convenience we present the explicit form of $GL_{r,q_{ij}}(n)$ invariant differential calculus in Table 1.1.

From these relations one can see that in one dimensional case it is natural to consider q-deformed differential calculus with the relation

$$\partial x = 1 + q^2 x \partial \ . \tag{1.6.19}$$

Actually, this relation follows also from the requirement of invariance with respect to the group transformations, this time non-semisimple, namely, the group of one-dimensional dilatations and translations. We postpone the discussion of this topic until Section 2.2. Here we note only that the relation (1.6.19) can be realized by the algebra of usual functions on $\mathbb{R}$ if ∂ is understood as the famous *Jackson derivative* D_q

$$D_q f(x) := \frac{f(q^2 x) - f(x)}{(q^2 - 1)x} \ , \tag{1.6.20}$$

which had been introduced in [202, 127]. One can easily check that this q-derivative (in fact, this is finite difference of the special form) satisfies the relation (1.6.19). The Jackson q-derivative is a corner stone of so-called basic

Table 1.1: Defining relations for $GL_{r,q_{ij}}(n)$ invariant differential calculus; in all the relations $i < k$

$x \leftrightarrow x$	$x^i x^k = q_{ik} x^k x^i$
$dx \leftrightarrow dx$	$dx^i dx^i = 0$ $dx^i dx^k = -\frac{q_{ik}}{r} dx^k dx^i$
$x \longleftrightarrow dx$	$x^i dx^i = r dx^i x^i$ $x^i dx^k = q_{ik} dx^k x^i + (r-1) dx^i x^k$ $x^k dx^i = \frac{r}{q_{ik}} dx^i x^k$
$\partial \leftrightarrow x$	$\partial_i x^i = 1 + r x^i \partial_i + (r-1) \sum_{a=i+1}^{n} x^a \partial_a$ $\partial_i x^k = \frac{r}{q_{ik}} x^k \partial_i$ $\partial_k x^i = q_{ik} x^i \partial_k$
$\partial \leftrightarrow dx$	$\partial_i dx^i = \frac{1}{r} dx^i \partial_i + (\frac{1}{r} - 1) \sum_{a=1}^{i-1} dx^a \partial_a$ $\partial_i dx^k = \frac{1}{q_{ik}} dx^k \partial_i$ $\partial_k dx^i = \frac{q_{ik}}{r} dx^i \partial_k$
$\partial \leftrightarrow \partial$	$\partial_i \partial_k = \frac{q_{ik}}{r} \partial_k \partial_i$
$\tilde{\partial} \leftrightarrow x$	$\tilde{\partial}_i x^i = 1 + \frac{1}{r} x^i \tilde{\partial}_i + (\frac{1}{r} - 1) \sum_{a=1}^{i-1} x^a \tilde{\partial}_a$ $\tilde{\partial}_i x^k = \frac{1}{q_{ik}} x^k \tilde{\partial}_i$ $\tilde{\partial}_k x^i = \frac{q_{ik}}{r} x^i \tilde{\partial}_k$
$\tilde{\partial} \leftrightarrow dx$	$\tilde{\partial}_i dx^i = r dx^i \tilde{\partial}_i + (r-1) \sum_{a=i+1}^{n} dx^a \tilde{\partial}_a$ $\tilde{\partial}_i dx^k = \frac{r}{q_{ik}} dx^k \tilde{\partial}_i$ $\tilde{\partial}_k dx^i = q_{ik} dx^i \tilde{\partial}_k$

analysis, the well developed branch of mathematics, which results in appearance of finite difference analogs of most special functions known in the usual analysis (see, e.g. [4, 99, 110]). These q-special functions are related to harmonic analysis on quantum groups in analogy with the case of non-deformed groups and special functions (see next Section).

The differential calculus on quantum Euclidian space $O_q^N[[x^i]]$ has been derived in [34] with the use of quite analogous method and the result for the one-parametric deformation is

$$\begin{aligned} x^i dx^j &= q\hat{R}^{ij}{}_{kl} dx^k x^l , \\ \partial^i x^j &= (C^{-1})^{ij} + q(\hat{R}^{-1})^{ij}{}_{kl} x^k \partial_l , \\ \partial^i \partial^j &= \frac{1}{q}\hat{R}^{ij}{}_{kl} \partial^k \partial^l , \\ \partial^i dx^j &= \frac{1}{q}\hat{R}^{ij}{}_{j'i'} dx^{j'} \partial_{i'} , \end{aligned} \tag{1.6.21}$$

where $R^{ij}{}_{kl}$ is the R-matrix (1.5.20) for q-orthogonal group; C_{ij} is the metric on O_q^N and $\partial^i := (C^{-1})^{ij}\partial_j$.

1.6.3 q-Deformation of integral calculus

Now we turn to integral calculus. This part of q-mathematics is not yet developed in the full completeness in both cases: of q-groups and q-spaces.

Invariant integration on $SU_q(2)$ group has been developed by Woronowicz [245]. We present the result using the technique of the differential operators Y [255] on a q-group which we considered in the first part of this Section.

The unitarity condition $T^\dagger = S(T)$ (cf. preceding Section), i.e.

$$\begin{pmatrix} a^* & c^* \\ b^* & d^* \end{pmatrix} = \begin{pmatrix} d & -q^{-1}b \\ -qc & a \end{pmatrix} , \tag{1.6.22}$$

and the determinant condition

$$ad - qbc = aa^* + bb^* = a^*a + q^{-2}b^*b = 1 ,$$

lead to the conjecture that a basis for this associative aalgebra as a linear space is provided by the monomials $a^k b^l c^m$ and $d^k b^l c^m$, where the exponents take all positive values $0, 1, 2, \ldots$.

To define invariant integral on a q-group means to associate to a function f on the group a real number $\langle f \rangle$, its invariant group average, i.e. a linear functional with a particular invariance properties. There are different ways to formulate the invariance. For algebraic manipulations a convenient way is to require that

$$\langle X^i{}_j f\rangle = \langle (Y^i{}_j - \delta^i{}_j) f\rangle = 0\ , \quad \forall i,j\ , \tag{1.6.23}$$

(see (1.6.7),(1.6.10)). This is the q-analog of usual condition of infinitesimal invariance. The explicit form of CR (1.6.8) for $SU_q(2)$ group is

$$\begin{aligned} y_+a &= ay_+ + q^{-1}\lambda by_2\ , & y_+b &= by_+\ , \\ y_-a &= ay_-\ , & y_-b &= by_- + q^{-1}\lambda ay_2\ , \\ y_2a &= q^{-1}ay_2\ , & y_2b &= qby_2\ , \end{aligned}$$

together with the relations obtained by replacing a by c and b by d. In these formulas the elements a and b are considered as multiplication operators in the linear basis, so these relations are q-analogue of the Leibniz rule. Here

$$Y =: \begin{pmatrix} y_1 & y_+ \\ y_- & y_2 \end{pmatrix}$$

and $\lambda := q - q^{-1}$.

Taking into account that $y_2 \cdot \mathbf{I} = \mathbf{I}$, one has

$$y_2 a^k b^l c^m = q^{-k+l-m} a^k b^l c^m\ . \tag{1.6.24}$$

So the corresponding relation

$$\langle (y_2 - \mathbf{I}) a^k b^l c^m\rangle = 0\ ,$$

from the set (1.6.23) gives that $\langle a^k b^l c^m\rangle$ must vanish unless $k + m - l = 0$. The condition

$$\langle y_- a^k b^l c^m\rangle = 0$$

strengthens this result and gives

$$\langle a^k b^l c^m\rangle = 0 \ \text{ unless both } k = 0 \text{ and } l = m\ .$$

Similarly,

$$\langle a^k b^l c^m\rangle = 0 \ \text{ unless both } k = 0 \text{ and } l = m\ .$$

To obtain the remaining nonvanishing average values $\langle b^k c^k\rangle$, we compute

$$y_+ ab^{k-1}c^k = \lambda q^{-1}[k;q^2] adc^{k-1}b^{k-1} + q^{-2}\lambda b^k c^k\ .$$

The average of lhs of this relation vanishes. Hence, using the determinant condition one arrives at the recursion relation

$$\langle b^k c^k \rangle = -q \frac{[k; q^2]}{[k+1; q^2]} \langle b^{k-1} c^{k-1} \rangle .$$

Choosing the normalization

$$\langle \mathbf{1} \rangle = 1 ,$$

one obtains finally

$$\langle b^k c^k \rangle = \frac{(-q)^k}{[k+1; q^2]} , \tag{1.6.25}$$

or, using the unitarity condition, one gets

$$\langle (c^*)^k c^k \rangle = \frac{1}{[k+1; q^2]} .$$

In the non-deformed limit this average, apparently, cannot be defined in terms of usual invariant integral over the group with Haar measure. This follows, in particular, from the fact that the result (1.6.25) is correct even without unitarity condition, i.e. for $SL_q(2)$ group. But in the classical limit this group is non-compact and corresponding integral does not exist.

In the final part of this Section we discuss integration on quantum spaces [15, 77, 140].

We start from the one dimensional case with CR for the coordinate and the derivative $\partial x - q^2 x \partial = 1$, $0 < q^2 < 1$, and indefinite integral which is understood as an inverse operation with respect to the derivative

$$\int^q := \partial^{-1} \equiv x x^{-1} \int^q = x(\partial x)^{-1} \equiv x \frac{1-q^2}{1-(1-(1-q^2)\partial x)} .$$

The operator $(1-(1-q^2)\partial x) = q^2 L$, where

$$L := (1-(1-q^2)x\partial) , \tag{1.6.26}$$

is diagonal in the basis of monomials

$$L x^n = q^{2n} x^n ,$$

with the eigenvalues of absolute values smaller than 1 and hence can be expanded in series to give

$$\int^q = (1-q^2)x \sum_{n=0}^{\infty} (1-(1-q^2)\partial x)^n = (1-q^2)x \sum_{n=0}^{\infty} q^{2n} L^n . \tag{1.6.27}$$

This q-integral can be considered as a q-analog of the *usual* integral

$$\int_0^x f(x')dx' \,. \tag{1.6.28}$$

Indeed, q-integration of the monomial x^m gives the result

$$\begin{aligned} {}^{q}\!\!\int x^m &= (1-q^2)x\sum_{n=0}^{\infty} q^{2n}L^n x^m = (1-q^2)x\sum_{n=0}^{\infty} q^{2n(m+1)}x^m \\ &= (1-q^2)\frac{x^{m+1}}{1-q^{2(m+1)}} = \frac{x^{m+1}}{[m+1;q^2]}\,, \end{aligned}$$

which in $q \longrightarrow 1$ limit coincides with that of usual integration (1.6.28).

As the operator L scales the arguments of functions on $\mathbb{R}$

$$Lf(x) = f(qx)\,, \tag{1.6.29}$$

and $0 < q^2 < 1$, it is natural to define the analog of the usual definite integral on $\mathbb{R}_+$, i.e. $\int_0^\infty f(x)dx$ as follows

$$ {}^{q}\!\!\int_0^\infty := \lim_{r\to\infty} L^{-r}\, {}^{q}\!\!\int = (1-q^2)x \sum_{n=-\infty}^{\infty} q^{2n}L^n \,.$$

Note, however, that the result of an action of this operator on any algebra element is another element but not a number as in non-deformed case. To obtain a number, one can use some concrete representation of CR (1.6.19) in a Hilbert space, then numbers appear as eigenvalues or average of operators.

In the case under consideration, the representation can be constructed [77, 140] starting from an eigenvector of the position operator x with the eigenvalue c

$$x \mid c\rangle = c \mid c\rangle \,. \tag{1.6.30}$$

Since

$$Lx = q^2 xL\,, \tag{1.6.31}$$

the vector

$$\mid q^{-2m}c\rangle = N(m)L^m \mid c\rangle\,, \tag{1.6.32}$$

has an eigenvalue $q^{2m}c$ ($N(m)$ is normalization factor). Also there exists an inverse operator

$$L^{-1} = \frac{1}{1-(1-q^2)x\partial} = \sum_{m=0}^{\infty}((1-q^2)x\partial)^m\,,$$

since in the space of functions (in monomial basis) the eigenvalues of $(1-q^2)x\partial$ are $1-q^{2m}$, i.e. smaller than 1. The inverse operator can be constructed by the use of the derivative $\bar{\partial}$ (see Section 3.2). One can thus let m in (1.6.32) to run over all the integers. Obviously, the scale c labels the representations.

A function f can be considered in the representation as a vector

$$\mid f\rangle := \sum_m f_m \mid q^{2m}c\rangle = f(x) \mid \mathbf{1}\rangle \,, \tag{1.6.33}$$

where

$$\mid \mathbf{1}\rangle = \sum_{m=-\infty}^{\infty} \mid q^{2m}c\rangle \,.$$

The position operator x can be written as the matrix

$$x = \sum_{m,s} \mid q^{2m}c\rangle c q^{2m} \delta_{ms} \langle q^{2s}c \mid \,.$$

The CR (1.6.19) and the condition $\partial \mid \mathbf{1}\rangle = 0$ then gives

$$\partial = \sum_{m,s} \mid q^{2m}c\rangle (\delta_{ms} - \delta_{(m+1)s})(q^{2m}c(1-q^2))^{-1} \langle q^{2s}c \mid \,, \tag{1.6.34}$$

and the normalization constants in (1.6.32) proves to be

$$N(m) = q^{-2m} \,.$$

Inverting (1.6.34), one obtains

$$\int^{q} = \partial^{-1} = c(1-q^2) \sum_{m=-\infty, t=0}^{\infty} q^{2(m+t)} \mid q^{2m}c\rangle \langle q^{2(m+t)}c \mid \,. \tag{1.6.35}$$

Then the definite q-integral from the "point" $\alpha = q^{2a}c$ to the point $\beta = q^{2b}c$ can be defined as follows

$$\int_{\alpha}^{q,\beta} f := \langle q^{2b}c \mid \int^{q} \mid f\rangle - \langle q^{2a}c \mid \int^{q} \mid f\rangle \,,$$

and by using (1.6.35) and (1.6.33) it can be rewritten [140] in the form of *Jackson integral* [4, 99]

$$\int_{\alpha}^{q,\beta} f = cq^{2b}(1-q^2)\sum_{t=0}^{\infty} q^{2t} f(cq^{2b}q^t) - cq^{2a}(1-q^2)\sum_{t=0}^{\infty} q^{2t} f(cq^{2a}q^t) \,. \tag{1.6.36}$$

Now consider the multidimensional generalization of this integral calculus covariant with respect to $GL_q(N)$ transformations. First of all we note that the last dimension in CR depicted in the Table 1.1 for $GL_q(n)$ invariant differential calculus on $\mathbb{C}_q^N[[x^i]]$ (recall that in the one-parametric deformation case $q_{ij} = \sqrt{r} \equiv q$) can be identified with the one dimensional case which we have considered so far.

The algebra is again naturally represented on the function algebra of power series in the generators x^i, obeying the commutation relations in Table (1.1).

As well as in the case of q-derivatives, we will omit in what follows the superscript "q", indicating deformation of an integral, if this will not cause a confusion.

Writing formally

$$\int_0^{x^i} := \partial_i^{-1} = x^i \frac{1-q^2}{1-(1-(1-q^2)\partial_i x^i)} , \tag{1.6.37}$$

we note that the last term is actually well defined as a power series. This is because the action of the operator $(1-(1-q^2)\partial_i x^i)$ in the basis of ordered monomials

$$(x^n)^{m_n}(x^{n-1})^{m_{n-1}} \cdots (x^1)^{m_1} ,$$

is diagonal

$$\begin{aligned}(1-(1-q^2)\partial_i x^i).(x^n)^{m_n} \cdots (x^i)^{m_i} \cdots (x^1)^{m_1} \\ = (1-q^{2(m_n+\ldots+m_{i+1})}(1-q^{2(m_i+1)})) \\ \times (x^n)^{m_n} \cdots (x^i)^{m_i} \cdots (x^1)^{m_1} ,\end{aligned} \tag{1.6.38}$$

and we can read off that its eigenvalues are all of absolute values smaller than 1. This allows its expansion as a geometrical series:

$$\int_0^{x^i} = (1-q^2)x^i \sum_{m=0}^{\infty} (1-(1-q^2)\partial_i x^i)^m . \tag{1.6.39}$$

We thus have the desired properties:

$$\partial_i \int_0^{x^i} f = f \qquad \forall f \in \mathbb{C}_q^N[[x^i]] , \tag{1.6.40}$$

$$\int_0^{x^i} \partial_i x^i f = x^i f \,' \qquad \forall f \in \mathbb{C}_q^N[[x^i]] . \tag{1.6.41}$$

The factor x^i prevents $\int_0^{x^i}$ from acting on zero which could otherwise occur through the annihilation of f by the action of ∂_i. It is not very difficult to prove these properties directly by using the above basis of ordered polynomials.

Actually we see that due to this choice of basis, those derivatives that arise from the term on the rhs of $\partial \leftrightarrow x$ CR in Table (1.1) do not find coordinates to act on. This means that for ordered polynomials in the evaluation of

$$\int_0^{x^n}\int_0^{x^{n-1}}\cdots\int_0^{x^1}(x^n)^{m_n}\cdots(x^1)^{m_1}\,, \tag{1.6.42}$$

one can neglect this term on the rhs of the CR. It is thus the special feature of these ordered polynomials that the operators $\int_0^{x^i}$ which act on them are built of operators which effectively obey the one dimensional commutation relation. Even then, the $\int_0^{x^i}$ are non-commuting with each other and with the coordinate functions.

Since the operator $\int_0^{x^i}$ is the inverse of the operator ∂_i we immediately get the commutation relations

$$\partial^j\int_0^{x^i} - q^{-1}\int_0^{x^i}\partial^j = 0\,, \quad \text{for} \quad i<j\,, \tag{1.6.43}$$

$$\partial^j\int_0^{x^i} - q\int_0^{x^i}\partial^j = 0\,, \quad \text{for} \quad i>j\,, \tag{1.6.44}$$

$$\int_0^{x^i}\int_0^{x^j} - q\int_0^{x^j}\int_0^{x^i} = 0\,, \quad \text{for} \quad i<j\,, \tag{1.6.45}$$

$$x^j\int_0^{x^i} - q\int_0^{x^i} x^j = 0\,, \quad \text{for} \quad i\neq j\,, \tag{1.6.46}$$

$$x^i\int_0^{x^i} - q^2\int_0^{x^i} x^i = \int_0^{x^i}\int_0^{x^i} + (q^2-1)\sum_{a=i+1}^{n} x^a\partial_a\int_0^{x^i}\int_0^{x^i}\,, \tag{1.6.47}$$

which could of course also be written in the R-matrix notation. Note that (1.6.47) describes integration by parts.

In complete analogy with the one dimensional case, one can also use the scaling operators L_i and the appropriate representations to define integrals resulting in numbers.

There is another approach [15],[140] for the definition of definite q-integrals. One can start from postulating the q-integral of some appropriate function on quantum space

$$I_0(g) := \int_{-\infty}^{\infty} g\,, \tag{1.6.48}$$

and using the postulate

$$\int_{-\infty}^{\infty}\partial(x^m g) = 0\,, \quad \forall m\in \mathbb{Z}_+\,, \tag{1.6.49}$$

(here the usual trick was used: a theorem in classical case - this time the Stokes one - becomes the defining property in deformed case) one can derive the recursive relation for

$$I_m := {}^{q}\!\!\int_{-\infty}^{\infty} x^m g \ , \tag{1.6.50}$$

and express it as some multiple of $I_0(g_\alpha)$. Multidimensional generalization is constructed quite straightforwardly. We shall not go into details of this approach in general here (see [140]). Instead, in Section 2.2 we will consider the construction of two-dimensional q-integral using this method with further application for derivation of Bargmann-Fock representation of the q-oscillator algebra.

1.7 Elements of quantum group representations

Most of group theoretical and algebraic methods and applications are connected with some type of representation theory. Considerable part of this book is devoted to different aspects of representations of quantum groups, universal enveloping algebras and q-deformed algebras of Heisenberg-Weyl type (q-oscillators).

For a Hopf algebra A associated with matrix quantum group, there are two important representation theories:

- irreducible corepresentations of A on (finite dimensional) Hilbert spaces (as a coalgebra);
- irreducible representations (irreps) of A on possibly infinite-dimensional Hilbert spaces (as algebra).

If $A = Fun(G)$ with G being a usual Lie group, then all irreducible representations of A are one-dimensional and correspond to point evaluations on the elements of G. By analogy, for non-commutative A, a set of irreducible representations of A may be considered as elements of the underlying quantum group. We have already mentioned in the Introduction about this aspect of quantum groups connected with general approach to non-commutative geometry and which originated in the Gel'fand-Naimark theorem. For physicists this problem is familiar from ordinary quantum mechanics: if a group manifold is considered as a phase space, then after quantization one is interested in possible representations of elements of $Fun_q(G)$ as operators in Hilbert spaces. In the simplest case of Heisenberg algebra in $\mathbb{R}^n$ all irreps are equivalent. This is the content of the famous von Neumann theorem (see, e.g., [17]). In other cases there are different inequivalent representations, e.g. for quantum top (spin) or a particle on a circle. Analogous situation appears in the theory of quantum groups. We will discuss this using the basic example of $SU_q(2)$ group.

The problem of construction of irreps of QUEA is quite analogous to the standard and well known problem of construction and classification of irreps of Lie algebras and UEA. Moreover, for generic value of the deformation parameter q, unitary irreps of quantum "Lie algebra" are in close correspondence with those of classical Lie algebras [197, 199, 188, 201] (see also [61, 60] and refs. therein). Essentially new phenomena appears for a root of unity value of q [188, 138, 204, 61, 60].

The general definition of a q-group corepresentation is a natural generalization of the definition of a matrix q-group itself (which are based on specific, namely, fundamental corepresentation; cf. Section 5). More generally, *a Hopf algebra corepresentation* is defined as follows.

Definition 1.31 *A matrix corepresentation of a Hopf algebra A is a matrix* $U = \{u^i{}_j\}_{i,j=1}^N$ *with* $u^i{}_j \in A$*, such that* $\Delta(u^i{}_j) = \sum_{k=1}^N u^i{}_k \otimes u^k{}_j$ *and* $\varepsilon(u^i{}_j) = \delta^i_j$.

Corepresentation theory deals with homomorphisms of the coalgebras

$$(End\ V)^* \longrightarrow Fun_q(G)\ ,$$

where $(End\ V)^*$ is the space dual to the space $End\ V$ of endomorphisms (i.e., homomorphic maps into itself) of a linear space V. For $q = 1$, it is equivalent to the usual representation theory in the vector space of a Lie group G in its *dual* formulation. In fact, the very definition of matrix quantum groups is based on such a corepresentation: in Section 1.5 we introduced a matrix q-group as the algebra dual to the q-deformed universal enveloping algebra in fundamental representation. The use of another representation of $\mathcal{U}_q(L)$ would lead to a different corepresentation of the Hopf algebra $Fun_q(G)$ (and to another representation for R–matrix). QUEA, being dual to $Fun_q(G)$ objects, also have corepresentations. In fact, we considered such corepresentation of $\mathcal{U}_q(s\ell(2))$ in Section 1.6: the matrices L^+ and L^- of the form (1.6.2) and (1.6.3) satisfy the Definition 1.31 (cf. (1.6.5)). Analogous corepresentations also exist for all simple q-groups [104].

In this concluding section of the first Chapter we give the basic definitions and present the most important properties of q-group corepresentations and representations.

1.7.1 Corepresentations of quantum groups

Here we will present basic definitions and facts concerning quantum group corepresentations and then consider the explicit construction of corepresentations of $SU_q(2)$.

The definition of a Hopf algebra corepresentation has been given already. The next definitions introduce the notions of equivalence, irreducibility and unitarity.

Definition 1.32 *Two corepresentations U and V of a Hopf algebra A are called equivalent if U and V are matrices of the same size $N \times N$ and if there is an invertible complex $N \times N$ matrix S (intertwiner) such that*

$$U = S^{-1} V S \ .$$

Definition 1.33 *A corepresentation U of a Hopf algebra A of dimension N is called irreducible if U is not equivalent to a corepresentation V of the form*

$$V = \begin{pmatrix} \alpha & \beta \\ 0 & \delta \end{pmatrix} ,$$

where α, β, δ are submatrices of dimensions smaller than N.

Definition 1.34 *A matrix corepresentation U of a Hopf algebra A is called unitary if*

$$(U^i{}_j)^* = S(U^j{}_i) \ , \tag{1.7.1}$$

where $*$ is an involution in the Hopf algebra A.

The relation (1.7.1) can be written in the form

$$\sum_k U^k{}_i{}^* U^k{}_j = \sum_k U^i{}_k U^j{}_k{}^* = \delta_{ij} \ .$$

One can note that the Definitions 1.31, 1.34 are analogous to those in Section 1.5 because as we noted already the very definition of a quantum matrix groups is based on the fundamental representation.

As is well known (see, e.g., [13, 142, 250]), finite dimensional *unitary* representations exist only for the compact groups. So we must define the corresponding notion of *a compact quantum group.*

Let G be a compact Lie group. It is isomorphic to a closed subgroup of $SU(n)$ as there is a faithful unitary matrix representation of G (*faithful* representation means that there exists *isomorphic* map to a set of matrices). Let $Pol(G)$ be the Hopf *-algebra of polynomials in matrix elements of this representation. Due to the group compactness, $Pol(G)$ can be made a normed *-algebra with a norm given by the supremum norm on G. After the norm completion of $Pol(G)$ as a normed linear space, it becomes the commutative C^*-algebra of continuous functions on G.

To introduce the notion of compactness for q-groups, one considers the existence of norm as a defining property (again the same method of transformation of a theorem in classical case into a defining postulate after deformation). In short, the norm of an element a of a Hopf algebra A is defined as follows

$$\|a\| := \sup_{\rho} \|\rho(a)\| \ ,$$

ρ running over all *-representations of A, where by a *-representation ρ of A in a Hilbert space $\mathcal{H}$ one means a homomorphism $\rho \ : \ A \longrightarrow B(\mathcal{H})$ of *-algebra A to the *-algebra $B(\mathcal{H})$ of all bounded operators on $\mathcal{H}$, the involution for the latter being defined by hermitian conjugation. If U is a unitary corepresentation, then

$$\|\rho(U^i{}_j)\| \leq 1 \ , \qquad \forall \rho, \ i, \ j \ ,$$

and, hence, $\|a\|$ is finite $\forall\, a \in A$, so $\|\cdot\|$ defines a norm on A (if A has a faithful *-representation so that $a = 0$, whenever $\|a\| = 0$). After an appropriate norm completion the algebra A becomes a C^*-algebra [76, 9], which is the cornerstone of Woronowicz's general theory of compact matrix quantum groups and we refer to the original papers [244]-[247] for further details.

One of the most important results of the construction is the following theorem [245]:

- Let A be a Hopf C^*-algebra. Then there exists a unique linear functional h on A such that:

    1. $h(\mathbf{1}) = 1$;
    2. $h(a^*a) \geq 0$, $\quad \forall a \in A$;
    3. $(h \otimes id)\Delta(a) = h(a)\mathbf{1} = (id \otimes h)\Delta(a)$, $\quad \forall\, a \in A$;
    4. if $h(a^*a) = 0$, then $a = 0$.

This functional h is the analog of a bi-invariant Haar integral on classical compact groups. Its explicit construction for $SU_q(2)$ group was considered in the preceding section.

Woronowicz has obtained also the analog of *Schur orthogonality relations*. Recall, that in the case of matrix representations $\{T^\sigma_{mn}\}^{d_\sigma}_{m,n=1}$ (of the dimension d_σ) of a classical group G, the relations read ([13, 142])

$$\int_G T^\sigma_{mk}(x)\overline{T^\tau_{nl}(x)}d\mu_x = d_\sigma^{-1}\delta_{\sigma\tau}\delta_{mn}\delta_{kl} , \tag{1.7.2}$$

where $d\mu_x$ is the invariant Haar measure on the group G.

The corresponding quantum relations have a more complicated form.

Choose for each equivalence class of irreducible unitary corepresentations of a Hopf *-algebra A a representative T^σ : $\{T^\sigma_{mn}\}^{d^\sigma}_{m,n=1}$. Then there exists a unique multiplicative linear functional f on A such that

$$\begin{aligned} h(T^\sigma_{mk}(T^\tau_{nl})^*) &= \frac{\delta_{\sigma\tau}\delta_{mn}f(T^\sigma_{lk})}{f(\sum_{l=1}^{d_\sigma} T^\sigma_{ll})} , \\ h((T^\sigma_{km})^*(T^\tau_{ln})) &= \frac{\delta_{\sigma\tau}\delta_{mn}\overline{f((T^\sigma_{lk})^*)}}{f(\sum_{l=1}^{d_\sigma} T^\sigma_{ll})} . \end{aligned} \tag{1.7.3}$$

Furthermore, the T^σ_{mn} form a basis of A.

The appearance of f in this quantum Schur theorem is a new phenomenon by which matrix elements belonging to the same representation are no longer orthogonal in a straightforward way. This phenomenon is related to the fact that the values of the Haar functional h depend , in general, on the order of elements in its argument, i.e. $h(ab) \neq h(ba)$.

For $SU_q(2)$, Woronowicz finds the explicit form [244]

$$f(a) = q^{-1} , \quad f(d) = q , \quad f(b) = f(c) = 0 ,$$

where a, b, c, d are the elements of the matrix (1.1.18).

Now we turn to the explicit construction of irreducible unitary corepresentations of $SU_q(2)$ [244, 231, 181, 147]).

Recall, at first, that unitary representation of $SU(2)$ group can be constructed in $(2l+1)$-dimensional spaces of homogeneous polynomials p^l_n of degree $2l$ in two complex variables (see, e.g. [233, 13])

$$p^l_n(\xi,\eta) = \begin{pmatrix} 2l \\ l-n \end{pmatrix}^{1/2} \xi^{l-n}\eta^{l+n} , \qquad n = -l, -l+1, \ldots, l ,$$

(l is integer or half-integer positive number). Matrix elements T^l_{mn} with respect to the basic vectors p^l_n are defined by

$$\begin{pmatrix} 2l \\ l-n \end{pmatrix}^{1/2} (a\xi + c\eta)^{l-n}(b\xi + d\eta)^{l+n}$$

$$=: \sum_{m=-l}^{l} T^l_{mn}(a,b,c,d) \binom{2l}{l-m}^{1/2} \xi^{l-m}\eta^{l+m}, \tag{1.7.4}$$

where a,b,c,d again are entries of the fundamental representation matrix $\begin{pmatrix} a & b \\ c & d \end{pmatrix}$ which acts on the two dimensional space $\begin{pmatrix} \xi \\ \eta \end{pmatrix}$.

As is well known many important applications of group representation theory are based on the fundamental connection of matrix elements of irreducible representations with the special functions of mathematical physics [233]. In particular, the T^l_{mn} are expressible in terms of Jacobi polynomials, the Schur relations (1.7.2) for them are equivalent to orthogonality relations for Jacobi polynomials.

To generalize the construction of irreps for the case of the quantum group $SU_q(2)$, one needs the generalization of binomial formula for the case of quantum space (see, e.g. [99]).

Let $xy = qyx$, $0 < q < 1$. Then

$$(x+y)^n = \sum_{k=0}^{n} \begin{bmatrix} n \\ k \end{bmatrix}_q y^{n-k}x^k = \sum_{k=0}^{n} \begin{bmatrix} n \\ k \end{bmatrix}_{q^{-1}} x^k y^{n-k}, \tag{1.7.5}$$

where q-binomial coefficients have the form

$$\begin{bmatrix} n \\ k \end{bmatrix}_q = \frac{[n;q]!}{[n-k;q]!\,[k;q]!}, \qquad [n;q]! = [1;q][2;q]\ldots[n;q].$$

These deformed coefficients are related to those in (1.3.23):

$$\begin{bmatrix} n \\ k \end{bmatrix}_{q^2} = q^{k(n-k)} \begin{pmatrix} n \\ k \end{pmatrix}_q .$$

Using this formula, the matrix elements of $SU_q(2)$ representations can be found from the analog of the relation (1.7.4)

$$\delta\left(\begin{bmatrix} 2l \\ l+n \end{bmatrix}_{q^2}^{1/2} x^{l-n}y^{l+n} \right) = \begin{bmatrix} 2l \\ l-n \end{bmatrix}_{q^{-2}}^{1/2} (a\otimes x + b\otimes y)^{l-n}(c\otimes x + d\otimes y)^{l+n}$$

$$=: \sum_{m=-l}^{l} T^l_{nm}(a,b,c,d)\otimes \begin{bmatrix} 2l \\ l-m \end{bmatrix}_{q^{-2}}^{1/2} x^{l-m}y^{l+m} . \tag{1.7.6}$$

Here δ is the coaction (1.5.4). Application of the coassociativity and counity axioms to the lhs and the rhs of (1.7.6) shows that $T^l = (T^l_{mn})$ is indeed a matrix corepresentation. The binomial coefficients in (1.7.6) are inserted as they make the corepresentation unitary: by direct inspection one finds the relations

$$T^l_{nm}(a,b,c,d) = T^l_{mn}(a,c,b,d) \; ,$$

which, together with the explicit form of *-involution in the algebra $SU_q(2)$ (see Definition 1.27, for $SU_q(2)$ which gives $b = -qc^*$, $d = a^*$) lead to unitarity.

Actually, any irreducible unitary matrix corepresentation of $SU_q(2)$ is equivalent to some T^l.

In analogy with the classical case, the matrix elements T^l_{nm} are expressible [231, 181] in terms of the little *q-Jacobi polynomials* (for a definition of q-Jacobi polynomials, independent of the q-group theory, see, e.g. [99, 4])

$$P_n^{(\alpha,\beta)}(z;q) := \sum_{m=0}^{\infty} \frac{(q^{-n};q)_m (q^{\alpha+\beta+n+1};q)_m}{(q;q)_m (q^{\alpha+1};q)_m} (qz)^m \; ,$$

where $(p;q)_m$ is one more useful q-expression

$$(p;q)_m := \prod_{k=0}^{m-1} (1 - pq^k) \; .$$

An explicit representation for the elements T^l_{mn} in terms of the little q-Jacobi polynomials reads as

- if $m+n \leq 0, \; m \geq n$,

$$\begin{aligned} T^l_{mn} &= q^{(l+n)(l-m)} \begin{bmatrix} l+m \\ l-n \end{bmatrix}^{1/2}_{q^2} \begin{bmatrix} l-n \\ m-n \end{bmatrix}^{1/2}_{q^2} \\ &\times \; a^{-m-n} c^{m-n} P^{m-n,-m-n}_{l+n}(c^* c; q^2) \; ; \end{aligned}$$

- if $m+n \leq 0, \; n \geq m$,

$$\begin{aligned} T^l_{mn} &= q^{(l+n)(m-m)} \begin{bmatrix} l-m \\ n-m \end{bmatrix}^{1/2}_{q^2} \begin{bmatrix} l+m \\ n-m \end{bmatrix}^{1/2}_{q^2} \\ &\times \; a^{-m-n} (c^*)^{n-m} P^{n-m,-m-n}_{l+n}(c^* c; q^2) \; ; \end{aligned}$$

- if $m+n \geq 0, \; n \geq m$,

$$\begin{aligned} T^l_{mn} &= q^{(n-m)(n-l)} \begin{bmatrix} l-m \\ n-m \end{bmatrix}^{1/2}_{q^2} \begin{bmatrix} l+n \\ n-m \end{bmatrix}^{1/2}_{q^2} \\ &\times \; P^{n-m,m+n}_{l-n}(-c^* c; q^2)(-c^*)^{n-m} (a^*)^{m+n} \; ; \end{aligned}$$

- if $m+n \geq 0,\ m \geq n$,

$$\begin{aligned} T^l_{mn} &= q^{(m-n)(m-l)} \begin{bmatrix} l+m \\ m-n \end{bmatrix}^{1/2}_{q^2} \begin{bmatrix} l-n \\ m-n \end{bmatrix}^{1/2}_{q^2} \\ &\times\ P^{n-m,-m-n}_{l+n}(c^*c;q^2)c^{m-n}(a^*)^{m+n} . \end{aligned}$$

The interpretations of other q-special functions in terms of harmonic analysis on q-groups are reviewed in [147, 61].

1.7.2 Representations of quantum universal enveloping algebras

The main aim of this subsection is to illustrate two most general and important facts:

- representations of QUEA for generic value (not a root of unity) of the deformation parameter q are similar (though smoothly deformed, of course) to those of the classical Lie algebras;
- properties of representations are changed drastically at a root of unity value of the deformation parameter ($q^N = 1,\ N \in \mathbb{Z}$).

It is the algebra structure that is important for the construction of a representation. The Hopf algebra structure (Δ, ε, S) enters into the picture when one considers *a set* of representations and relations between them (in particular their tensor product).

1) Generic q.

As usual we consider at first the example of $su_q(2)$ algebra. It is convenient to use the basis $\{\widetilde{X}^{\pm}, k\}$ and CR (1.3.19) and to introduce for compactness the notations

$$e := \widetilde{X}^+, \qquad f := \widetilde{X}^- .$$

The CR (1.3.19) for $su_q(2)$ take the form

$$\begin{aligned} kek^{-1} &= q^2 e , \\ kfk^{-1} &= q^{-2} f , \\ [e,f] &= \frac{k-k^{-1}}{q-q^{-1}} \end{aligned} \tag{1.7.7}$$

The real form is defined by the *-involution: $e^* = f$, $k^* = k$ for $q \in \mathbb{R}$ and $k^* = k^{-1}$ for $q \in \mathbb{C}$, $|q| = 1$. Let (ρ, V) be a linear representation of $su_q(2)$ consisting of a vector space V and a homomorphism $\rho :\ su_q(2) \longrightarrow End\,V$.

Assume that $v \in V$ is the nonzero eigenvector of $\rho(e)$

$$\rho(e)v = cv , \qquad c \in \mathbb{C} .$$

Since k is invertible,

$$v^{(\pm n)} := k^{\pm n} v , \qquad \forall n \in \mathbb{Z} ,$$

are also eigenvectors of $\rho(e)$. For generic q (not a root of unity) the eigenvectors and corresponding eigenvalues are different. So if V is *finite* dimensional, the action of the generator e must be nilpotent and the eigenvalue $c = 0$. The generator f also acts nilpotently and if v_0 is the highest weight vector, $v_0 \in V_{high}$, where $V_{high} := \{v \in V \mid ev = 0\}$, then there is a smallest $l \in \mathbb{Z}_+$ satisfying

$$\begin{aligned} f^n v_0 &\neq 0 , \qquad 0 \leq n \leq l , \\ f^{l+1} v_0 &= 0 . \end{aligned}$$

Since $kV_{high} = V_{high}$, one can find $v_0 \in V_{high}$ such that

$$kv_0 = \xi v_0 , \qquad \xi \in \mathbb{C} , \xi \neq 0 .$$

Let $v_n := f^n v_0$. Then from the equalities

$$0 = \frac{1}{[l+1]_q!}[e, f^{l+1}]v_0 = \frac{q^{-l}\xi - q^l \xi^{-1}}{q - q^{-1}} v_l ,$$

one gets

$$\xi = \pm q^l$$

and the use of (1.7.7) gives

$$\begin{aligned} \rho_l^\sigma(e)v_n^{(l)} &= \sigma[l-n+1]_q v_{n-1}^{(l)} , \\ \rho_l^\sigma(f)v_n^{(l)} &= \sigma[n+1]_q v_{n+1}^{(l)} , \\ \rho_l^\sigma(k)v_n^{(l)} &= \sigma q^{l-2n} v_n^{(l)} , \end{aligned} \tag{1.7.8}$$

where $\sigma = \pm$, $v_n^{(l)} = 0$ if $n > l$ or $n < 0$. Therefore, the representation $(\rho_l^\sigma, V_l^\sigma)$ is a representation of dimension $(l+1)$. Thus (ρ_l^+, V_l^+) is a q-analog of the spin $l/2$ representation of $su(2)$. But contrary to this, (ρ_l^-, V_l^-) has not a non-deformed counterpart. This is related to the existence of the $su_q(2)$ automorphism κ

$$\kappa(e) = -e , \quad \kappa(f) = f , \quad \kappa(k) = -k ,$$

so that (ρ_l^-, V_l^-) can be obtained by the composition

$$\rho_l^- = \rho_l^+ \circ \kappa \,. \tag{1.7.9}$$

Standard arguments show that

- all representations $(\rho_l^\sigma, V_l^\sigma)$, $(l = 0, 1, 2, \ldots;\ \sigma = \pm)$ are irreducible and inequivalent to each other;
- any finite dimensional irreps of $su_q(2)$ is equivalent to one of $(\rho_l^\sigma, V_l^\sigma)$;
- a finite dimensional representation of $su_q(2)$ is completely reducible.

Thus *all* finite dimensional representations of $su_q(2)$ are *direct sum* of some copies of the $(\rho_l^\sigma, V_l^\sigma)$.

After appropriate normalization and transition to generators $\{X^\pm, H\}$ matrices of the representations $(\rho_l^\sigma, V_l^\sigma)$ coincide with the standard ones presented in (1.1.40) (note that j in the basis (1.1.40) is equal to $l/2$).

As in the case of classical Lie algebra $su(2)$, the representations are characterized by eigenvalues of the Casimir operator (1.1.39)

$$C_2^q \mid j, m\rangle = [j + \frac{1}{2}]_q^2 \mid j, m\rangle \,, \qquad \forall m \,. \tag{1.7.10}$$

Action of the generators in tensor product of irreps is defined by comultiplication

$$su_q(2) \xrightarrow{\Delta} su_q(2) \otimes su_q(2) \xrightarrow{\rho_1 \otimes \rho_2} End(V_1) \otimes End(V_2) \subset End(V_1 \otimes V_2) \,.$$

Let

$$w^{(l)} := \sum_{j=0}^{s} a_j v_j^{(m)} \otimes v_{s-j}^{(n)} \in V_m^+ \otimes V_n^+ \,,$$

where $s = 0, 1, \ldots, \min(m, n)$; $l = m + n - 2s$. A highest weight vector is defined by the condition

$$\Delta(e) w_0^{(l)} = 0 \,,$$

and has the form

$$w_0^{(l)} = \sum_{j=0}^{s} a_0 (-1)^j q^{j(m+1-j)} \frac{[n - s + j]_q! [m - j]_q!}{[m]_q! [n - s]_q!} v_j^{(m)} \otimes v_{s-j}^{(n)} \,.$$

Acting by $\Delta(f^k)$, one obtains the general weight vectors $w_k^{(l)}$ in $V_l \subset V_m \otimes V_n$ and the Clebsch-Gordan rule [143, 187, 184]:

- $$V_m^+ \otimes V_n^+ = V_{m+n}^+ \oplus \ldots \oplus V_{|m-n|}^+ \ , \quad \forall m,n \in \mathbb{Z}_+ \ .$$

We will consider a construction of the Clebsch-Gordan coefficients in the Section 2.4 using the Jordan-Schwinger realization of the algebra generators in terms of q-deformed oscillators.

Representations of arbitrary simple quantum algebra $\mathcal{U}_q(L)$ have the same general properties for generic q as the particular case of $\mathcal{U}_q(su(2))$:

- generators $X_i^\pm$ (or $\widetilde{X}_i^\pm$), $i = 1, \ldots, r := rank\ L$, are nilpotent;
- generators k_i, $i = 1, \ldots, r$ are diagonalizable and a representation space $V = \oplus V_\mu$, where
$$V_\mu := \{ v \in V \mid k_i v = q_i^{\mu_i} v \ \ \forall i \} \ ,$$
and $\mu = (\mu_1, \mu_2, \ldots, \mu_r)$ is the weight of the representation;
- finite dimensional representations of $\mathcal{U}(L)$ are completely reducible;
- if $\tilde{\mu}$ is the dominant weight ($\mu_i \in \mathbb{Z}_+$) , $V_{\tilde{\mu}}$ is irreducible and finite dimensional;
- any finite dimensional irreps of $\mathcal{U}_q(L)$ is equivalent to $V_{\tilde{\mu}}$ up to the sign twist, analogous to (1.7.9).

2) Representations for a root of unity value of q.

Let $q^p = 1$, i.e. $q = e^{\frac{2\pi i}{p}}$. In fact, all the features of quantum group representations at these specific values of the deformation parameter originate from the fundamental property of q-numbers

$$q^p = 1 \ \Rightarrow \ \left[\frac{p}{2}\right]_q = \left[\frac{p}{2}; q^2\right] = 0 \ . \tag{1.7.11}$$

The first obvious consequence of (1.7.11) is the appearance of additional zero entries in the representation matrices (1.7.8) or (1.1.40) for dimensions $l \geq p-1$ (or $j \geq (p-1)/2$). This implies that some representations being irreducible for generic value of q, may become *reducible* for a root of unity.

For example, the four-dimensional representation of $su_q(2)$ in the basis (1.1.36) has the following explicit form (cf. (1.1.40))

$$X^+ = \begin{pmatrix} 0 & \sqrt{[3]_q} & 0 & 0 \\ 0 & 0 & [2]_q & 0 \\ 0 & 0 & 0 & \sqrt{[3]_q} \\ 0 & 0 & 0 & 0 \end{pmatrix} ,$$

$$H = \begin{pmatrix} 3 & 0 & 0 & 0 \\ 0 & 1 & 0 & 0 \\ 0 & 0 & -1 & 0 \\ 0 & 0 & 0 & -3 \end{pmatrix} ,$$

$X^- = (X^+)^\dagger$. If $q^3 = \pm 1$ and, hence, $[3]_q = 0$, there is an obvious decomposition: $V_3 = V_0 \oplus V_1 \oplus V_0$, i.e. to the two one-dimensional and one two-dimensional irreps. This implies, in particular, that $(X^\pm)^2 = 0$. In general, one can check that $(X^\pm)^p = 0$ if $q^p = 1$.

Another new fact is that the Casimir operator (1.1.39) takes the same eigenvalues for the representations

$$|\, j,m\rangle \;, \quad |\, j+np,m'\rangle \;\text{ and }\; |\, p-1-j+np,m''\rangle \;, \quad \forall\, 2j,n \in \mathbb{Z} \;. \tag{1.7.12}$$

Thus, C_2^q does not separate such representations. An important consequence of this fact is the existence of *reducible* but *indecomposable representations.* Let us consider a concrete example again: product of the representations $V_1 \otimes V_2$. Action of the generators in the six-dimensional tensor product space is defined by the comultiplication

$$\Delta(X^+) = \begin{pmatrix} 0 & \sqrt{[2]_q q^{-1}} & 0 & q & 0 & 0 \\ 0 & 0 & \sqrt{[2]_q q^{-1}} & 0 & 1 & 0 \\ 0 & 0 & 0 & 0 & 0 & q^{-1} \\ 0 & 0 & 0 & 0 & \sqrt{[2]_q q} & 0 \\ 0 & 0 & 0 & 0 & 0 & \sqrt{[2]_q q} \\ 0 & 0 & 0 & 0 & 0 & 0 \end{pmatrix} . \tag{1.7.13}$$

In the case of generic q the representation can be decomposed into the sum of the representations V_3 and V_1: $V_1 \otimes V_2 \sim V_3 \oplus V_1$

$$\Delta_q(X^+) \to \mathcal{G}_q^{-1} \Delta_q(X^+) \mathcal{G}_q = \left(\begin{array}{cccc|cc} 0 & \sqrt{[3]_q} & 0 & 0 & 0 & 0 \\ 0 & 0 & [2]_q & 0 & 0 & 0 \\ 0 & 0 & 0 & \sqrt{[3]_q} & 0 & 0 \\ 0 & 0 & 0 & 0 & 0 & 0 \\ \hline 0 & 0 & 0 & 0 & 0 & 1 \\ 0 & 0 & 0 & 0 & 0 & 0 \end{array}\right) ,$$

by the use of the Clebsch-Gordan coefficients

$$\mathcal{G} = \begin{pmatrix} 1 & 0 & 0 & 0 & 0 & 0 \\ 0 & \sqrt{[2]_q(q[3]_q)^{-1}} & 0 & -q\sqrt{[3]_q^{-1}} & 0 & 0 \\ 0 & 0 & (q\sqrt{[3]_q})^{-1} & 0 & -\sqrt{q[2]_q[3]_q^{-1}} & 0 \\ 0 & q\sqrt{[3]_q^{-1}} & 0 & \sqrt{[2]_q(q[3]_q)^{-1}} & 0 & 0 \\ 0 & 0 & \sqrt{q[2]_q[3]_q^{-1}} & 0 & (q\sqrt{[3]_q})^{-1} & 0 \\ 0 & 0 & 0 & 0 & 0 & 1 \end{pmatrix} .$$

However, if $q^3 = \pm 1$, the entries of the $\mathcal{G}_q$ become singular and the decomposition is impossible. Note that the Casimir operator C_2^q has the same eigenvalues for V_3 and V_1 in this case (cf. (1.7.12))

$$j = \frac{3}{2} \leftrightarrow j' = p - 1 - j = \frac{1}{2} .$$

So the representations are mixed up to reducible, but not fully reducible representations, because of the presence of zero norm vectors. We can introduce the following orthogonal basis in the $V_1 \otimes V_2$ [251]

$$\begin{aligned} a^{\mathsf{T}} &= (0,0,0,0,0,1) , \\ d^{\mathsf{T}} &= (-1,0,0,0,0,0) , \\ b &= X^+ a , \\ c &= X^- d , \\ b'^{\mathsf{T}} &= q(0,0,1,0,-i,0) , \\ c'^{\mathsf{T}} &= q^2(0,i,0,0,-1,0) . \end{aligned} \tag{1.7.14}$$

Using the relation

$$\sum_{i=0}^{p} q^i = \frac{1-q^{p+1}}{1-q} = 1 \quad \text{if} \quad q^p = 1 ,$$

and (1.7.13), one finds that the vectors b and c have zero norm

$$b^{\mathsf{T}} b = c^{\mathsf{T}} c = 0 .$$

Besides,

$$X^+ b = c , \quad X^+ c = 0 ; \quad X^- c = b , \quad X^- b = 0 ,$$

so that these vectors form an invariant subspace. However, this subspace is not isolated from the remaining vectors (by construction, see (1.7.14)). Moreover, the vectors a and d can also be obtained from the others

$$a = X^- b' , \qquad d = X^+ c' .$$

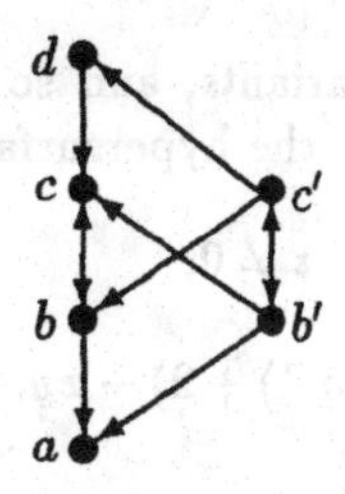

Figure 1.7: Structure of $V_3 \otimes V_1$ for $q^3 = 1$.

Complete structure of the indecomposable representation $V_3 \otimes V_1$ is depicted in Fig. 1.7 (arrows which have a projection up (down) correspond to X^+ (X^-)).

An appropriate characteristic, which distinguishes indecomposable representations from decomposable ones, is proved to be so-called *q-dimension*

$$\dim_q V_l := [2j+1]_q \,, \qquad l = 2j \,.$$

In the example above

$$\dim_q V_3 + \dim_q V_1 = [4]_q + [2]_q = 0 \quad \text{if} \quad q^3 = 1 \,.$$

The general situation is as follows

- type-I representations have q-dimension zero and are indecomposable;
- type-II representations
  - have a nonzero q-dimension,
  - have restricted *usual* dimension $0 \le j \le \frac{1}{2}p - 1$,
  - are described by their highest weight vector and are in correspondence with the representations of non-deformed $su(2)$.

This is not the whole story. The point is that in the case of root of unity value of q, there appear *new* central elements: in the basis of generators (1.7.7) these are ($q^p = \pm 1$)

$$e^p \,, \quad f^p \,, \quad k^p \,,$$

(note that H does not commute with e^p, f^p; thus different bases of $su_q(2)$ are not equivalent in this case). The representations considered already correspond to *zero* eigenvalue of e^p and f^p. The Casimir operator C_2^q is not algebraically independent on these new invariants, and so the general representations of $su_q(2)$ are marked by points on the hypersurface

$$\begin{aligned} Spec\ Z = \{(x,y,z,C) \in \mathbb{C} \mid\ z \neq 0, \\ \prod_{i=0}^{p-1}(C(q-q^{-1})^2 \mp (q^i+q^{-i})+2) - xy - (\pm z - 1)(\pm 1 - z^{-1}) = 0\}\ , \end{aligned}$$

where C is an eigenvalue of C_2^q; x, y, z are eigenvalues of the operators $((q - q^{-1})e)^p$, $(q-q^{-1})f)^p$, k^p, respectively.

The hypersurface $Spec\ Z$ has singularities, given by the set Sg of $p-1$ points

$$Sg := \left\{0, 0, \pm 1, \frac{\pm(q^j + q^{-j}) - 2}{(q-q^{-1})^2};\ \ j = 1, 2, \ldots, \frac{p-1}{2}\right\}\ .$$

The following results have been obtained by De Concini and Kac [73]:

- Every point of $Spec\ Z$ corresponds to irrep of $\mathcal{U}_q(su(2))$.
- Any non-singular point of $Spec\ Z$ marks one irreducible finite dimensional representation of dimension p. Define the $p \times p$ matrices

$$X = \begin{pmatrix} 0 & 0 & \cdots & 0 & 1 \\ 1 & 0 & \cdots & 0 & 0 \\ 0 & 1 & \cdots & 0 & 0 \\ \vdots & \vdots & \ddots & \vdots & \vdots \\ 0 & 0 & \cdots & 1 & 0 \end{pmatrix}, \qquad Z = \begin{pmatrix} 1 & 0 & 0 & \cdots & 0 \\ 0 & q & 0 & \cdots & 0 \\ 0 & 0 & q^2 & \cdots & 0 \\ \vdots & \vdots & \vdots & \ddots & \vdots \\ 0 & 0 & 0 & \cdots & q^{p-1} \end{pmatrix},$$

which satisfy

$$ZX = qXZ\ , \qquad Z^p = X^p = 1\ .$$

The representation is given by

$$\begin{aligned} e &= (q-q^{-1})^{-1}x_1(a_1 Z - a_1^{-1}Z^{-1})X\ , \\ f &= (q-q^{-1})^{-1}x_1^{-1}(a_2 Z^{-1} - a_2^{-1}Z)X^{-1}\ , \\ k &= \frac{a_1}{a_2}Z^2\ , \end{aligned}$$

where either $a_1^{2p} \neq 1$ or $a_2^{2p} \neq 1$; $a_1, a_2, x_1 \in \mathbb{C}\backslash 0$ and

$$x = (a_1^p - a_1^{-p})x_1^p\ , \quad y = (a_2^p - a_2^{-p})x_1^{-p}\ , \quad z = \left(\frac{a_1}{a_2}\right)^p\ ,$$

$$C = (q - q^{-1})^{-2}(qa_1a_2 + (qa_1a_2)^{-1} - 2)\ .$$

From the explicit form of the matrices X, Z, it follows that there is no highest weight vectors in this case and the operators e and f shift basic vectors of the representation spaces *cyclically*. These representations are called *cyclic representations*.

- To any singular point from the set Sg, there correspond exactly two usual highest weight irreducible finite dimensional representations $\rho^{\pm}_{j-1}, \rho^{\pm}_{p-j-1}$ (see (1.7.8)).

Representations of general quantum algebras $\mathcal{U}_q(L)$ for a root of unity value of q require technically more involved description but have essentially the same properties as in the case of $\mathcal{U}_q(su(2))$ [73, 170, 61].

1.7.3 Representations of quantized algebras of functions

The construction of representations of a non-commutative algebra with CR which in the classical limit correspond to Poisson brackets of functions on classical (phase) space, is the standard problem in quantum mechanics. In the simplest case of usual Heisenberg algebra, there exists only one (up to unitary equivalence) such representation (von Neumann theorem). Other, more complicated, algebras have a set of inequivalent representations which are in correspondence with *symplectic leaves* of the corresponding Poisson manifold (see Definition 1.5 and 1.6 in Section 1.2 and the discussion after them). For example, the space $su(2)^*$ dual to Lie algebra $su(2)$ is a union of spheres $S_r^{(2)}$ of radius r, $su(2)^* = \bigcup_r S_r^{(2)}$, which are symplectic leaves with respect to Lie-Kirillov-Kostant brackets (1.2.32) specified to this case

$$\{j_m, j_n\}_P = \varepsilon_{mnk} j_k\ , \tag{1.7.15}$$

where j_m ($m = 1, 2, 3$) are the generators of $Fun(su(2)^*)$. This bracket is non-degenerate [19] on the spheres $S_r^{(2)}$, defined by the condition

$$\sum_{m=1}^{3} (j_m)^2 = r^2\ . \tag{1.7.16}$$

The spheres $S_r^{(2)}$ are in correspondence with representations of the Lie algebra $su(2)$ generated by J_m ($m = 1, 2, 3$) with the CR

$$[J_m, J_n] = i\chi\varepsilon_{mnk} J_k\ . \tag{1.7.17}$$

The Poisson brackets on the very group $SU(2)$ satisfying the condition (1.2.33) (or,equivalently, (1.2.34)) are defined by (1.2.49) and in the standard parameterization

$$T = \begin{pmatrix} a & -\bar{c} \\ c & \bar{a} \end{pmatrix} \in SU(2) ,$$

have the following explicit form

$$\begin{aligned} \{a, c\}_P &= -iac , & \{a, \bar{c}\}_P &= -ia\bar{c} , \\ \{a, \bar{a}\}_P &= i2\bar{c}c , & \{c, \bar{c}\}_P &= 0 . \end{aligned} \tag{1.7.18}$$

Note, that since we know already the quantum CR (1.1.17), the technically simplest way to derive (1.7.18) is to use the classical limit

$$\{\cdot,\cdot\}_P = \lim_{\chi \to 0} \frac{1}{i\chi}[\cdot,\cdot] ;$$

in this subsection we consider real $q = e^{\chi}$.

From (1.7.18) one obtains that $\bar{c}c^{-1}$ has zero Poisson brackets with functions on $SU(2)$. Thus the points of $SU(2)$ with different values of $(\arg c)$ belong to different symplectic leaves. This gives the following classification (for the rigorous proof see [231]):

- first series: zero-dimensional symplectic leaves of the form

$$L^{(1)}_{\phi} = \begin{pmatrix} e^{i\phi} & 0 \\ 0 & e^{-i\phi} \end{pmatrix} , \qquad 0 \le \phi \le 2\pi ;$$

- second series: two-dimensional symplectic leaves of the form

$$L^{(2)}_{\phi} = \begin{pmatrix} a & -\rho e^{-i\phi} \\ \rho e^{i\phi} & \bar{a} \end{pmatrix} ,$$

$$a \in \mathbb{C} , \quad \rho \in \mathbb{R} , \quad 0 \le \phi \le 2\pi .$$

It is clear, that the second series defines two-dimensional spheres due to the unimodularity condition

$$\bar{a}a + \rho^2 = 1 .$$

Thus both sets of the Poisson brackets (1.7.15) and (1.7.18) define the Hamiltonian dynamics of the two-dimensional spheres and both cases were derived from the requirements of consistency with $SU(2)$ transformations! It is instructive to understand distinctly the difference between the consistency requirement in orbit method and in quantum group approach.

The Poisson relations (1.7.15) are invariant with respect to $SU(2)$ transformations with group parameters being considered as numbers or, in other words, having *trivial* Poisson brackets among themselves. So the consistency condition has the same general form (1.2.33) as in the case of Poisson-Lie group, but one of the terms in the definition (1.2.35) of Poisson brackets on the rhs of the (1.2.33) disappears. Contrary to this, the Poisson-Lie groups are defined by (1.2.33) where functions are considered on a tensor product of two copies of a group, both having *non-trivial* Poisson brackets and both terms on the rhs of (1.2.35) come to the play.

This circumstance leads to quite different explicit forms (1.7.15) and (1.7.18) of the Poisson brackets on two-dimensional spheres and to general distinction of the invariance conditions in the orbit method and Poisson-Lie group construction.

Analogous consideration is applicable to a quantum case and explains the difference in the commutation relations in orbit method and on quantum groups, in particular, (1.7.17) and (1.1.17). Unitary representations of (1.7.17) (i.e. irreps of $su(2)$) are well known and all are finite dimensional (this is general feature of compact Lie algebras [13]). Representations of CR (1.1.17) with the involution (1.6.22) have been found in [231]. There are two series of irreducible representations of $Fun_q(SU(2))$, each one is parameterized by $\phi \in [0, 2\pi]$:

- one-dimensional representation ρ_ϕ,

$$\begin{aligned} \rho_\phi(a) &= e^{i\phi}\,, \qquad \rho_\phi(c) = 0\,, \\ \rho_\phi(d) &= \rho_\phi(a^*) = \rho_\phi(a)^\dagger = e^{-i\phi}\,, \\ \rho_\phi(b) &= \rho_\phi(-qc^*) = -q\rho_\phi(c)^\dagger = 0\,, \end{aligned}$$

  (recall that in the $SU_q(2)$ case: $d = a^*$, $b = -qc^*$);

- infinite dimensional representations π_ϕ in the Hilbert space $\mathcal{H} = l_2(\mathbb{Z}_+)$; in the standard basis $\{|\, n\rangle\}_{n=0}^{\infty}$ one has

$$\begin{aligned} \pi_\phi(a)\,|\,n\rangle &= \sqrt{1-q^{2n}}\,|\,n-1\rangle\,, \qquad \pi_\phi(a)\,|\,0\rangle = 0\,, \\ \pi_\phi(c)\,|\,n\rangle &= e^{i\phi}q^n\,|\,n\rangle \\ \pi_\phi(d) &= \pi_\phi(a)^\dagger\,, \qquad \pi_\phi(b) = -q\pi_\phi(c)^\dagger\,. \end{aligned} \tag{1.7.19}$$

It is easy to see close relation of the representation classification with the pattern of symplectic leaves in $SU(2)$.

To understand the origin of infinite dimensionality of the $Fun_q(SU(2))$ representations, it is convenient to express the quantum matrix (1.1.18) as *an exponential* of a matrix with suitable non-commuting entries

$$T = \begin{pmatrix} a & b \\ c & d \end{pmatrix} = \exp\{\chi M\} , \qquad (1.7.20)$$

where

$$M = \begin{pmatrix} D & P_1 \\ P_2 & -D \end{pmatrix} ,$$

and

$$\begin{aligned} [D, P_1] &= P_1 , \\ [D, P_2] &= P_2 , \\ [P_1, P_2] &= 0 . \end{aligned} \qquad (1.7.21)$$

The expression (1.7.20) can hardly be interpreted as an analog of the exponential map for classical groups. It is rather the generalization of a transition from Heisenberg CR (1.1.1) to Weyl CR (1.1.8) in quantum mechanics. The operators D, P_1, P_2 with the CR (1.7.21) generate non-semisimple, non-compact solvable Lie algebra of infinitesimal translations and dilatations of a two-dimensional space. Its unitary representations are either trivial or infinite dimensional (because of non-compactness of the Lie algebra, see, e.g. [13]) and are characterized by eigenvalues $\phi \in [0, 2\pi]$ of the Casimir operator

$$C = P_1 P_2^{-1} .$$

This explains the properties of representations of $Fun_q(SU(2))$.

Representations of $Fun_q(SL(2,\mathbb{C}))$ serve as a building blocks for constructions of representations of other quantum groups [222]. This is based on the map: $Fun_q(G) \longrightarrow Fun_q(SL(2,\mathbb{C})_i)$, where $Fun_q(SL(2,\mathbb{C})_i)$ corresponds to the subalgebra $\mathcal{U}_q(s\ell(2)_i)$ generated by a standard triple $\{X_i^{\pm}, H_i\}$ in Chevalley basis of $\mathcal{U}_q(L)$. For details of this construction see [222, 223, 231, 162].

This concludes the Chapter 1 devoted to the general constructions and properties of quantum groups and algebras. In the following Chapters along with further developments of general aspects, we shall consider different applications of q-groups and non-commutative geometry.

Chapter 2

q-Deformation of Harmonic Oscillators, Quantum Symmetry and All That

Harmonic oscillator plays in physics the role analogous to that of the $SL(2)$ in group theory. It is the starting point for construction of more complicated systems, basic object for development of new methods in mathematical physics, excellent example for demonstrations of different physical phenomena etc. Besides, and perhaps most important, it has numerous direct applications for description of real physical systems. Finally, algebra of oscillator operators plays essential role in mathematics being connected with certain class of special functions and representations of semisimple Lie algebras.

In this chapter we will study q-deformations of oscillator algebras both from physical point of view as simplest deformed dynamical system and from pure mathematical point of view as a tool for construction of representations of the q-deformed Lie algebras.

Actually, the generalization of commutation relations for the canonical operators (coordinate-momentum or creation-annihilation operators) was suggested, long before the discovery of quantum groups, by Heisenberg [120] in attempts to achieve regularization for his (nonrenormalizable) nonlinear spinor field theory (see also Wigner [241]). In the beginning of seventies such commutation relations were introduced in the form of q-deformation

$$aa^{+} - qa^{+}a = 1 ,$$

in the works [66, 6] devoted to investigations of dual models with nonlinear spectrum. Some modifications of the relations for oscillator algebra and their possible physical interpretations as spectrum generating algebras or algebras of nonstandard statistics were considered in [7, 158, 128].

After the development of quantum group theory there appeared a huge number of works devoted to different aspects and applications of q-oscillators. Here we mention only a few papers which opened this "new wave" of interest to q-oscillators: [172, 23, 118, 38, 157].

2.1 q-Deformation of single harmonic oscillator

There are a few different forms of q-deformed oscillator algebra. In most modern works the basic definition of a q-oscillator is the following:

Definition 2.1 *A q-oscillator is an associative algebra $\mathcal{A}(q)$ with three generators a, a^+, N which satisfy the relations*

$$aa^+ - qa^+a = q^{-N} , \tag{2.1.1}$$

$$[N, a] = -a , \quad [N, a^+] = a^+ , \tag{2.1.2}$$

$$(a^+)^\dagger = a , \quad N^\dagger = N ,$$

$$q \in \mathbb{R} \quad \text{or} \quad q \in \mathbb{C}, \ |q| = 1 .$$

Note that in this definition we do not postulate any relation among a^+, a and N. This means that a q-oscillator must be considered as the deformation of the solvable oscillator algebra

$$[a_0, a_0^+] = 1 , \quad [N_0, a_0] = -a_0 , \quad [N_0, a_0^+] = a_0^+ \tag{2.1.3}$$

(no relations between N_0 and a_0^+, a_0) and not of the nilpotent algebra of usual harmonic oscillator

$$[b_0, b_0^+] = 1 , \quad N = b_0^+ b_0 ,$$

(in this relation the unity is considered as one of the Lie algebra generators commuting with all others).

To analyze irreducible representations we need an analog of the Casimir operator, i.e. a central element of the algebra (2.1.1),(2.1.2). With use of the identities

$$aa^+a = (qa^+a + q^{-N})a ,$$

and

$$a(q^{-N}a^+a) = (q^{-N}a^+a + q^{-1}q^{-2N})a \ ,$$

one finds that the central element has the form [157]

$$\zeta = q^{1-N}([N]_q - a^+a) \ . \qquad (2.1.4)$$

In the $q \to 1$ limit, the central element becomes trivial

$$\zeta \longrightarrow \zeta_0 = N_0 - a_0^+ a_0 \ ,$$

since any of its eigenvalues can be shifted to zero by the algebra (2.1.3) automorphism

$$N_0 \longrightarrow N_0 + \gamma \ , \qquad \gamma \in \mathbb{R} \ .$$

This corresponds to the famous von Neumann theorem (see, e.g. [17, 195]) which states that all representations of the oscillator or the Heisenberg (coor dinate-momentum) algebras are unitary equivalent.

This is not the case for a q-deformed oscillator. Values of the central element label different representations. Before we will proceed to their classification, we mention that in the case of real q-parameter there are other forms of the q-algebra related to the initial one by invertible transformations

$$\alpha = q^{-\frac{N}{2}}a \ , \quad \alpha^+ = a^+q^{-\frac{N}{2}} \ , \qquad (2.1.5)$$

$$\alpha\alpha^+ - \alpha^+\alpha = q^{-2N} \ , \qquad (2.1.6)$$

$$[N, \alpha] = -\alpha \ , \quad [N, \alpha^+] = \alpha^+$$

and

$$b = q^{\frac{N}{2}}a \ , \quad b^+ = a^+q^{\frac{N}{2}} \ , \qquad (2.1.7)$$

$$bb^+ - q^2b^+b = 1 \ , \qquad (2.1.8)$$

$$[N, b] = -b \ , \quad [N, b^+] = b^+ \ .$$

For complex values of q these transformations also exist but the new creation and annihilation operators prove to be non-conjugate to each other. The central element in terms of these operators takes the form

$$\begin{aligned} \zeta &= [N; q^{-2}] - \alpha^+\alpha \\ &= q^{2(1-N)}([N; q^2] - b^+b) \ . \end{aligned} \qquad (2.1.9)$$

Irreducible representations for *real* q have been classified in [41, 157, 51]. We present here the classification in terms of the generators b^+, b, (2.1.7),(2.1.8), following the last paper:

- **(A)** the Fock representation for any $q > 0$ with a non-degenerate spectrum of b^+b given by the formula

$$\lambda_k = \frac{1-q^{2k}}{1-q^2} =: [k;q^2], \quad k = 0,1,2,\ldots, \qquad \zeta = 0 \,. \tag{2.1.10}$$

  If $0 < q < 1$, the operators b and b^+ are bounded;

- **(B)** the non-Fock representations for $0 < q < 1$ with the non-degenerate spectrum of b^+b, given as

$$\lambda_k = \frac{1+q^{2k+2\gamma}}{1-q^2} =: \{k+\gamma;q^2\}, \quad k \in \mathbb{Z}\,, \qquad \zeta = -q^2\{\gamma;q^2\}\,, \tag{2.1.11}$$

  and the representations are classified by $\gamma \in [0,1)$;

- **(C)** the degenerate representation for $0 < q < 1$ with

$$b^+b = bb^+ = (1-q^2)^{-1}\,, \qquad \zeta = -q^2(1-q^2)^{-1}\,. \tag{2.1.12}$$

To prove the classification, we define the self-adjoint operator

$$K = bb^+ - b^+b. \tag{2.1.13}$$

From eq. (2.1.8) it follows that K has the property

$$Kb = q^{-2}bK, \quad Kb^+ = q^2b^+K. \tag{2.1.14}$$

This equation gives

$$b^+Kb = q^{-2}b^+bK, \quad bKb^+ = q^2bb^+K\,. \tag{2.1.15}$$

Since b^+b and bb^+ are positive operators commuting with K, we see that K is definite on a subspace $\mathcal{H}$ spanned by vectors $(b^+)^m|k\rangle$, $b^m|k\rangle$, $m = 0,1,2,\ldots$, where $|k\rangle$ is some eigenstate of K. Thus, the representations are classified by the sign of the commutator K.

The case $K = 0$ just corresponds to the case **(C)**, and (2.1.12) immediately follows.

If $K \neq 0$ we put $|K| = q^M$. From (2.1.14) we obtain

$$[M,b] = -b, \quad [M,b^+] = b^+\,. \tag{2.1.16}$$

The operator $\exp(2\pi i M)$ then commutes with b and b^+ and has in any irreducible representation the fixed value

$$\exp(2\pi i M) = \exp(2\pi i \gamma)\,, \quad \gamma \in [0,1)\,. \tag{2.1.17}$$

Consequently, M has a discrete spectrum containing points of the form $k+\gamma$, with k being an integer. Let $|k\rangle$ be some normalized eigenstate

$$M|k\rangle = (k+\gamma)|k\rangle. \tag{2.1.18}$$

Since N commutes with M, it has the same eigenstates and due to the algebra automorphism $N \longrightarrow N+\omega$, $\forall \omega \in \mathbb{R}$, we can put

$$N \mid k\rangle = k \mid k\rangle\,,$$

so that

$$M = N + \gamma\,.$$

Using (2.1.16), one can show that $b^+|k\rangle$ and $b|k\rangle$ are the eigenstates of M (provided that they are non-vanishing) with the eigenvalues $k+1+\gamma$ and $k-1+\gamma$, respectively. Simultaneously, they are the eigenstates of N with the eigenvalues $k+1$ and $k-1$.

Let $K = q^M > 0$. Then from (2.1.8) and (2.1.13) we obtain

$$b^+b = \frac{1-q^{2M}}{1-q^2} = [M;q^2]\,. \tag{2.1.19}$$

As $[k+\gamma] < 0$ for $k+\gamma < 0$, we see that in this case only $\gamma = 0$ is allowed. Then there exists a system of normalized eigenstates $|k\rangle, k = 0,1,2,\ldots$, such that

$$M|k\rangle = k|k\rangle,\ k = 0,1,2,\ldots\,, \quad N = M\,, \tag{2.1.20}$$

$$b^+|k\rangle = [k+1;q^2]^{1/2}|k+1\rangle,\ b|k\rangle = [k;q^2]^{1/2}|k-1\rangle\,, \tag{2.1.21}$$

and then the formula (2.1.10) directly follows.

We see that all states $|k\rangle, k = 1,2,\ldots$, can be obtained by the repeated action of b^+ on the vacuum state $|0\rangle$, satisfying $b|0\rangle = 0$. We refer to this as to *the Fock case* (**A**).

If $K = -q^M < 0$, then

$$b^+b = \frac{1+q^{2M}}{1-q^2} = \{M;q^2\}\,. \tag{2.1.22}$$

As b^+b is positive, only $0 < q < 1$ is allowed. Then the normalized eigenstates $|k\rangle, k \in \mathbf{Z}$, of M satisfy

$$M|k\rangle = (k+\gamma)|k\rangle,\ k \in \mathbb{Z}\ , \quad N = M - \gamma\ . \tag{2.1.23}$$

The formula (2.1.11) follows immediately. In this case

$$b^+|k\rangle = \{k+\gamma+1\}^{1/2}|k+1\rangle\ , b|k\rangle = \{k+\gamma\}^{1/2}|k-1\rangle\ . \tag{2.1.24}$$

Since now there is no vacuum state, we refer to this as to *the non-Fock case* **(B)**.

This completes the classification for real parameter q.

For transformations which do not change the values of the central element ζ, there exists an analog of the von Neumann theorem [54]

- Let b_i and b_i^+, $i = 1,2$, be two irreducible representations of the q-oscillator algebra with the same eigenvalue of the central element ζ. If in each of the spaces V_i there exists a vector $|0_i>$: $N_i|0_i>= 0$, then there exists a unitary (or quasi-unitary) operator U, such that

$$b_2 = Ub_1U^{-1},\ \ b_2^+ = Ub_2U^{-1},\ \ N_2 = UN_1U^{-1}\ . \tag{2.1.25}$$

By definition, quasi-unitary operator is an operator satisfying the relations: $UU^\dagger = U^\dagger U = -1$. It is sufficient to consider only the case when $\langle 0_1|0_1\rangle = \langle 0_2|0_2\rangle$, since the proof in the case $\langle 0_1|0_1\rangle = -\langle 0_2|0_2\rangle$ is similar. If $\langle 0_1|0_1\rangle = \langle 0_2|0_2\rangle$, then U is a unitary operator; if $\langle 0_1|0_1\rangle = -\langle 0_2|0_2\rangle$, then U is a quasi-unitary one.

For the proof, we notice that to any arbitrary vector $|\alpha_1\rangle$ of the space V_1, one can put into correspondence a vector $|\alpha_2\rangle$ of the space V_2, which is expressed in terms of the vectors $|k_2\rangle$ in the same way as $|\alpha_1\rangle$ is expressed in terms of $|k_1\rangle$. Let us prove, that the above-mentioned correspondence preserves the scalar product, i.e. for arbitrary vectors we have

$$\langle\alpha_1|\beta_1\rangle = \langle\alpha_2|\beta_2\rangle\ . \tag{2.1.26}$$

To prove (2.1.26), we observe that for an arbitrary function $\psi(\mathrm{N})$ the following equation is valid:

$$\langle 0_1|\psi(\mathrm{N}_1)|0_1\rangle = \langle 0_2|\psi(\mathrm{N}_2)|0_2\rangle.$$

Using the commutation relations (2.1.2) and (2.1.9) we obtain that $\langle\alpha_i|\beta_i\rangle = \langle 0_i|\psi(\mathrm{N}_i)|0_i\rangle$, where the functional form of $\psi(\mathrm{N}_i)$ does not depend on i, i.e. eq. (2.1.26) is proven.

Introduce now the operator U: $|\alpha_2\rangle = U|\alpha_1\rangle$. According to (2.1.26),

$$\langle\alpha_1|U^\dagger U|\beta_1\rangle = \langle\alpha_1|\beta_1\rangle .$$

We prove that $U^\dagger U = 1$, i.e. U is an isometric operator. Since the operator U provides with a one-to-one correspondence between the spaces V_1 and V_2, then there exists its inverse U^{-1}. In such a case, as easily seen, from the equality $U^\dagger U$=1 follows that $UU^\dagger$=1, i.e. U is a unitary operator. The required eq. (2.1.25) is then a direct consequence of unitarity of U and of the equation (2.1.26). Indeed, let $|\beta_i\rangle = a_i|\beta_i'\rangle$. According to previously proven: $\langle\alpha_1|\beta_1\rangle = \langle\alpha_2|\beta_2\rangle$, but $|\alpha_2\rangle = U|\alpha_1\rangle$, $|\beta_2'\rangle = U|\beta_1'\rangle$, and consequently

$$\langle\alpha_1|a_1|\beta_1'\rangle = \langle\alpha_1|U^+ a_2 U|\beta_1'\rangle ,$$

and (2.1.25) is derived easily.

In the case of $q^2 < 0$ (i.e. pure *imaginary* q), the intervals $(-1,0)$ and $(-\infty,-1)$ are equivalent due to the substitution $b \longrightarrow \sqrt{-q^2}b$. There are two representations with negative q^2 [157]

- **(D)** the degenerate representation of the type (C)

$$b^+b = bb^+ = (1-q^2)^{-1} , \qquad \zeta = -\frac{q^2}{1-q^2};$$

- **(E)** the Fock representation of the type (A), $\zeta = 0$.

Note that the representations **(B)**,**(C)** are singular at $q^2 \longrightarrow 1$, **(D)** is singular at $q^2 \longrightarrow -1$ and the representation **(E)** in the limit $q^2 \longrightarrow 1$ corresponds to a fermionic oscillator.

In the case of the $*$-algebra of a, a^+, N and complex $q \in \mathbb{C}$, there is the only possibility $|q| = 1$, and after conjugation of (2.1.1) an additional relation appears [172, 23]:

$$aa^+ - q^{-1}a^+a = q^N \tag{2.1.27}$$

The algebra (2.1.1), (2.1.2) and (2.1.27) is denoted by $\mathcal{A}(q, q^{-1})$. The relations (2.1.1) and (2.1.27) immediately give

$$a^+a = [N]_q \, , \; aa^+ = [N+1]_q \, , \tag{2.1.28}$$

and thus there exist the representations of Fock type only in this case.

Similar to the quantum Lie algebras, properties of the oscillator algebra representations are changed drastically for special values of deformation parameter, i.e. root of unity: $q^p = 1$ for some integer p. In this case the algebra $\mathcal{A}(q, q^{-1})$ also has Fock type representations of the form (2.1.20),(2.1.21), but Fock spaces prove to be finite dimensional, the dimension being dependent on the value of q, or rather p. This follows directly from (1.7.11). Thus the algebras $\mathcal{A}(q, q^{-1})$ with root of unity parameter q, have finite dimensional Fock representations of dimension p for odd values of p and of dimension $p/2$ for even p.

As we mentioned above, if q is complex the central charge ζ is equal to zero but in roots of unity case there appear two new central elements of the q-boson algebra: a^{+^p} and a^p. This follows, in particular, from the general relation

$$a(a^+)^i = (qa^+)^i a + [i]_q q^N (a^+)^{i-1} , \qquad \forall i \in \mathbb{Z} .$$

One can simultaneously diagonalize these central elements and the q^N, since the latter commute with them. The representations for which a^{+^p}, a^p have nonzero eigenvalues are called *cyclic representations*, as all vectors in the representation (of dimension p) can be generated from any nonzero vector (hence these representations are not of highest weight type). In some basis $\xi_0, ..., \xi_{p-1}$, the representations are labeled by $l \in \mathbb{C} - \{0\}$, $\mu, \nu \in \mathbb{C}$ and have the form

$$\begin{aligned} q^N \xi_n &= l q^n \xi_n , \qquad n = 0, ...p-1 , \\ a^+ \xi_n &= \xi_{n+1} , \qquad n = 0, ...p-2 , \\ a\xi_{p-1} &= \mu \xi_0 , \\ a\xi_n &= (l^{-1}[n]_q + q^n \mu\nu)\xi_{n-1} , \qquad n = 1, ..., p-1 , \\ a\xi_0 &= \nu \xi_{p-1} . \end{aligned}$$

For general values of parameters l, μ, ν, the operators a^+, a are not hermitian conjugate to each other. So, in general, the algebra (2.1.1), (2.1.2) is not equivalent to the restricted algebra with zero central element even for complex values of parameter q (in the derivation of expression for the central charge, we assumed that the creation and annihilation operators are conjugate). The restricted algebra

$$a^+ a = [N]_q , \qquad aa^+ = [N+1]_q ,$$

corresponds to an additional relation for the parameters μ, ν, l

$$\mu\nu = \frac{l - l^{-1}}{q - q^{-1}} .$$

Further restrictions on the parameters and use of rescaling

$$a^+ \longrightarrow \omega a^+ \,, \quad a \longrightarrow \omega^{-1} a \,, \omega \in \mathbb{C} - \{0\} \,,$$

which leaves the q-algebra unchanged, permit to obtain representation with the hermitian conjugate operators. The properties of $\mathcal{A}(q, q^{-1})$ have close relations with unusual properties of representations of quantum groups with root of unity parameter q (cf. Subsection 1.7.2 and Section 2.5).

Note that because of the exponent in the rhs of (2.1.1), the algebra of a^+, a operators is non-quadratic. But if parameter q is equal to root of unity, the relation (2.1.1) in non-cyclic irreducible representations is polynomial due to nilpotency of a^+, a (finite dimensionality of the Fock space). One can check easily that, e.g. for $q^2 = -1$ the relation (2.1.1) is equivalent to anticommutation relation for Pauli matrices $\sigma^\pm$

$$\sigma^+\sigma^- + \sigma^-\sigma^+ = 1 \,. \tag{2.1.29}$$

Let us consider the next even value of integer parameter, $p = 6$, for the q-oscillators, i.e. $q^3 = -1$. In this case the Fock space is 3-dimensional and operator q^{-N} can be written in the form

$$q^{-N} = C_0 + C_1 a^+ a + C_2 a^{+^2} a^2 \,. \tag{2.1.30}$$

Comparing the action of the lhs and rhs of (2.1.30) on arbitrary vector of Fock space, one finds

$$q^{-N} = 1 - q a^+ a - a^{+^2} a^2 \,. \tag{2.1.31}$$

Substitution of this relation into (2.1.1), gives

$$a a^+ + a^{+^2} a^2 = 1 \,.$$

This looks as a very natural generalization of relation (2.1.29) for Pauli matrices.

Representation (2.1.21),(2.1.7) of a^+, a in this case has a very simple form

$$a^+ = \begin{pmatrix} 0 & 1 & 0 \\ 0 & 0 & 1 \\ 0 & 0 & 0 \end{pmatrix} \,, \qquad a = \begin{pmatrix} 0 & 0 & 0 \\ 1 & 0 & 0 \\ 0 & 1 & 0 \end{pmatrix} \,. \tag{2.1.32}$$

We immediately recognize in (2.1.32) the generators $J_\pm$ of usual non-deformed $su(2)$ in the spin-1 representation up to a trivial rescaling. Now one can find N in the same way as q^{-N} in (2.1.30),(2.1.31); the result is

$$N = a^+ a + a^{+^2} a^2 \,,$$

so that

$$[a,a^+] = 1 - N\ .$$

Thus we can identify

$$J_+^{(1)} = a^+/\sqrt{2}\ , \quad J_-^{(1)} = a/\sqrt{2}\ , \quad J_3^{(1)} = (N-1)/2\ ,$$

where the superscript index indicates the representation (spin 1).

As $J_\pm$ coincide with a^+, a for $K = 2,3$, one can expect that analogous relations hold for all values of K. However, this is not the case. Indeed, considering just the next value $K = 8$, one finds

$$aa^+ = 1 + (\sqrt{2}-1)a^+a - (\sqrt{2}-1)a^{+^2}a^2 + (\sqrt{2}-2)a^{3^+}a^3\ .$$

But $[a,a^+] \neq A + BN$, for any constants A, B. Instead, inspired by the $q^2 = -1$ and $q^3 = -1$ cases one can assume that the algebras with generators c^+, c and CR [58]

$$cc^+ + (c^+)^p c^p = 1\ , \tag{2.1.33}$$

have Fock representations of dimensions $p+1$. This is indeed the case, as (2.1.33) have the representations

$$(c^+)_{ij} = \delta_{i,j-1}\ , \quad (c)_{ij} = \delta_{i-1,j}\ , \quad i,j = 1,...,p+1\ ,$$

which is the natural generalization of Pauli matrices $\sigma^\pm$. The matrices of the representation are extremely simple and, in a sense, basic as any operator in finite dimensional Fock space can be expressed in terms of c^+, c.

Now we return to the *general* values of the parameter q and notice that the q-oscillator algebra can be obtained [38] as the result of the contraction of the quantum Lie algebra $su_q(2)$. For this aim one has to redefine the generators of $su_q(2)$

$$H \longrightarrow N := j\mathbf{1} - H\ , \qquad X^\pm \longrightarrow Y^\pm := X^\pm [2j]_q^{-\frac{1}{2}}\ ,$$

and to consider the vectors of the representations (1.1.40) which satisfy

$$\lim_{j\to\infty} \frac{j-m}{j} = 0\ .$$

Then the operators $\alpha^+ = \lim_{j\to\infty} Y^+$, $\alpha = \lim_{j\to\infty} Y^-$ generate q-oscillator algebra in terms of α-operators with the CR (2.1.6), which is equivalent to the relations (2.1.1),(2.1.2) for a and a^+ due to (2.1.5).

In contrast to deformed Lie algebras, a Hopf structure for the q-oscillator (2.1.1),(2.1.2) (or (2.1.5)-(2.1.8)) is not known. However, there is another deformed oscillator algebra with known and nontrivial Hopf structure. This algebra also appears as the result of appropriate contraction of the q-Lie algebra $u_q(2)$ [35]. Recall that the usual oscillator algebra can be obtained by the contraction of Lie algebra $u(2)$. For this aim, one rescales the $u(2)$ generators as follows

$$\begin{aligned} a_0 &:= \varepsilon J_+ , \qquad & a_0^+ &:= \varepsilon J_- , \\ h &:= 2K , \qquad & N &:= \varepsilon^{-2} K - J_0 , \end{aligned}$$

and takes the limit $\varepsilon \to 0$; the operator K generates $u(1)$ in $u(2) = u(1)\otimes su(2)$ and h, being a central element, plays the role of the Planck constant.

New feature of q-algebra contraction considered in [35] is that the parameter q takes part in the contraction also. q-Deformed oscillator algebra is obtained after the following transformation of $u(1)\otimes su_q(2)$ generators and $q = e^{\chi}$

$$\begin{aligned} A &:= \varepsilon X^+ , \qquad & A^+ &:= \varepsilon X^- , \\ h &:= 2K , \qquad & N &:= \varepsilon^{-2} K - H , \\ \omega &:= \varepsilon^{-2} \chi . \end{aligned}$$

Taking the limit $\varepsilon \to 0$, one gets

$$\begin{aligned} [A, A^+] &= \frac{\sinh(\omega h/2)}{\omega/2} , \\ [N, A] &= -A , \qquad [N, A^+] = A^+ , \\ [h, A] &= [h, A^+] = [h, N] = 0 . \end{aligned}$$

It is clear that these CR are equivalent to those of a usual oscillator (the only difference is redefinition of the central element $h \longrightarrow 2\sinh(\omega h/2)/\omega$). But the comultiplication remains, after the contraction, nontrivial (cf. (1.3.5))

$$\begin{aligned} \Delta A &= A \otimes e^{-\omega h/4} + e^{\omega h/4} \otimes A , \\ \Delta A^+ &= A^+ \otimes e^{-\omega h/4} + e^{\omega h/4} \otimes A^+ , \\ \Delta N &= \mathbf{1} \otimes N + N \otimes \mathbf{1} , \\ \Delta h &= \mathbf{1} \otimes h + h \otimes \mathbf{1} . \end{aligned}$$

Thus the only non-trivially deformed part of this construction is the coalgebra. Another rescaling of $su_q(2)$ generators

$$\begin{aligned} h &:= \varepsilon^2 (X^+ + X^-)/2 , \qquad & x &:= -i\varepsilon (X^+ - X^-)/2 , \\ p &:= \varepsilon H , \qquad & \omega &:= \varepsilon^{-1} \chi , \end{aligned}$$

gives after the contraction $\varepsilon \to 0$, the Heisenberg algebra

$$[x,p] = ih\ , \quad [h,x] = [h,p] = 0\ ,$$

with nontrivial coalgebra structure

$$\begin{aligned}
\Delta h &= e^{-\omega p/2} \otimes h + h \otimes e^{\omega p/2}\ , \\
\Delta x &= e^{-\omega p/2} \otimes x + x \otimes e^{\omega p/2}\ , \\
\Delta p &= \mathbf{1} \otimes p + p \otimes \mathbf{1}\ , \\
S(x) &= -x + i\frac{\omega}{2}h\ , \quad S(p) = -p\ , \quad S(h) = -h\ .
\end{aligned}$$

Another contraction of $s\ell_q(2)$ leads to the q-oscillator (2.1.1) [38, 157]. Notice also that appropriate contractions of simple quantum algebras give inhomogeneous ones, see Section 3.5.3.

Notice that there does not exist a two-parametric deformation of the q-oscillator algebra. An attempt to introduce a q-oscillator algebra, e.g. of the form

$$\tilde{a}\tilde{a}^+ - qs^{-1}\tilde{a}^+\tilde{a} = (qs)^{-N}\ ,$$

$$[N, \tilde{a}] = -\tilde{a}\ , \quad [N, \tilde{a}^+] = \tilde{a}^+\ ,$$

with seemingly two deformation parameters s and q is nothing but a simple redefinition of the generators

$$\tilde{a} = s^{-\frac{N}{2}}a\ , \qquad \tilde{a}^+ = a^+ s^{-\frac{N}{2}}\ ,$$

which brings back the above algebra to a one-parametric q-oscillator algebra (2.1.1),(2.1.2).

In conclusion of this section we consider two types of realizations of the q-oscillator algebra (2.1.8) in terms of non-polynomial operator functions of usual coordinates and derivatives.

1) Bargmann (holomorphic) realization of the Fock representation.

The Hilbert space $\mathcal{H}$ in this case is spanned by functions of the complex variable z:

$$\phi_k(z) = ([k;q^2]!)^{-1/2} z^k, \quad k = 0,1,2,\dots\ . \tag{2.1.34}$$

They are eigenfunctions of the operator $M = z\partial_z$ with the eigenvalue k, where ∂_z here is *the usual non-deformed* derivative . The operators b^+ and b have the form

$$b^+ = z, \quad b = \frac{1}{z}[M;q^2]\ . \tag{2.1.35}$$

In $\mathcal{H}$ the scalar product is defined by $(\phi_n, \phi_m) = \delta_{nm}$, and can be expressed in terms of the Jackson integral.

The functions (2.1.34) can be reinterpreted as coefficients of q -coherent states in the basis $|k\rangle$, $k = 0, 1, 2, \ldots$,(see [23, 131, 196]):

$$\langle k|z\rangle = \phi_k^*(y), \; k = 0, 1, 2, \ldots . \tag{2.1.36}$$

For $q > 1$ all $z \in \mathbb{C}$ are admissible, whereas $|z| < (1-q^2)^{-1}$ for $0 < q < 1$.

2) Macfarlane realization.

The standard Macfarlane q-oscillator representation [172] is defined by the operators

$$\tilde{b} = \frac{1}{1-q^2}[e^{-2is\varphi} - e^{-is\varphi}e^{is\partial}],$$

$$\tilde{b}^+ = \frac{1}{1-q^2}[e^{2is\varphi} - e^{is\partial}e^{is\varphi}], \tag{2.1.37}$$

where $\partial = \partial_\varphi$ is again *non-deformed* derivative and $0 < q = \exp(-s^2) < 1$. The operators (2.1.37) formally act on a suitable subset of the functions defined on the interval $J = (-\frac{\pi}{2s}, +\frac{\pi}{2s})$ and belonging to the Hilbert space $\mathcal{L}_2(J, |h(\varphi)|^2 d\varphi)$ where

$$h(\varphi) = \prod_{m=0}^{\infty} [1 + e^{-s^2(2m+1)+2is\varphi}] . \tag{2.1.38}$$

We shall use slightly different Macfarlane representations [51] given by the operators

$$b = \frac{1}{1-q^2}[e^{-2is\varphi} \mp (e^{-2is\varphi} + e^{-s^2})e^{is\partial}],$$

$$b^+ = \frac{1}{1-q^2}[e^{2is\varphi} \mp e^{is\partial}(e^{2is\varphi} + e^{-s^2})]. \tag{2.1.39}$$

They act in a suitable subspace $\mathcal{H}$ of functions from $\mathcal{L}_2(J, d\varphi)$ specified below. Their commutator (2.1.13) is

$$K = \pm(1 + e^{s^2 - 2is\varphi})(e^{is\partial} \mp e^{2is\partial})(1 + e^{s^2 + 2is\varphi}). \tag{2.1.40}$$

The upper signs in (2.1.39), (2.1.40) corresponds to the Fock representation in which the operators b and b^+ are related to $\tilde{b}$ and $\tilde{b}^+$as

$$b = A(\varphi)\tilde{b}A^{-1}(\varphi), \quad b^+ = A(\varphi)\tilde{b}^+A^{-1}(\varphi),$$

where $A(\varphi) = h(\varphi)\exp(\frac{i}{2}s\varphi)$.

In this case the operators b and b^+ act in the space $\mathcal{H}$ spanned by functions $\psi_k(\varphi) = \exp(2isk\varphi)$, $k = 0, 1, 2, \ldots$. The operator $\exp\{is\partial\}$ acts in $\mathcal{H}$ as

$$e^{is\partial}\psi_k(\varphi) = e^{-ks^2}\psi_k(\varphi). \tag{2.1.41}$$

Using this, one can show that the (unnormalized) vacuum state $\phi_0(\varphi)$ satisfying

$$K\phi_0(\varphi) = \phi_0(\varphi) , \tag{2.1.42}$$

is

$$\phi_0(\varphi) = h(\varphi) , \tag{2.1.43}$$

with $h(\varphi)$ given in (2.1.38). This follows from the relation

$$h(\varphi + is) = (1 + e^{-s^2+2is\varphi})^{-1}h(\varphi). \tag{2.1.44}$$

The eigenstates $\phi_k(\varphi)$, $k = 1, 2, \ldots$ are given according to (2.1.21) as

$$\phi_k(\varphi) = ([k; q^2]!)^{-1/2}(b^+)^k\phi_0(\varphi) , \tag{2.1.45}$$

where b^+ is given by (2.1.39) with the upper signs.

The lower signs in (2.1.39), (2.1.40) correspond to the non-Fock representation in the subspace $\mathcal{H}$ of $\mathcal{L}_2(J, d\varphi)$ spanned by the functions $\psi_k(\varphi) = \exp[2is(k+\gamma)\varphi]$, $k \in \mathbb{Z}$, where $\gamma \in [0, 1)$ is fixed. In this space the operator $\exp(is\partial)$ acts as

$$e^{is\partial}\psi_k(\varphi) = e^{-2(k+\gamma)s^2}\psi_k(\varphi), k \in \mathbb{Z}. \tag{2.1.46}$$

Obviously, the operator K (given in (2.1.40) with lower signs) is negative, so that we are really dealing with the non-Fock case.

The (unnormalized) eigenfunction $\phi_0^\gamma(\varphi)$ satisfying

$$-K\ \phi_0^\gamma(\gamma) = e^{-2\gamma s^2}\phi_0^\gamma(\varphi) , \tag{2.1.47}$$

is given by the formula

$$\phi_0^\gamma(\varphi) = \frac{1}{k(\varphi)}\sum_{m=0}^{\infty} a_m e^{2is(m+\gamma)\varphi} , \tag{2.1.48}$$

where

$$a_m = e^{-m^2s^2}\prod_{n=1}^{m}\frac{1 + e^{-2(n+\gamma)s^2}}{1 - e^{-2ns^2}} , \tag{2.1.49}$$

$$k(\varphi) = \prod_{n=1}^{\infty} [1 + e^{-(2n-1)s^2 - 2is\varphi}]. \tag{2.1.50}$$

Eq. (2.1.47) follows from the relations

$$a_{m-1} = e^{(2m-1)s^2} \frac{1 - e^{-2ms^2}}{1 + e^{-2(m+\gamma)s^2}} a_m, \tag{2.1.51}$$

$$k(\varphi + is) = (1 + e^{s^2 - 2is\varphi}) k(\varphi), \tag{2.1.52}$$

and the formula (2.1.40) for K with lower signs.

The eigenstates $\phi^{\gamma}_{\pm k}(\varphi), k = 1, 2, \ldots$, (with the same norm as $\phi^{\gamma}_0(\varphi)$) are given according to (2.1.24) as

$$\phi^{\gamma}_{-k} = (\gamma\{\gamma + k - 1\}!)^{-1/2}\, b^k \phi^{\gamma}_0(\varphi),$$

$$\phi^{\gamma}_{k} = (\{\gamma + k\}!)^{-1/2}\, (b^+)^k \phi^{\gamma}_0(\varphi), \tag{2.1.53}$$

where $\{\gamma + k\}! = \{\gamma + 1\} \ldots \{\gamma + k\}$, and b and b^+ are defined in (2.1.39) with lower signs. We stress that the parameter γ introduced in (2.1.46) is identical to the one introduced in (2.1.17).

The discussed Bargmann and Macfarlane representations can be considered as the realizations of the q-oscillator operators in terms of *usual* coordinate and momentum operators. For example, the Macfarlane realization obviously corresponds to a quantum mechanical particle on a circle with the coordinate φ and momentum $p := i\partial_\varphi$ with Weyl-like commutation relations (cf. Sections 1.1 and 1.7.3). The existence of nonequivalent representations on a circle is a well-known fact. It is clear that Hamiltonians which have natural form in terms of q-oscillators are highly non-polynomial in terms of the coordinates and momenta. The basic CR in this case are the canonical one and CR (2.1.1),(2.1.2) have an auxiliary meaning with the parameter q being some function of the usual Planck constant $\hbar$, with $q \to 1$ if $\hbar \to 0$.

In another case, one considers CR (2.1.1),(2.1.2) as the basic relation and represents the raising and lowering q-operators in terms of differential operators and coordinates on a quantum plane. The latter possibility is considered in details in the next Section.

2.2 Bargmann-Fock representation for q-oscillator algebra in terms of operators on quantum planes

If one tries to construct the *Bargmann-Fock representation* (BFR) of (2.1.8) or its multioscillator generalization, in terms of coordinates $\bar{z}_i$ and derivatives $\bar{\partial}_i$ on a q-hyperplane, i.e. to maintain the correspondence $\mathbf{b}_i^+ \rightarrow \bar{z}_i$, $\mathbf{b}_i \rightarrow \bar{\partial}_i$ (here $\bar{z}_i$, $\bar{\partial}_i$ are coordinates and *q-deformed* derivatives on q-hyperplane), one needs CR between coordinates and derivatives, in other words, a q-differential calculus. Usually, one develops the $GL(N)_q$-invariant differential calculus as we considered in Section 1.6, but this is not the case in BFR construction.

Let us consider the basic construction for just one (bosonic) oscillator. In this case, we have to use a 2-dimensional q-plane with coordinates $z, \bar{z}$, satisfying CR

$$z\bar{z} - p\bar{z}z = 0 , \tag{2.2.1}$$

and the corresponding derivatives ∂, $\bar{\partial}$. The reason for denoting the deformation parameter in (2.2.1) by p, and not by standard q, will be clear soon. As is well known, BFR is constructed in the space of antianalytic functions $f(\bar{z})$ ($\bar{z}$ is conjugate to z: $z^* = \bar{z}$). Hence, the group of invariance cannot mix z and $\bar{z}$ (and thus it cannot be $GL(2)_q$) and the admissible homogeneous transformations have the following matrix form:

$$\begin{pmatrix} z' \\ \bar{z}' \end{pmatrix} = \begin{pmatrix} u & 0 \\ 0 & \bar{u} \end{pmatrix} \begin{pmatrix} z \\ \bar{z} \end{pmatrix} . \tag{2.2.2}$$

It is easy to see that such transformations do not fix differential calculus on the q-plane. On the other hand, derivatives are connected with translations and, hence, differential calculi are connected with inhomogeneous groups. So it is natural to require an invariance of the calculus with respect to *inhomogeneous* generalization of (2.2.2), i.e. with respect to deformed two-parametric Euclidean group $E(2)_{r,p}$. General consideration of q-deformed inhomogeneous groups will be given in the next Chapter. Fortunately, the 2-dimensional case is the simplest one and its algebraic structure can be introduced easily [230].

$E(2)_{r,p}$-transformations have the form

$$\begin{aligned} z &\longrightarrow z' = uz + t , \\ \bar{z} &\longrightarrow \bar{z}' = \bar{u}\bar{z} + \bar{t} . \end{aligned} \tag{2.2.3}$$

The multiparametric CR for the q-group generators are

$$\begin{aligned} \bar{u}u &= u\bar{u} = 1\,, \\ t\bar{t} &= p\bar{t}t\,, \\ ut &= rtu\,, \\ \bar{u}t &= p^{-1}t\bar{u}\,. \end{aligned} \tag{2.2.4}$$

Other CR can be obtained after involution of (2.2.4): $u^* = \bar{u}$, $t^* = \bar{t}$. Comultiplication Δ for group elements is defined as follows

$$\begin{aligned} \Delta(u) &= u \otimes u\,, & \Delta(\bar{u}) &= \bar{u} \otimes \bar{u}\,, \\ \Delta(t) &= u \otimes t + t \otimes \mathbf{1}\,, & \Delta(\bar{t}) &= \bar{u} \otimes \bar{t} + \bar{t} \otimes \mathbf{1}\,. \end{aligned} \tag{2.2.5}$$

This comultiplication can be expressed in the usual matrix form

$$\Delta(M_{ij}) = M_{ik} \otimes M_{kj}\,,$$

if one writes the transformations (2.2.3) in the matrix projective form

$$\begin{pmatrix} z' \\ \bar{z}' \\ 1 \end{pmatrix} = M \begin{pmatrix} z \\ \bar{z} \\ 1 \end{pmatrix} \equiv \begin{pmatrix} u & 0 & t \\ 0 & \bar{u} & \bar{t} \\ 0 & 0 & 1 \end{pmatrix} \begin{pmatrix} z \\ \bar{z} \\ 1 \end{pmatrix}. \tag{2.2.6}$$

This form is also convenient for writing the antipode:

$$S\left(\begin{pmatrix} u & 0 & t \\ 0 & \bar{u} & \bar{t} \\ 0 & 0 & 1 \end{pmatrix}\right) = \begin{pmatrix} \bar{u} & 0 & -\bar{u}t \\ 0 & u & -u\bar{t} \\ 0 & 0 & 1 \end{pmatrix},$$

and counity: $\varepsilon(M_{ij}) = \delta_{ij}$. It is easy to see that the quantum translations do not form a q-subgroup of the $E(2)_{r,p}$ because the relations $u = \bar{u} = 1$ contradict both the CR (2.2.4) and the homomorphism condition for the comultiplication (2.2.5). This illustrates the remark about q-subgroups in the end of Section 1.5.

To develop the differential calculus, we introduce the q-differentials dz, $d\bar{z}$ which are transformed homogeneously: $dz \to u dz$, $d\bar{z} \to \bar{u} d\bar{z}$ and require that CR for the differentials and coordinates be invariant with respect to $E(2)_{r,p}$-transformations (cf. Section 1.1 and 1.6). Let us consider as an example the CR for z and $d\bar{z}$. The general form of CR is the following:

$$z d\bar{z} = A_1(d\bar{z})z + A_2(dz)z + A_3(d\bar{z})\bar{z} + A_4(dz)\bar{z}\,, \tag{2.2.7}$$

where A_i are constants to be defined from the invariance condition. After $E(2)_{r,p}$-transformation of (2.2.7), one has

$$\begin{aligned} zd\bar{z} + t\bar{u}d\bar{z} = A_1(d\bar{z})z + A_2u^2(dz)z + A_3\bar{u}^2(d\bar{z})\bar{z} + A_4(dz)\bar{z} \\ + A_1\bar{u}td\bar{z} + A_2utdz + A_3\bar{u}\bar{t}d\bar{z} + A_4u\bar{t}dz \ . \end{aligned}$$

Comparing both sides of the equation, one obtains

$$A_1 = p \ ,$$

$$A_2 = A_3 = A_4 = 0 \ .$$

Hence

$$zd\bar{z} = pd\bar{z}z \ . \tag{2.2.8}$$

Introducing external differential d with the standard properties (i.e. satisfying the Leibniz rule and the condition $d^2 = 0$), one immediately derives from (2.2.8) the CR for differentials:

$$dzd\bar{z} = -pd\bar{z}dz \ .$$

The CR for other pairs of coordinates and differentials are obtained analogously. Now we introduce q-derivatives ∂, $\bar{\partial}$ through the relation (see Section 1.1 and 1.6)

$$d = d\bar{z}\bar{\partial} + dz\partial \ ,$$

and derive the complete set of CR for coordinates, differentials and derivatives [15],[149],[53]:

$$\begin{aligned}
z\bar{z} &= p\bar{z}z \ , & \partial\bar{\partial} &= p\bar{\partial}\partial \ , \\
dzd\bar{z} &= -pd\bar{z}dz \ , & (dz)^2 &= (d\bar{z})^2 = 0 \ , \\
zd\bar{z} &= p(d\bar{z})z \ , & \bar{z}dz &= p^{-1}(dz)\bar{z} \ , \\
zdz &= r^{-1}(dz)z \ , & \bar{z}d\bar{z} &= r(d\bar{z})\bar{z} \ , \\
\partial z &= 1 + r^{-1}z\partial \ , & \bar{\partial}\bar{z} &= 1 + r\bar{z}\bar{\partial} \ , \\
\partial\bar{z} &= p^{-1}\bar{z}\partial \ , & \bar{\partial}z &= pz\bar{\partial} \ , \\
dz\bar{\partial} &= p^{-1}\bar{\partial}dz \ , & d\bar{z}\partial &= p\partial d\bar{z} \ , \\
d\bar{z}\bar{\partial} &= r\bar{\partial}d\bar{z} \ , & dz\partial &= r^{-1}\partial dz \ .
\end{aligned} \tag{2.2.9}$$

Thus the differential calculus is not arbitrary and it is not fixed by $GL(2)_q$-*group but is* defined by inhomogeneous Euclidean quantum group. Note that it considerably differs from $GL(2)_{q,r}$-invariant calculus developed in Section 1.6.

In the standard BF representation the derivative $\bar{\partial}$ plays the role of annihilation operator and a multiplication by the coordinate $\bar{z}$ plays the role of creation operator. As is seen from (2.2.9), q-deformed $\bar{\partial}$ and $\bar{z}$ satisfy the CR (2.1.8) for q-deformed oscillator operators if one puts $r = q^2$. The second parameter p can take arbitrary value: this does not influence the CR (2.1.8). From the other hand properties of the quantum plane, which supposedly plays the role of a classical phase space for the q-oscillator, strongly depends on the value of parameter p: if $p = 1$ the plane has commutative coordinates, if $p \neq 1$ the "classical phase space" proves to be non-commutative. We postpone the discussion of the subtle point concerning classical limits until the next Section 2.3. And now continue the construction of BF representation in two cases:

1. non-commutative plane with $p = r = q^2$;
2. commutative q-plane with $p = 1$, $r = q^2$.

Consider at first **non-commutative space**, $p = r = q^2$. To define a scalar product in BF Hilbert space we need a q-analog of integration. In the case of usual non-deformed BF representation the scalar product of two antiholomorphic functions is defined with the help of exponential measure (see, e.g. [17])

$$< g, f >= \int d\bar{v}dv \, e^{-\bar{v}v}\overline{g(\bar{v})}f(\bar{v}) \, , \tag{2.2.10}$$

($\bar{v}, v$ are commuting variables on ordinary complex plane). Thus to construct q-deformed analog of the scalar product, it is natural to use the definition of q-integral mentioned in the end of Section 1.6. As a basic function g_a in the formulas (1.6.48)-(1.6.49), one must take the *q-deformed exponent*

$$e_q^x \equiv \exp_q x := \sum_{n=0}^{\infty} \frac{x^n}{[n;q]!} \, , \tag{2.2.11}$$

with the defining property

$$\partial e_{q^2}^{ax} = a e_{q^2}^{ax} \, , \quad \text{if} \quad \partial x - q^2 x \partial = 1, \quad a \in \mathbb{C} \, .$$

Then the following postulates define the integral calculus on a q-plane [15, 149, 53]:

1. Normalization
$$I_{00} := \int d\bar{z}dz\ e_{q^2}^{-\bar{z}z} = 1;$$

2. Analog of the Stokes formula
$$\int d\bar{z}dz\ \bar{\partial}f(\bar{z},z) = \int d\bar{z}dz\ \partial f(\bar{z},z) = 0\ ;$$

3. Change of variables
$$\int d\bar{z}dz\ e_q^{-a^2\bar{z}z} f(\bar{z},z) = a^{-2}\int d\bar{z}dz\ e_q^{-\bar{z}z} f(a^{-1}\bar{z},a^{-1}z)\ .$$

These postulates allow to prove the following result,
$$I_{mn} = \int d\bar{z}dz\ e_{q^2}^{-\bar{z}z} z^n\bar{z}^m = \delta_{mn}[n;q^2]!\ , \tag{2.2.12}$$

which is sufficient to compute the integral of any function which has a power series expansion.

Indeed, let us calculate $\partial\,(e_r^{-\bar{z}z}z^n\bar{z}^m)$ (we use r instead of q^2 to simplify the notation of the q-exponent; recall that $r = q^2$ in the cases under consideration) using the formulas
$$\begin{aligned}\bar{\partial}e_r^{a\bar{z}z} &= aze_r^{a\bar{z}z}\ ,\\ \partial e_r^{a\bar{z}z} &= aq^{-2}\bar{z}e_r^{aq^{-2}\bar{z}z}\ ,\end{aligned} \tag{2.2.13}$$

which can be obtained from (2.2.11), (2.2.9):
$$\partial\left(e_r^{-\bar{z}z}z^n\bar{z}^m\right) = -q^{-2}\bar{z}e_r^{-q^{-2}\bar{z}z}z^n\bar{z}^m + e_r^{-q^{-4}\bar{z}z}[n;q^{-2}]z^{n-1}\bar{z}^m\ .$$

Then the above postulate 2 for the q-integral gives
$$q^{-2n-2}\int d\bar{z}dz\ e_r^{-q^2\bar{z}z}z^n\bar{z}^{m+1} = [n;q^{-2}]\int d\bar{z}dz\ e_r^{-q^{-4}\bar{z}z}z^{n-1}\bar{z}^m\ ,$$

and the third postulate permits to rewrite this in the form
$$I_{n\ m+1} = [n;q^2]I_{n-1\ m}\ . \tag{2.2.14}$$

The Stokes formula (the postulate 2) leads to the equality
$$I_{n0} = I_{0n} = 0\ ,\quad \forall n\ ,$$

and by induction with use of (2.2.14), one obtains

$$I_{mn} = 0 \ , \text{if } m \neq n \ .$$

For $n = m + 1$, the relation (2.2.14) gives

$$I_{nn} = [n; q^2] I_{n-1\ n-1} \ ,$$

from which one obtains easily (2.2.12) by induction, and the normalization for I_{00}. The Stokes formula for the derivative $\bar{\partial}$ is also consistent with (2.2.12).

The BF representation is constructed in the Hilbert space $\mathcal{H}$ of antianalytic functions $f(\bar{z})$ on the q-plane with scalar product of the form

$$< g, f > = \int d\bar{z}dz \ e_r^{-\bar{z}z} \overline{g(\bar{z})} f(\bar{z}) \ , \tag{2.2.15}$$

so that the monomials

$$\psi^n(\bar{z}) = \frac{\bar{z}^n}{\sqrt{[n; q^2]!}} \tag{2.2.16}$$

form the orthonormal complete set of vectors in $\mathcal{H}$. The creation and annihilation operators are represented as coordinate and derivative

$$b^+ = \bar{z} \ , \qquad b = \bar{\partial} \ . \tag{2.2.17}$$

Using (2.2.13), one can check that b^+ and b are hermitian conjugate of each other with respect to the scalar product (2.2.15).

Now we will consider another possibility: **plane with commuting coordinates,** $p = 1$, $r = q^2$. The BF representation of CR (2.1.8) with commuting $z, \bar{z}$ variables and q-deformed differential and integral calculus is constructed in the space of antiholomorphic functions of the form

$$\psi_n = \frac{\bar{z}^n}{\sqrt{[n; q^2]!}} \ ,$$

which are orthonormalized with respect to scalar product

$$< g, f > = \int d\bar{z}dz \ e_{1/r}^{-q^2\bar{z}z} \overline{g(\bar{z})} f(\bar{z}) \ ,$$

where the integral is defined by the relation

$$\int d\bar{z}dz \ e_{1/r}^{-q^2\bar{z}z} z^n \bar{z}^m = \delta_{mn} [n; q^2]! \ .$$

This integral is defined by the same postulate as in the non-commutative case. Here we have used the so-called second q-deformed (basic) exponential function [99],

$$\exp_{1/q^2}\{x\} := \sum_{n=0}^{\infty} \frac{x^n}{[n; q^{-2}]!} = \sum_{n=0}^{\infty} q^{n(n-1)} \frac{x^n}{[n; q^2]!} \ .$$

2.3 Quasi-classical limit of q-oscillators and q-deformed path integrals

In this section we consider the problem of quasi-classical limit and the path integral representation of the quantum mechanical evolution operator kernel for q-deformed oscillator. Besides the case of real parameter of deformation, path integral for root of unity value of q is considered briefly in the final part of this section. In all cases we will discuss two possibilities: non-commutative and commutative coordinates.

Let us write the CR for q-oscillator algebra in the form (with recovered $\hbar$)

$$bb^+ - q^2 b^+ b = \hbar \ , \tag{2.3.1}$$

$$bb^+ - b^+ b = \hbar C q^{2N} \ , \tag{2.3.2}$$

where C is the central element.

There are two possibilities:

1. In the limit $\hbar \to 0$ and fixed C the CR are incompatible if $q \neq 1$. This means that in this case q must be a function of $\hbar$ so that $q(\hbar) \longrightarrow 1$ as $\hbar \to 0$. This leads to *commutative dynamics* in the classical limit. If, e.g. $q = 1 + \hbar\gamma$, the CR (2.3.1) becomes

$$[b, b^+] = \hbar(1 + \gamma b^+ b) \ , \tag{2.3.3}$$

and we expect to obtain in this limit classical dynamics with commutative observables, curved phase space and the symplectic form

$$\omega(\bar{z}, z) = i(1 + \gamma\bar{z}z)^{-1} dz \wedge d\bar{z} \ . \tag{2.3.4}$$

2. In the limit $\hbar \longrightarrow 0$ with fixed $\hbar C$ there is no contradiction in the case of nontrivial $q \neq 1$. In this case one can expect to obtain q-classical dynamics with q-commuting variables: after a simple rescaling, $\bar{z} = \sqrt{(q^2-1)/\hbar C}\ b^+$; $z = \sqrt{(q^2-1)/\hbar C}\ b$, one gets in this limit

$$\begin{aligned} z\bar{z} &= q^2 \bar{z} z \ , \\ \bar{z} z &= q^{2N} \ . \end{aligned}$$

The aim of this Section is to obtain quasi-classical approximations for both cases and to discuss the possibility of construction of a path integral representation for the q-oscillator evolution operator. The last subsection is devoted to the path integrals in the case of root of unity value of the deformation parameter.

2.3.1 Quasi-classical limit of q-oscillators (with real parameter of deformation)

As we discussed in Section 1.1, the adequate formalism for the study of quasi-classical limit is a symbol calculus [182, 17]. In the case of oscillator-like systems, the most convenient symbol map is the *normal* one, which put in correspondence to a regular function $\mathcal{N}_A(\bar{z}, z) = \sum_{m,n} c_{mn}\bar{z}^m z^n$ the operator $A = \sum_{m,n} c_{mn}(b^+)^m b^n$, so that in each monomial of the operators, creation operators are on the left of annihilation ones.

In the Bargmann-Fock representations this map can be made explicit with the use of *operator kernels* $\mathcal{A}$ of an operator A defined by the relation

$$(Af)(\bar{z}_1) = \frac{1}{\hbar}\int d\bar{z}_2 dz_2\; \mathcal{A}(\bar{z}_1, z_2)\mathcal{E}(\bar{z}_2, z_2)f(\bar{z}_2)\;,$$

where the integration is understood in an appropriate sense (usual integration for non-deformed oscillator or the integrations considered in the preceding Section in q-deformed case) and $\mathcal{E}(\bar{z}, z)$ is the function defined by Bargmann-Fock scalar products, e.g.

$$\begin{aligned} \mathcal{E}(\bar{z}, z) &= e^{-\bar{z}z/\hbar}\,, && \text{in non-deformed case;} \\ \mathcal{E}(\bar{z}, z) &= e_{1/q^2}^{-q^2\bar{z}z/\hbar}\,, && \text{deformed BFR with } p=1,\ r=q^2\,; \\ \mathcal{E}(\bar{z}, z) &= e_{q^2}^{-\bar{z}z/\hbar}\,, && \text{deformed BFR with } p=r=q^2\,. \end{aligned}$$

Of course, in cases of non-commutative variables one has to define CR between different pairs and care about the order of factors in the integrand. In non-deformed case (see, e.g. [17]) the kernel of an operator is expressed through the normal symbol by the relation

$$\mathcal{A}(\bar{z}, z) = e^{\bar{z}z/\hbar}\mathcal{N}(\bar{z}, z)\;,$$

and the product of two operators corresponds to the convolution of their kernels. This allows to find the explicit form of star-product of normal symbols and its quasi-classical limit.

Now let us turn to the q-deformed cases. At first we will consider the limit:$\hbar \to 0$, $C = const$. As we expect to obtain classical dynamics with commutative observables, we express operators in terms of integral kernels with commuting variables (but q-deformed differential and integral calculi with $p=1$, $r=q^2$).

Manipulations analogous to those in the non-deformed case, give that the kernel of product A_1A_2 of two operators A_1, A_2 with kernels $\mathcal{A}_1, \mathcal{A}_2$, equals to the convolution

$$\mathcal{A}_1 * \mathcal{A}_2(\bar{z}, z) := \frac{1}{\hbar} \oint^{q} d\bar{\xi}\xi e_{1/r}^{-q^2\bar{\xi}\xi/\hbar} \mathcal{A}_1(\bar{z}, \xi)\mathcal{A}_2(\bar{\xi}, z) ,$$

and the relation between the normal symbol $\mathcal{N}$ and the kernel $\mathcal{A}$ of an operator A is

$$\mathcal{A}(\bar{z}, z) = e_r^{\bar{z}z/\hbar}\mathcal{N}(\bar{z}, z) .$$

So to understand how nontrivial (nonzero) $\hbar$ modifies the classical (commutative) multiplication we must calculate in the quasi-classical limit $\hbar \sim 0$ the expression

$$\mathcal{N}_{\mathcal{A}_1} \star \mathcal{N}_{\mathcal{A}_2} = \mathcal{N}_{\mathcal{A}_1 * \mathcal{A}_2} = \oint^{q} d\bar{\xi}\xi\, \mathcal{N}_1(\bar{z}, \xi)\mathcal{N}_2(\bar{\xi}, z) e_r^{\bar{z}\xi/\hbar} e_{1/r}^{-q^2\bar{\xi}\xi/\hbar} e_r^{\bar{\xi}z/\hbar} e_{1/r}^{-\bar{z}z/\hbar} .$$

In the usual classical case of harmonic oscillator one would combine the exponents, then shift the variables of integration and use, e.g., the steepest descent method to evaluate the integral in $\hbar \to 0$ limit. None of these methods can be used in q-deformed case. Instead, we can use the q-exponent expansion and the q-integral property (2.2.12).

Consider operators of the form $P_m = f_m(b^+)b^m$, where $m \in \mathbb{Z}_+$ and f_m is an arbitrary polynomial. Any operator can be represented as a sum (possibly infinite) of P_m. For the convolution of two such operators one has (after trivial change of integration variables $\bar{\xi}, \xi \longrightarrow \bar{\xi}/\sqrt{\hbar}, \xi/\sqrt{\hbar}$)

$$\mathcal{N}_{\mathcal{P}_m * \mathcal{P}_l} = \oint^{q} d\bar{\xi}\xi\, f_m(\bar{z})\xi^m g_l(z)\bar{\xi}^l(\sqrt{\hbar})^{m+l} e_r^{\bar{z}\xi/\sqrt{\hbar}} e_{1/r}^{-q^2\bar{\xi}\xi} e_r^{\bar{\xi}z/\sqrt{\hbar}} e_{1/r}^{-\bar{z}z/\hbar}$$

$$= \sum_{s,n=0}^{\infty} \oint^{q} d^2\xi f_m(\bar{z})\xi^m g_l(z)\bar{\xi}^l(\sqrt{\hbar})^{m+l-s-n} \frac{\bar{z}^s\xi^s}{[s;q^2]!} \frac{\bar{\xi}^n z^n}{[n;q^2]!} e_{1/r}^{-q^2\bar{\xi}\xi} e_{1/r}^{-\bar{z}z/\hbar} . \quad (2.3.5)$$

For definiteness consider the case $l \leq m$ (opposite case is dealt quite similarly). Then from (2.2.12) we have

$$\mathcal{N}_{\mathcal{P}_m * \mathcal{P}_l} = f_m(\bar{z})g_l(z)e_{1/r}^{-\bar{z}z/\hbar} \sum_s \hbar^{l-s}\bar{z}^s z^{m+s-l} \frac{[m+s;q^2]!}{[s;q^2]![s+m-l;q^2]!}$$

$$= f_m(\bar{z})g_l(z)e_{1/r}^{-\bar{z}z/\hbar} \sum_s \hbar^{l-s}\bar{z}^s z^{m+s-l} \frac{[s+m-(l-1);q^2]...[s+m;q^2]}{[s;q^2]!} . \quad (2.3.6)$$

To go further, let us note that the sum of the form

$$\sum_{s=0}^{\infty} \bar{z}^{s} z^{s-l} \hbar^{l-s} \frac{1}{[s-k;q^2]!},$$

where $k < l-1$, can be rewritten up to $O(\hbar)$ terms as

$$\sum_{s=k}^{\infty} \bar{z}^{s} z^{s-l} \hbar^{l-s} \frac{1}{[s-k;q^2]!} = \sum_{s=0}^{\infty} \bar{z}^{s+k} z^{s+k-l} \hbar^{l-k} \hbar^{-s} \frac{1}{[s;q^2]!}$$

$$= \bar{z}^{k} z^{k-l} \hbar^{l-k} e_{r}^{\bar{z}z/\hbar}. \tag{2.3.7}$$

Use of the *summation theorem* [99] for q-exponentials with commuting arguments

$$e_q^A e_{1/q}^B = \sum_{n=0}^{\infty} \frac{(A+B)_q^{(n)}}{[n;q]!}, \qquad AB = BA, \tag{2.3.8}$$

where

$$(A+B)_q^{(n)} := (A+B)(A+qB)\dots(A+q^{n-1}B),$$

gives that the last factor is cancelled by the last exponent in (2.3.6) so that the resulting expression of the form (2.3.6) would be of the order $\sim O(\hbar^2)$. So in (2.3.6) we must take into account the parts of the last factor of the form $\sim 1/[r-l;q^2]!$ and $\sim 1/[r-l+1;q^2]!$ only.

The identities

$$[i+j;q^2] = q^j([i;q^2] - [-j;q^2]) = q^j[i;q^2] + [j;q^2], \tag{2.3.9}$$

allow to present the ratio

$$\frac{[s-l+1+m;q^2]\dots[s+m;q^2]}{[s-l+1;q^2]\dots[s;q^2]}, \tag{2.3.10}$$

in the form

$$q^{ml} + q^l[m;q^2]\frac{1}{[s-l+1;q^2]}\sum_{i=0}^{l-1} q^{-i} \tag{2.3.11}$$

$$+ \text{ terms with higher orders of } [s;q^2] \text{ in denominators,}$$

so that for (2.3.6) we have

$$f_m(\bar{z})g_l(z)e_{1/r}^{-\bar{z}z/\hbar} \sum_{s} \hbar^{l-s} \bar{z}^{s} z^{m+s-l} \frac{[s+m-(l-1);q^2]\dots[s+m;q^2]}{[s;q^2]!}$$

$$\begin{aligned}
&= f_m(\bar{z})g_l(z)e_{1/r}^{-\bar{z}z/\hbar}\sum_s \hbar^{l-s}\bar{z}^s z^{m+s-l}\frac{q^{ml}}{[s-l;q^2]!} \\
&+ f_m(\bar{z})g_l(z)e_{1/r}^{-\bar{z}z/\hbar}q^l[m;q^2] \\
&\times \left(\sum_{i=0}^{l-1} q^{-i}\right)\sum_s \hbar^{l-s}\bar{z}^s z^{m+s-l}\frac{1}{[s-l+1;q^2]!} + O(\hbar^2)\,. \qquad (2.3.12)
\end{aligned}$$

The series over s in this expression can be summed up to $e_r^{\bar{z}z/\hbar}$, so that we finally obtain for the convolution, up to $\sim O(\hbar)$ terms

$$\mathcal{N}_{\mathcal{P}_m*\mathcal{P}_l} = q^{ml}\mathcal{N}_{\mathcal{P}_m}\mathcal{N}_{\mathcal{P}_l} + \hbar f_m(\bar{z})z^{m-1}\bar{z}^{l-1}g_l(z)q^l[m;q^2]\sum_{i=0}^{l-1}q^{-i} + O(\hbar^2)\,. \quad (2.3.13)$$

The factor q^{ml} in the first term shows that in the limit $\hbar \to 0$ the convolution does not reduce to usual commutative multiplication of functions on"classical" phase space. This means that there is no such a classical limit for q-deformed oscillator with $q \neq 1$. From the other hand, if $q = 1 + \gamma\hbar + O(\hbar^2)$, we have for (2.3.13) in the same order in $\hbar$

$$\begin{aligned}
\mathcal{N}_{\mathcal{P}_m*\mathcal{P}_l} &= \mathcal{N}_{\mathcal{P}_m}\mathcal{N}_{\mathcal{P}_l}(1 + ml\gamma\hbar) + \hbar ml f_m(\bar{z})z^{m-1}\bar{z}^{l-1}g_l(z) + O(\hbar^2) \\
&= \mathcal{N}_{\mathcal{P}_m}\mathcal{N}_{\mathcal{P}_l} + \hbar(1+\gamma\hbar)\partial\mathcal{N}_{\mathcal{P}_m}\bar{\partial}\mathcal{N}_{\mathcal{P}_l} + O(\hbar^2)\,. \qquad (2.3.14)
\end{aligned}$$

Here $\partial, \bar{\partial}$ denote, obviously, the usual (non-deformed) derivatives.

The Poisson bracket derived from (2.3.14) has the form

$$\begin{aligned}
\{\mathcal{N}_1, \mathcal{N}_2\}_P &= \lim_{\hbar\to 0}\frac{i}{\hbar}(\mathcal{N}_1 * \mathcal{N}_2 - \mathcal{N}_2 * \mathcal{N}_1) \\
&= i(1+\gamma\bar{z}z)\left(\partial\mathcal{N}_1\bar{\partial}\mathcal{N}_2 - \bar{\partial}\mathcal{N}_2\partial\mathcal{N}_1\right)\,, \qquad (2.3.15)
\end{aligned}$$

and, as we expected, it corresponds to the symplectic form (2.3.4).

Now consider the second possibility for quasi-classical limit mentioned in the beginning of this Section which leads to non-commutative "classical" variables: $\hbar \to 0$, $\hbar C = const$. Here we use the q-deformed BF representation with $p = r = q^2$ (see the preceding Section). Again the action of any operator A in the BF Hilbert space $\mathcal{H}$ based on non-commutative variables, can be represented by its kernel $\mathcal{A}(\bar{z}, z)$

$$(Af)(\bar{z}_1) = \frac{1}{\hbar}\int d\bar{z}_2 dz_2\; \mathcal{A}(\bar{z}_1, z_2)e_r^{-\bar{z}_2 z_2/\hbar}f(\bar{z}_2)\,, \qquad (2.3.16)$$

where

$$\mathcal{A}(\bar{z}_1, z_2) = \sum_{m,n} A_{mn}\frac{\bar{z}_1^m}{\sqrt{\hbar^m[m;q^2]!}}\frac{z_2^n}{\sqrt{\hbar^n[n;q^2]!}}\,. \qquad (2.3.17)$$

The special order of operator kernel and "the integration measure" (q-exponent) is convenient for the subsequent definition of convolution of operators in the case of q-commuting "classical" variables. Another new feature in this definition is that one more pair of q-commuting coordinates is introduced. So we have to define the CR for coordinates on different copies of q-planes. The choice of the CR depends on the concrete meaning of different planes. In our case these q-planes will correspond to different time slices in the process of time evolution. In the continuous limit they become infinitesimally close to each other and it would be quite unnatural if coordinates on different time slices would commute. More formally this argument can be expressed as follows. We assume that the classical analog of oscillator operators b^+, b, i.e. the variables $\bar{z}, z$, obey some Hamiltonian-like equation of motion of the general form

$$\frac{dz}{dt} = F(\bar{z}(t), z(t)) ,$$

$$\frac{d\bar{z}}{dt} = \bar{F}(\bar{z}(t), z(t)) .$$

rhs of these equations have definite CR with $\bar{z}(t)$, $z(t)$. Hence lhs must have the same CR. It means that $\bar{z}(t + \Delta t)$, $z(t + \Delta t)$ have the same CR with $\bar{z}(t)$, $z(t)$ as the latter have CR with each other. As a result, we postulate that any copies of coordinates $\bar{z}_i$, z_i $(i = 1, 2, ...)$ on q-planes have the following CR:

$$z_i\bar{z}_j = q^2\bar{z}_j z_i , \qquad \bar{z}_i\bar{z}_j = \bar{z}_j\bar{z}_i , \qquad z_i z_j = z_j z_i , \tag{2.3.18}$$

i.e. they do not depend on the indices which distinguish the copies.

Now we have to derive CR for differentials, derivatives and coordinates for *different* pairs of variables. As usual we can do this using the consistency requirement. For example, assume that CR for a derivative ∂_z and coordinate ξ from another copy of variables is a homogeneous one: $\partial_z\xi = a\xi\partial_z$, where a is a constant to be defined. Then, acting by both sides of this relation on a function $f(z)$, we obtain $a = 1$. Proceeding in this way, one comes to the following CR for any two different pairs of variables $\{\bar{z}, z\}$ and $\{\bar{\xi}, \xi\}$:

$$\begin{aligned}
\partial_z\bar{\xi} &= q^{-2}\bar{\xi}\partial_z\,, & \bar{\partial}_z\xi &= q^2\xi\bar{\partial}_z\,,\\
\partial_z\xi &= \xi\partial_z\,, & \bar{\partial}_z\bar{\xi} &= \bar{\xi}\bar{\partial}_z\,,\\
dz\bar{\xi} &= q^2\bar{\xi}dz\,, & d\bar{z}\xi &= q^{-2}\xi d\bar{z}\,,\\
dz\xi &= \xi dz\,, & d\bar{z}\bar{\xi} &= \bar{\xi}d\bar{z}\,,\\
dzd\bar{\xi} &= q^2d\bar{\xi}dz\,, & d\bar{z}d\xi &= q^{-2}d\xi d\bar{z}\,,\\
dzd\xi &= d\xi dz\,, & d\bar{z}d\bar{\xi} &= d\bar{\xi}d\bar{z}\,,\\
\partial_z\bar{\partial}_\xi &= q^2\bar{\partial}_\xi\partial_z\,, & \bar{\partial}_z\partial_\xi &= q^{-2}\partial_\xi\bar{\partial}_z\,,\\
\partial_z\partial_\xi &= \partial_\xi\partial_z\,, & \bar{\partial}_z\bar{\partial}_\xi &= \bar{\partial}_\xi\bar{\partial}_z\,,\\
\partial_z d\bar{\xi} &= q^{-2}d\bar{\xi}\partial_z\,, & \bar{\partial}_z d\xi &= q^2 d\xi\bar{\partial}_z\,,\\
\partial_z d\xi &= d\xi\partial_z\,, & \bar{\partial}_z d\bar{\xi} &= d\bar{\xi}\bar{\partial}_z\,.
\end{aligned}$$

One can check that the definition of q-integral is consistent with these CR for different variables if requires that CR of the symbol $\int d\bar{z}dz$ (map from q-plane to $\mathbb{C}$) are defined by (i.e. coincide with) the CR for $d\bar{z}dz$. This gives sense to the notation for the functional ("definite integral") on a q-plane.

Coefficients A_{mn} in (2.3.17) can be expressed through the scalar product

$$A_{mn} = q^{2m(n+1)-2m} < \psi_m \mid \mathbf{A} \mid \psi_n > \,. \tag{2.3.19}$$

Consider an action of two operators A_1 and A_2 on arbitrary wave function $f(\bar{z})$

$$A_2A_1 = \frac{1}{\hbar}\int d\bar{z}_1dz_1\, A_2(\bar{z}_2, z_1)e_r^{-\bar{z}_1z_1/\hbar}\frac{1}{\hbar}\int d\bar{z}_0dz_0e_r^{-\bar{z}_0z_0/\hbar}A_1(\bar{z}_1, z_0)f(\bar{z}_0)\,. \tag{2.3.20}$$

Using the CR for different pairs of q-variables, we obtain the formula for the convolution of operator kernels

$$\mathcal{A}_2 * \mathcal{A}_1(\bar{z}, z) := \frac{1}{\hbar}\int d\bar{\xi}d\xi\, \mathcal{A}_2(q^{-2}\bar{z}, q^2\xi)e_r^{-\bar{\xi}\xi/\hbar}\mathcal{A}_1(\bar{\xi}, z)\,. \tag{2.3.21}$$

The appearance of another copy of variables implies that in general we have to consider q-commuting constants ("q-constants") in addition to (dynamical) q-variables. Consistency requires that after physical quantization (i.e. with $\hbar$ as a deformation parameter) this q-numbers must q-commute with q-oscillator operators homogeneously. But in introducing such quantities we have to be very careful to avoid any contradiction. For example, it would be tempting to introduce a pair of complex q-numbers $\bar{\xi}, \xi$ which q-commute with operators b^+, b

$$\begin{aligned} b\bar{\xi} &= q^2\bar{\xi}b\,, & \xi b^+ &= q^2 b^+\xi\,, \\ b\xi &= \xi b\,, & b^+\bar{\xi} &= \bar{\xi}b^+\,. \end{aligned}$$

In this case the CR (2.3.1) seem to be invariant with respect to the q-analog of classical shifting $b^+ \to b^+ + \bar{\xi}$, $b \to b + \xi$. But q-constants with such CR definitely contradict the very (2.3.1) as lhs does not commute with, e.g. ξ, while rhs (usual number, $\hbar$ or unity) does. However, we can introduce a real q-constant $\omega = \omega^\dagger$ with the CR

$$b\omega = q^{-2}\omega b\,, \qquad b^+\omega = q^2\omega b^+\,, \tag{2.3.22}$$

and with analogous CR for any copy of "classical" (dynamical) variables

$$z\omega = q^{-2}\omega z\;; \qquad \bar{z}\omega = q^2\omega\bar{z}\,. \tag{2.3.23}$$

This q-constant will play the role of q-deformed "frequency parameter" of corresponding q-oscillator Hamiltonian. As is clear from (2.3.22), the operator ω is diagonal in the standard basis $\mid n\rangle$, $n = 0, 1, \ldots$ of q-oscillator Hilbert space (simultaneously with the number operator N) and if the vacuum vector satisfies

$$\omega \mid 0; \omega_0\rangle = \omega_0 \mid 0; \omega_0\rangle\,, \quad \omega_0 \in \mathbb{R}\,,$$

then for arbitrary vector of the basis one has

$$\omega \mid n; \omega_0\rangle = q^{-2n}\omega_0 \mid 0; \omega_0\rangle\,. \tag{2.3.24}$$

Let us consider a normal monomial

$$N^p_{rs} := \omega^p (b^+)^r b^s\,, \tag{2.3.25}$$

and the corresponding normal symbol

$$\mathcal{N}^p_{rs} := \omega^p \bar{z}^r z^s\,. \tag{2.3.26}$$

From (2.3.17) and (2.3.19) we can derive the expression for the corresponding integral kernel

$$\mathcal{A}^p_{rs} = \omega_0^p q^{2(s(s+1)-r(p+1))}\bar{z}^r z^s \exp_{1/q^2}\{q^{2(s-p+1)}\bar{z}z/\hbar\}\,, \tag{2.3.27}$$

(note that in the kernel q-constants become usual c-numbers and q-oscillator operators become q-commuting variables).

Now we are ready to calculate the normal symbol corresponding to the product of two monomials in ω, b^+ and b in quasi-classical approximation in analogy to the case of commuting variables. Consider the convolution of two operator kernels

$$\begin{aligned} \mathcal{A}^p_{ab} * \mathcal{A}^t_{cd} &= \frac{\omega_0^{p+t}}{\hbar} \int d\bar{\xi} d\xi\, q^{2(b(b+1)-a(p+1))} q^{-2a+2b} \bar{z}^a \xi^b \exp_{1/q^2}\{q^{2(b-p+1)} \bar{z}\xi/\hbar\} \\ &\times \exp_{q^2}\{-\bar{\xi}\xi/\hbar\} q^{2(d(d+1)-c(t+1))} \bar{\xi}^c z^d \\ &\times \exp_{1/q^2}\{q^{2(d-t+1)} \bar{\xi} z/\hbar\} . \end{aligned} \tag{2.3.28}$$

Long but straightforward calculations of this convolution up to $O(\hbar)$ terms are analogous to those in the case of commuting variables. However, one has to care about the order of all factors in the expressions. The result is

$$\begin{aligned} \mathcal{A}^p_{ab} * \mathcal{A}^t_{cd} &= \omega_0^{p+t} q^{(b+d)(b+d+1)-(a+c)(p+t+1)} q^{2bc} q^{2t(a-b)} \bar{z}^{a+c} z^{b+d} \\ &\times \exp_{1/q^2}\{q^{2(b+d-p-t+1)} \bar{z} z/\hbar\} \\ &+ \omega_0^{p+t} q^{(b+d-1)(b+d)-(a+c-1)(p+t+1)} q^{2(b-1)(c-1)} q^{2t(a-b)} [b; q^2][c; q^2] \\ &\times \bar{z}^{a+c-1} z^{b+d-1} \exp_{1/q^2}\{q^{2(b+d-p-t)} \bar{z} z/\hbar\} + O(\hbar^2) . \end{aligned} \tag{2.3.29}$$

This convolution corresponds to the normal symbol

$$\begin{aligned} \mathcal{N}_{\mathcal{A}^p_{ab} * \mathcal{A}^t_{cd}} &= q^{2bc} q^{2t(a-b)} \omega^{p+t} \bar{z}^{a+c} z^{b+d} + \hbar q^{2(b-1)(c-1)} q^{2t(a-b)} [b; q^2][c; q^2] \\ &\times \omega^{p+t} \bar{z}^{a+c-1} z^{b+d-1} + O(\hbar^2) . \end{aligned} \tag{2.3.30}$$

Introducing the star-product for normal symbols

$$\mathcal{N}^p_{ab} \star \mathcal{N}^t_{cd} := \mathcal{N}_{\mathcal{A}^p_{ab} * \mathcal{A}^t_{cd}} , \tag{2.3.31}$$

we can present it in the form

$$\mathcal{N}^p_{ab} \star \mathcal{N}^t_{cd} = \mathcal{N}^p_{ab} \mathcal{N}^t_{cd} + \hbar (\mathcal{N}^p_{ab} \partial_R)(\bar{\partial} \mathcal{N}^t_{cd}) + O(\hbar^2) , \tag{2.3.32}$$

where for convenience we have introduced right derivative ∂_R which has the same CR as the left derivative ∂ but acts on functions from the right. The same concerns its complex conjugate $\bar{\partial}_R$. In particular,

$$z^n \partial_R = [n; q^2] z^{n-1} , \qquad \bar{z}^n \bar{\partial}_R = q^{-2(n-1)} \bar{z}^{n-1} .$$

Using (2.3.32), one can easily obtain the expression for q-commutator of star-products of normal symbols in quasi-classical limit

$$\left(\mathcal{N}^p_{ab} \star \mathcal{N}^t_{cd} - q^{2(bc-ad+p(d-c)+t(a-b))}\mathcal{N}^t_{cd} \star \mathcal{N}^t_{cd}\right)$$
$$= \hbar\mathcal{N}^p_{ab}\left(\partial_R\bar\partial - q^{2(b+c-1)}\bar\partial_R\partial\right)\mathcal{N}^t_{cd} + O(\hbar^2)\,. \qquad (2.3.33)$$

An analogy with usual definition and procedure of physical quantization inspires to define a q-deformed "Poisson bracket" as the factor in front of $\hbar/i$ in the rhs of (2.3.33) in the limit $\hbar \to 0$

$$\left\{\mathcal{N}^p_{ab}, \mathcal{N}^t_{cd}\right\}_q := i\mathcal{N}^p_{ab}\left(\partial_R\bar\partial - q^{2(b+c-1)}\bar\partial_R\partial\right)\mathcal{N}^t_{cd}\,. \qquad (2.3.34)$$

This means that in the limit $\hbar \to 0$, the normal symbol of q-commutator

$$[N^p_{ab}, N^t_{cd}]_q := N^p_{ab}N^t_{cd} - q^{2(bc-ad+p(d-c)+t(a-b))}N^t_{cd}N^p_{ab}\,, \qquad (2.3.35)$$

of two monomials of the form (2.3.25) divided by $\hbar/i$ is equal to the q-Poisson bracket

$$\lim_{\hbar\to 0}\frac{i}{\hbar}\mathcal{N}_{[.,.]_q} = \{.,.\}_q\,. \qquad (2.3.36)$$

There is a special class of monomials, namely N^a_{aa}, $\forall a \in \mathbb{Z}$, for which the q-commutator (2.3.35) becomes a usual commutator for any second monomials entering the q-commutator

$$[N^a_{aa}, N^t_{cd}]_q = N^a_{aa}N^t_{cd} - N^t_{cd}N^a_{aa}\,, \quad \forall a, b, d, t\,.$$

If we want to consider the Heisenberg equations of motion for q-oscillator operators with *usual* time variables and, hence, with time derivative satisfying *usual* Leibniz rule, we have to use the Hamiltonians (time shift generators) of the type N^a_{aa}, since only the commutator acts on products of operators according to the Leibniz rule. (Of course, this is true as we want to consider classical limit of the above described type. If we would not care about any (quasi)classical limit at all, we could consider usual Heisenberg equations for any operators constructed from b^+, b.) Thus the natural choice for Hamiltonian is

$$H_{qh} := N^1_{11} = \omega b^+ b\,. \qquad (2.3.37)$$

In the q-oscillator basis (2.3.24) this Hamiltonian is diagonal

$$H_{qh} \mid n, \omega_0\rangle = E_n \mid n, \omega_0\rangle\,,$$

where

$$E_n = \frac{\omega_0}{q^2}[n; q^{-2}]\,. \qquad (2.3.38)$$

Essentially the new feature of the Hamiltonian (2.3.37) with q-deformed "frequency" ω is that the corresponding Heisenberg equations of motion prove to be very simple

$$\frac{\partial b}{\partial t} = \frac{i}{\hbar}[H_{qh}, b] = -\frac{i}{\hbar}q^{-2}\omega b = -\frac{i}{\hbar}b\omega \ , \tag{2.3.39}$$

$$\frac{\partial b^+}{\partial t} = \frac{i}{\hbar}\left[H_{qh}, b^+\right] = \frac{i}{\hbar}\omega b^+ = \frac{i}{\hbar}q^{-2}b^+\omega \ . \tag{2.3.40}$$

This operator equations have the obvious harmonic solution

$$b(t) = e^{-iq^{-2}\omega t/\hbar}b(0) = b(0)e^{-i\omega t/\hbar} \ , \tag{2.3.41}$$

$$b^+(t) = e^{i\omega t/\hbar}b^+(0) = b^+(0)e^{iq^{-2}\omega t/\hbar} \ , \tag{2.3.42}$$

in spite of non-diagonal, non-equidistant and exponentially growing (for $q < 1$) spectrum (2.3.38). Time evolution (2.3.41),(2.3.42) of the operators inspires to call the system "q-deformed *harmonic* oscillators". Note that often considered in literature Hamiltonian of the form

$$H_0 = \omega_0 b^+ b \ , \tag{2.3.43}$$

with c-number frequency ω_0 leads to complicated non-harmonic time evolution.

"q-Classical" equations of motion are defined by the q-deformed Poisson brackets

$$\frac{\partial z}{\partial t} = \{H_{cl,qh}, z\}_q = -iq^{-2}\omega z = -iz\omega \ , \tag{2.3.44}$$

$$\frac{\partial \bar{z}}{\partial t} = \{H_{cl,qh}, \bar{z}\}_q = i\omega z = iq^{-2}\bar{z}\omega \ . \tag{2.3.45}$$

with the solutions of the form (2.3.41),(2.3.42) as in the quantum case. Here we used the normal symbol of H_{qh} as classical Hamiltonian

$$H_{cl,qh} = \mathcal{N}_{H_{qh}} = \omega\bar{z}z \ . \tag{2.3.46}$$

Of course, the variables $\bar{z}, z$ and the "constant" ω are operators and one has to construct their representation in some Hilbert space $\mathcal{H}_{cl}$ and to extract all information about the non-commutative classical system using the technique of standard quantum mechanics in the Heisenberg picture. However, we call this theory a q-*classical* one because an evolution of the dynamical variables is defined by q-Poisson brackets but not by a commutator (note that commutators of monomials in $\bar{z}, z$ are quite different from their q-Poisson brackets).

A very important question concerns the possible applications of the non-commutative classical mechanics. This is outside the scope of the present book. We note only that it is by now a well known fact that the notion of trajectories in classical physics becomes too restrictive for a description of complicated phenomena, e.g. dynamical chaos and irreversible (classical) processes. It is tempting to find some relations between non-commutative classical mechanics and these important phenomena.

2.3.2 Path integral for q-oscillators (real q)

Another interesting question is the possibility to construct q-deformed path integral for q-oscillator systems using as the underlying classical system the commutative (case 1 in the beginning of this Section) and non-commutative (case 2) limits. Let us start from the latter.

We consider the usual Schrödinger equation

$$i\frac{d}{dt}\Psi(\bar{z},t) = H(b^+,b)\Psi(\bar{z},t) , \tag{2.3.47}$$

and at first take as a Hamiltonian $H_0 = \omega_0 b^+ b$ with c-number frequency parameter. Here and in what follows we again put $\hbar = 1$ for simplicity. As we discussed already, the time evolution of an operator in Heisenberg picture is defined by usual commutator with a Hamiltonian and classical limit is connected with q-commutator (2.3.35). It is instructive to see how this contradiction reveals itself in corresponding path integral. The integral kernel for infinitesimal operator

$$U \approx 1 - iH\Delta t = 1 - i\omega_0 b^+ b\Delta t ,$$

takes the form

$$\mathcal{U}(\bar{z}z) = e_{1/r}^{q^4\bar{z}z}\left(1 - q^2 e_r^{-q^4\bar{z}z} e_{1/r}^{q^6\bar{z}z} i\omega\bar{z}z\Delta t\right) \approx e_{1/r}^{q^4\bar{z}z} exp\{-iH_{eff}\Delta t\} , \tag{2.3.48}$$

where

$$H_{eff} = q^2 e_r^{-q^4\bar{z}z} e_{1/r}^{q^6\bar{z}z}\omega_0\bar{z}z = \frac{q^2\omega\bar{z}z}{1+q^4(1-q^2)\bar{z}z}$$

(we have used here (2.3.8)).

Now we can write the convolution of K infinitesimal evolution operator

$$\begin{aligned}
&\mathcal{U}(\bar{z}_K z_{K-1}) * \mathcal{U}(\bar{z}_{K-1}z_{K-2}) * ... * \mathcal{U}(\bar{z}_1 z_0) \\
&\quad = \int d\bar{z}_{K-1}dz_{K-1}...d\bar{z}_1 dz_1 e_r^{-\bar{z}_{K-1}z_{K-1}}...e_r^{-\bar{z}_1 z_1} e_{1/r}^{-q^4\bar{z}_K z_{K-1}}...e_{1/r}^{-q^4\bar{z}_1 z_0} \\
&\quad \times\ e^{-iH_{eff}(\bar{z}_K z_{K-1})\Delta t}...e^{-iH_{eff}(\bar{z}_1 z_0)\Delta t} .
\end{aligned} \tag{2.3.49}$$

Using the product representation for the second q-exponent [99],

$$e^{x}_{1/r} = \prod_{r=0}^{\infty}\{1 + xq^{2r}(1-q^2)\} \ ,$$

and introducing Δz_{K-l} as

$$z_{K-l-1} = z_{K-l} - \Delta z_{K-l} \ ,$$

one can write

$$e_{1/r}^{q^4 \bar{z}_K z_{K-1}} \ldots e_{1/r}^{q^4 \bar{z}_1 z_0} = \exp\left\{\sum_{l=1}^{K-1} \ln \ e_{1/r}^{q^4 \bar{z}_{K-l} z_{K-l-1}}\right\}$$

$$= \exp\left\{\sum_{l=1}^{K-1}\sum_{r=0}^{\infty} \ln(1 + q^{2r+4}(1-q^2)\bar{z}_{K-l} z_{K-l-1}\right\}$$

$$\approx e_{1/r}^{q^4 \bar{z}_{K-1} z_{K-1}} \ldots e_{1/r}^{q^4 \bar{z}_1 z_1} \exp\left\{\sum_{l=1}^{K-1}\left(q^4 \sum_{r=0}^{\infty} \frac{q^{2r}(1-q^2)}{1 + q^{2r+4}(1-q^2)\bar{z}_{K-l} z_{K-l}}\right)\bar{z}_{K-l}\Delta z_{K-l}\right\} \tag{2.3.50}$$

Substituting (2.3.50) into (2.3.49) and taking the continuous limit $\Delta t \to 0$ as usual, we finally obtain the path integral representation for the evolution operator [53]:

$$U(\bar{z}, z; t'' - t') = \int \left(\prod_t \frac{d\bar{z}(t)dz(t)}{(1 + (1-q^2)\bar{z}(t)z(t))(1 + q^2(1-q^2)\bar{z}(t)z(t))}\right)$$
$$\times e_{1/r}^{q^4 \bar{z}(t'')z(t'')} \exp\left\{-\int_{t'}^{t''} (\phi(\bar{z}(t)z(t))\bar{z}(t)\dot{z}(t) + iH_{eff}(\bar{z}(t)z(t)))\, dt\right\} \ , \tag{2.3.51}$$

where the dot on variable z means time derivative and

$$\phi(\bar{z}(t)z(t)) = \sum_{r=0}^{\infty} \frac{q^{2r}}{(q^4(1-q^2))^{-1} + q^{2r}\bar{z}(t)z(t)} \ .$$

Note that in the $q^2 \to 1$ limit

$$\phi \to 1 \ ,$$

$$H_{eff} \to H_{cl} \equiv \omega_0 \bar{z} z \ ,$$

so that in this limit one obtains the usual expression for the harmonic oscillator path integral [18].

Thus we see that the above mentioned contradiction between time shifting by commutator with the Hamiltonian H_0 and q-commutator needed for definition of q-classical limit leads, to the appearance of the "effective Hamiltonian" in the path integral-like representation of the evolution operator.

Consider analogous representation in the case of the Hamiltonian H_{qh} (2.3.37). As the q-commutator (2.3.35) for it coincides with the usual commutator and the relation between its normal symbol and kernel is the same as for unity operator, there are no effective Hamiltonians in this case and the kernel of infinitesimal evolution operator takes the form

$$\mathcal{U}(\bar{z}z) = e_{1/r}^{q^2\bar{z}z}\,(1 - i\omega_0\bar{z}z\Delta t) \approx e_{1/r}^{q^2\bar{z}z} exp\{-i\omega_0\bar{z}z\Delta t\}\,. \tag{2.3.52}$$

Proceeding as in the case of the Hamiltonian H_0, we obtain

$$U(\bar{z}, z; t'' - t') = \int \left(\prod_t \frac{d\bar{z}(t)dz(t)}{(1 + (1 - q^2)\bar{z}(t)z(t))}\right) e_{1/r}^{q^2\bar{z}(t'')z(t'')}$$
$$\times \exp\left\{-\int_{t'}^{t''} \left(\phi(\bar{z}(t)z(t))\bar{z}(t)\dot{z}(t) + i\omega_0\bar{z}(t)z(t)\right) dt\right\}\,, \tag{2.3.53}$$

and the function ϕ is almost the same as in the preceding case

$$\phi(\bar{z}(t)z(t)) = \sum_{r=0}^{\infty} \frac{q^{2r}}{(q^2(1 - q^2))^{-1} + q^{2r}\bar{z}(t)z(t)}\,.$$

This result shows that even for the Hamiltonian H_{qh} which has well defined and simple dynamics in q-classical limit, the q-path integral (2.3.53) (defined as the convolution of infinitesimal operator kernels) has a complicated structure with nontrivial (non-flat) integration measure and with the "action" (argument of the exponential in the integrand) which does not lead to correct equations of motion (2.3.44),(2.3.45) through the principle of least action. Technically, this follows from the fact that for q-oscillator the kernel of identity operator is not equal to the inverse measure of scalar product in q-deformed Bargamann-Fock representation. From a more general point of view, the obvious reason is the absence of the very "paths" in a non-commutative space. Notice that if q-oscillator operators are constructed from ordinary canonical coordinates and momenta with CR $[p, x] = \hbar/i$ (as in the Macfarlane representation (2.1.37)), the limit $\hbar \to 0$ does lead to classical limit (though with some peculiarities due to topologically nontrivial configuration space of the system: the operator of coordinate is defined on a circle). We will not describe this construction and refer the reader to [220].

Now consider the first possibility indicated in the beginning of this Section: simultaneous classical limit $\hbar \to 0$, $q(\hbar) \to 1$ which leads to commutative classical case, and we derive the corresponding path integral. The method of the q-path integral constructions is quite analogous to that in the case of non-commuting variables and the calculation is even simpler due to the commutativity of the coordinates [55]. So we present here only the result. For the evolution operator one obtains

$$U(t''-t') = \int \left(\prod_t d\bar{z}(t) dz(t) \omega(t) \right) e_r^{\bar{z}(t'')z(t'')}$$
$$\times \exp \left\{ - \int_{t'}^{t''} \left(\phi(\bar{z}(t)z(t))\bar{z}(t)\dot{z}(t) + iH(\bar{z}(t)z(t)) \right) dt \right\} ,$$

where

$$\omega(t) = (1 + \varepsilon \bar{z}(t)z(t))^{-1}$$

and the function $\phi(\bar{z}z)$ is expressed through the Jackson integral (1.6.36)

$$\phi(\bar{z}z) = \frac{q^2}{\bar{z}z} {}^{q^{-2}}\!\!\int_0^{\bar{z}z/q^2} \omega(y) \, .$$

This permits us to write the q-path integral in the covariant form, e.g. for partition function Z one has

$$Z = \int \left(\prod_t d\bar{z}(t) dz(t) \omega(t) \right) exp\{-S_q\} \, ,$$

where

$$S_q = \int_0^T [\theta_z \dot{z} + \theta_{\bar{z}} \dot{\bar{z}} + iH(\bar{z}z)] \, dt \, ,$$

$$\theta_z = \bar{z}\phi(\bar{z}z) \, , \qquad \theta_{\bar{z}} = z\phi(\bar{z}z) \, , \qquad d\theta = \omega \, .$$

Thus one obtains the expected covariant form for the path integral, the phase space being the plane with nontrivial q-deformed differential calculus and the integrals for fixed time slices being q-deformed Jackson-like integrals over the q-plane.

2.3.3 Path integral for q-oscillators with root of unity value of deformation parameter

Finally we consider the root of unity case of the deformation parameter [56].

For definiteness we shall consider the even roots, so that $q^{k+1} = -1$ for some integer k. The usual starting CR for q-oscillator operators in this case, i.e. (2.1.1),(2.1.2), are not suitable for the construction of BF representation. Hence we introduce the operators $\bar{b}, b$

$$b = q^{N/2}a , \qquad \bar{b} = a^{+}q^{N/2}$$

with CR

$$b\bar{b} - q^2\bar{b}b = 1 , \qquad b\bar{b} - \bar{b}b = q^{2N} , \qquad \bar{b}b = [N] \equiv \frac{q^{2N}-1}{q^2-1} . \qquad (2.3.54)$$

But for complex values of q the operator $\bar{b}$ is not hermitian conjugate to b

$$b^{\dagger} = \bar{b}q^{-N}, \qquad \bar{b}^{\dagger} = q^{-N}b .$$

The Fock space representation of these operators is the following

$$b|n\rangle = q^{n-1}\sqrt{[n]_q}|n-1\rangle , \quad \bar{b}|n\rangle = \sqrt{[n+1]_q}|n+1\rangle , \quad N|n\rangle = n|n\rangle . \quad (2.3.55)$$

Since $[k+1]_q = 0$, this Fock space is k-dimensional.

As concerns the explicit realization of (2.3.55), again there exist two possibilities. One can construct BF representation in the space of antiholomorphic functions with basis

$$\psi_n = \frac{\bar{z}^n}{\sqrt{[n]_q!}} , \qquad (2.3.56)$$

with either commuting $\bar{z}, z$ variables or non-commuting ones. In both cases the variables must satisfy the condition of nilpotency $z^{k+1} = \bar{z}^{k+1} = 0$ to provide the finiteness of the Fock space. We shall illustrate the both possibilities on the example of Grassmann-like $q^2 = -1$ case. In the case of commuting variables, one has

$$z\bar{z} = \bar{z}z , \qquad \bar{\partial}\bar{z} + \bar{z}\bar{\partial} = 1 , \qquad z^2 = \bar{z}^2 = 0 .$$

The creation $\bar{b} = \bar{z}$ and destruction $b = \bar{\partial}$ operators act in the BF space with the basis $\{\psi_0 = 1, \psi_1 = \bar{z}\}$ and the scalar product

$$\int d\bar{z}dz e^{\bar{z}z}\bar{\psi}_n\psi_m = \delta_{nm} ,$$

where the integral is defined by the usual Berezin rules

$$\int d\bar{z}\,\bar{z} = \int dz\,z = 1 , \qquad \int d\bar{z} = \int dz = 0 .$$

As a result, the evolution operator takes the form

$$U(t''-t') = \int \left(\prod_t \frac{d\bar{z}(t)dz(t)}{1-2\bar{z}(t)z(t)} \right)$$
$$\times \exp\left\{ \bar{z}(t'')z(t'') - \int_{t'}^{t''} (\bar{z}(t)\dot{z}(t) + iH(\bar{z}(t)z(t)))\, dt \right\} .$$

Note that the integral measure reproduces the form of non-canonical Poisson bracket

$$\{z, \bar{z}\}_p = i(1 - 2\bar{z}z) ,$$

which corresponds to CR $[b, \bar{b}] = \hbar(1 - 2\bar{b}b)$, the latter being equivalent to (2.3.54) after putting $\hbar = 1$.

The case of non-commuting (actually anticommuting in the $q^2 = -1$ case) BF variables seems to be identical to the usual Grassmann path integral. But there is one subtlety. In the usual construction not only $z(t_i)$ and $\bar{z}(t_j)$ anticommute for any time slices t_i, t_j but the same variables on different time slices anticommute also, e.g. $z(t_i)z(t_j) + z(t_j)z(t_i) = 0$. Such CR cannot be generalized to other roots of unity values of parameter q. So we introduce CR on different time slices as in the case of real q

$$\begin{aligned} z(t_i)\bar{z}(t_j) + \bar{z}(t_j)z(t_i) = 0 \ , \qquad & z^2(t_i) = \bar{z}^2(t_i) = 0 \ , \\ \bar{z}(t_i)\bar{z}(t_j) = \bar{z}(t_j)\bar{z}(t_i) \ , \qquad & z(t_i)z(t_j) = z(t_j)z(t_i) \ . \end{aligned} \tag{2.3.57}$$

One can check that for such CR all ingredients of path integral construction (scalar product measure, relation between normal symbol and kernel of operators, etc.) remain the same as in the case of usual Grassmann path integral (see, e.g. [18]). So the result proves to be the same also.

The CR (2.3.57) can be easily generalized to other values of deformation parameter, for example, for $q^3 = -1$ one has

$$\begin{aligned} z(t_i)\bar{z}(t_j) = q^2\bar{z}(t_j)z(t_i) \ , \qquad & z^3(t_i) = \bar{z}^3(t_i) = 0 \ , \\ \bar{z}(t_i)\bar{z}(t_j) = \bar{z}(t_j)\bar{z}(t_i) \ , \qquad & z(t_i)z(t_j) = z(t_j)z(t_i) \ . \end{aligned} \tag{2.3.58}$$

In this case the BF representation is defined in the space of functions with the basis

$$\psi_0 = 1 \ , \quad \psi_1 = \bar{z} \ , \quad \psi_2 = \bar{z}^2 \ ,$$

which is orthonormal with respect to the scalar product

$$\int d\bar{z}dz\ \overline{\psi_n(\bar{z})}\mu(\bar{z}z)\psi_m(\bar{z}) = \delta_{nm} \ , \tag{2.3.59}$$

where

$$\mu(\bar{z}z) := (1 + z\bar{z} + z^2\bar{z}^2) .$$

Here the integral is defined by the relation [15, 53]

$$\int d\bar{z}dz\; z^n \bar{z}^m = \delta_{nk}\delta_{mk}[k]_q! \,. \tag{2.3.60}$$

Because of nilpotency, a general Hamiltonian has a form

$$H = \Omega(u\bar{b}b + v\bar{b}^2 b^2) , \tag{2.3.61}$$

where the constants u, v are restricted by the hermiticity condition $H^\dagger = H$ and can take three sets of values:
i) $u = 1,\ \ v = -q;$
ii) $u = 1,\ \ v = 1 - 2q;$
iii) $u = 0,\ \ v = -q^2.$
In this case, and in the cases of higher roots of unity, the measure of scalar product in BF spaces and relation between the normal symbols and the kernels of operators cannot be expressed in terms of appropriated q-deformed exponents. This prevents from writing the general expression of path integral for arbitrary root of unity. But due to nilpotency, for any given value of parameter q it can be done. In the case under consideration ($q^3 = -1$), the path integral takes the form

$$U(t'' - t') = \int \left(\prod_t d\bar{z}(t)dz(t)(1 + q\bar{z}(t)z(t)) \right) (1 + \bar{z}(t'')z(t'') + \bar{z}^2(t'')z^2(t''))$$

$$\times \exp\left\{ - \int_{t'}^{t''} [(1 + (1 + 2q)\bar{z}z)\bar{z}(t)\dot{z}(t) + iH_{eff}(\bar{z}(t)z(t))]\, dt \right\} . \tag{2.3.62}$$

As in the case of real parameter q (2.3.51), the path integral contains effective Hamiltonian H_{eff} instead of the initial one

$$H_{eff}(\bar{z}(t)z(t)) = \Omega(u\bar{z}z + q(u + v - qu)\bar{z}^2 z^2) .$$

Let us consider for $q^3 = -1$, another possibility using commuting variables, with CR (2.2.9) at $p = 1$, $r = q^2$, in analogy with Grassmann-like case. The basis of BF representation has the same form (2.3.56) as in the case of noncommuting variables and is orthonormal with respect to scalar product (2.3.59) with the measure $\mu = 1 + \bar{z}z + \bar{z}^2 z^2$, the integral being defined by (2.3.60) (but now with commuting variables). The derivation of the path integral is essentially the same as in the case of noncommuting variables and the result is

$$U(t''-t') = \int \left(\prod_t d\bar{z}(t)dz(t)(1+2\bar{z}(t)z(t)+3\bar{z}^2(t)z^2(t))\right)$$
$$\times\,(1+\bar{z}(t'')z(t'')+\bar{z}^2(t'')z^2(t''))$$
$$\times\exp\left\{-\int_{t'}^{t''}[(1+\bar{z}z)\bar{z}(t)\dot{z}(t)+iH_{eff}(\bar{z}(t)z(t))]\,dt\right\} \quad (2.3.63)$$

The effective Hamiltonian H_{eff}, for all hermitian Hamiltonians of the form (2.3.61) listed above, prove to be real functions:

i) $H_{eff}=\omega\bar{z}z,$ for $u=1, v=-q;$
ii) $H_{eff}=\omega(\bar{z}z-\bar{z}^2z^2),$ for $u=1, v=1-2q;$
iii) $H_{eff}=\omega\bar{z}^2z^2,$ for $u=0, v=-q^2.$

After restoring the Planck constant $\hbar$ the expression (2.3.63) takes the form (for definiteness we consider the Hamiltonian (2.3.61) with $u=1, v=-q$)

$$U(t''-t') = \int\left[\prod_t \frac{d\bar{z}(t)dz(t)}{\hbar}\left(1+2\frac{\bar{z}(t)z(t)}{\hbar}+3\frac{\bar{z}^2(t)z^2(t)}{\hbar^2}\right)\right]$$
$$\times\left(1+\frac{\bar{z}(t'')z(t'')}{\hbar}+\frac{\bar{z}^2(t'')z^2(t'')}{\hbar^2}\right)$$
$$\times\exp\left\{-\frac{1}{\hbar}\int_{t'}^{t''}\left[\left(1+\frac{\bar{z}z}{\hbar}\right)\bar{z}(t)\dot{z}(t)+i\omega\bar{z}(t)z(t)\right]dt\right\}\,. \quad (2.3.64)$$

One can see that there is simply no quasi-classical approximation corresponding to the limit $\hbar\to 0$ (since the terms in the measure and in the action diverge). This is an expected result because, as is known, the quasi-classical approximation for spin-like systems considered here corresponds to the limit $S\to\infty$, where S is the spin of the system [19]. Systems with fixed spins (numbers of states) have no (quasi)classical limit, as is confirmed once again by the expression (2.3.64). Recall that spin systems with two-dimensional Hilbert space of states can be described either as particular representation of a usual quantum mechanical system with spherical phase space [19] or as a quantization of "classical" system with Grassmann variables [21].

Above we have shown that such a system can be considered as a result of quantization of a "classical" system with commuting but also nilpotent variables. Further options concern different possibilities of commutation relations between variables on different time slices. The resulting path integrals have quite different forms. The freedom in the choice of "classical" variables prove to be crucial for the generalization to the case of systems with higher dimensional Hilbert spaces. In this case the variables can be either q-commuting or commuting in ordinary sense. The first choice leads to the path integrals with operator-valued complex action functionals. The second gives the expression which is similar to the case of Grassmann-like variables and with real action. Thus the quantum plane with commuting nilpotent variables can be considered as a "classical" phase space for a system with finite dimensional Hilbert space. Of course, classical limit $\hbar \to 0$ in the usual sense does not exist for such systems as for any system with finite value of spin.

2.4 q-Oscillators and representations of QUEA

Quantum groups, quantum universal enveloping algebras (QUEA) and deformed oscillator algebras are the main objects of q-deformation theory. Actually, there is close connection between them. In particular

(*i*) irreducible representations of QUEA can be constructed via q-analog of Jordan-Schwinger (JS) realization from QUEA to an algebra of q-oscillator system;

(*ii*) in many applications it is necessary to consider q-oscillators which are covariant with respect to action of $SU_q(N)$ group (N is number of oscillators under consideration).

2.4.1 q-Deformed Jordan-Schwinger realization

We start from the q-deformation of very important JS construction [217]. Let us recall this approach on the example of $sl(2,\mathbb{C})$ or, equivalently, $su(2)$. Using an algebra $\mathcal{A}^{(2)}$ of a pair of independent oscillators β_i, β_i^+, $i = 1,2$ with the CR

$$[\beta_i, \beta_j^+] = \delta_{ij} ,$$

(all other CR vanish) one can define the map $su(2) \longrightarrow \mathcal{A}^{(2)}$:

$$\begin{aligned} J_+ &= \beta_1^+\beta_2 , \qquad J_- = \beta_2^+\beta_1 , \\ J_0 &= \frac{1}{2}[J_+, J_-] = \frac{1}{2}(N_1 - N_2) , \\ N_i &:= \beta_i^+\beta_i , \quad i = 1,2 , \end{aligned} \tag{2.4.1}$$

where N_i are number operators for usual harmonic oscillators. Fock representation for the oscillators permits to construct all finite dimensional irreps of $su(2)$.

In the deformed case this construction was given in [172, 23, 118, 155]. The mapping (2.4.1) is replaced by

$$\begin{aligned} &X^+ = a_1^+ a_2 \ , \qquad X^- = a_2^+ a_1 \ , \\ &H = \tfrac{1}{2}(N_1 - N_2) \ , \end{aligned} \tag{2.4.2}$$

where a_i^+, a_i $(i = 1,2)$ are q-oscillators in the Fock representations with the CR of the type (2.1.1),(2.1.2)

$$\begin{aligned} a_i a_i^+ - q a_i^+ a_i &= q^{-N_i} \ , \\ a_i a_i^+ - q^{-1} a_i^+ a_i &= q^{N_i} \ , \\ \left[N_i, a_i^+\right] = a_i^+ \ , &\qquad [N_i, a_i] = -a_i \ , \\ [a_i, a_j] = \left[a_i^+, a_j^+\right] = \left[a_i^+, a_j\right] &= 0 \ , \qquad i \neq j \ , \end{aligned} \tag{2.4.3}$$

(the second relation follows from the condition $\zeta = 0$ for the central element (2.1.4) in the Fock representation). Introduce the eigenstates which are analogous to undeformed angular momentum states

$$|j,m\rangle = ([j+m]_q![j-m]_q!)^{-1/2} (a_1^+)^{j+m} (a_2)^{j-m} \ |0\rangle \ ,$$

$$j = 0, \frac{1}{2}, 1, ... \ ; \quad m = -j, -j+1, ..., j \ .$$

Then the map (2.4.2) gives the representations (1.1.40)

$$X^\pm |jm\rangle = \sqrt{[j \mp m]_q [j \pm m + 1]_q} \ |j, m \pm 1\rangle \ ,$$

$$H|jm\rangle = m|jm\rangle \ .$$

This realization reproduces $s\ell_q(2)$ algebra provided $a_i^+ a_i = [N_i]$. For $\mathcal{A}(q)$ algebra this relation is valid only in the Fock representation, but not on the algebraic level. The latter aim can be achieved using the generators of the q-Heisenberg algebra

$$W_+ = \tau(b^+b)^{1/2} b^+ \ , \ W_- = \tau b(b^+b)^{1/2}$$

$$W_0 = (2q^2 b^+ b + 1)/(q + q^{-1})$$

with

$$b = q^{N/2}a \ , \ r = \frac{q}{(q+q^{-1})^{1/2}} \ .$$

These elements, due to the CR $bb^{+} - q^2 b^{+} b = 1$ satisfy the relations

$$\begin{aligned} [W_0, W_-]_{q,\frac{1}{q}} &= -W_- \ , \\ [W_0, W_+]_{\frac{1}{q},q} &= W_+ \ , \quad [A,B]_{p,r} := pAB - rBA \ , \\ [W_+, W_-]_{q^2,\frac{1}{q^2}} &= W_0 \ , \end{aligned}$$

and give another set of generators of the quantum algebra $s\ell_q(2)$, which is connected with the generators (2.4.2) by invertible map [69]. Such a form of $s\ell_q(2)$ algebra was suggested by E.Witten [242].

The Jordan-Schwinger construction for $su(2)_q$ confirms the important result: for all $q \in \mathbb{R}_+$, the unitary irreps of $su(2)_q$ are in correspondence with the unitary irreps of $su(2)$ and have the same dimensions (Section 1.7).

Construction of JS map for $s\ell_q(n)$ in Chevalley basis is the straightforward generalization of that for $s\ell_q(2)$ [118]. One defines n sets of q-oscillator operators with CR (2.4.3) (where now $i = 1, \ldots, n$) and in the basis $\{X_i^{\pm}, h_i\}_{i=1}^{n-1}$ (see (1.3.13)) makes the assignments

$$X_i^{+} = a_i^{+} a_{i+1} \ , \qquad X_i^{-} = a_{i+1}^{+} a_i \ ,$$
$$H = \tfrac{1}{2}(N_i - N_{i+1}) \ .$$

It is very important that this realization satisfies the q-deformed Serre relations and thus satisfies all the defining relations for $s\ell_q(n)$.

The special properties of Lie algebra representations at root of unity value of parameter q (see Section 1.7.2) also can be reproduced using the JS map and the corresponding representations (Sections 2.1) of q-oscillator algebras.

2.4.2 Quantum Clebsch-Gordan coefficients and Wigner-Eckart theorem

Next problem of the representation theory which we shall consider using the JS map is the decomposition of direct product of representations in direct sum of irreps. In the theory of Lie groups and algebras the problem is solved with the use of Clebsch-Gordan (CG) coefficients. q-Deformed analog of the latter for $su_q(2)$ was derived in [143]. We consider a construction of deformed CG coefficients for $su_q(2)$ algebra with real parameter $0 < q < 1$ in the basis of the monomials

$$|jm\rangle := \frac{s^{j+m}t^{j-m}}{\sqrt{[j+m]_q![j-m]_q!}} =: Q^j_{(q)m} \cdot \tag{2.4.4}$$

To avoid confusion, we remark that these monomials remind the basis (1.7.6) but the variables s and t in (2.4.4) are *commutative*; notice that in Section 1.7.1 we constructed *the corepresentations* of $SL_q(2)$ and here we are interested in *representations* of $su_q(2)$.

Combining BF-type representation with JS map, one can express $su_q(2)$ generators as follows

$$X^+ = sD_t^{(q)}, \qquad X^- = tD_s^{(q)}, \tag{2.4.5}$$

$$\tag{2.4.6}$$

$$H = \frac{1}{2}(s\partial_s - t\partial_t), \tag{2.4.7}$$

where $D_s^{(q)}, D_t^{(q)}$ are *the symmetrical* Jackson derivatives

$$D_x^{(q)} f(x) = \frac{f(qx) - f(q^{-1}x)}{(q-q^{-1})x} \tag{2.4.8}$$

and ∂_s, ∂_t are the usual (*nondeformed*) derivatives. Note that the symmetric Jackson derivative is necessary for construction of BF type representation of the q-oscillator algebra (2.1.1),(2.1.2) in terms of operators a^+, a in contrast to the case of operators b^+, b which are represented by the Jackson derivative D_x of the form (1.6.20). One can check that representation (2.4.6) indeed satisfies the CR (1.1.36) for $su_q(2)$ algebra.

In this monomial basis CG coefficients were constructed [207] via appropriate generalization of Van der Waerden's method [237]. The method is based on a possibility of explicit construction of invariants of $su(2)$ Lie algebra in terms of variables s, t. Indeed, consider three pairs of variables $(s_1, t_1), (s_2, t_2)$ and $(\bar{s}_1, \bar{t}_1)$ and with generators of the form

$$J_+ = s\partial_t, \qquad J_- = t\partial_s,$$

$$J_0 = \frac{1}{2}(s\partial_s - t\partial_t),$$

for the first and second pairs of variables and

$$J_+ = \bar{s}\partial_{\bar{t}}, \qquad J_- = \bar{t}\partial_{\bar{s}},$$

$$J_0 = \frac{1}{2}(\bar{s}\partial_{\bar{s}} - \bar{t}\partial_{\bar{t}}),$$

for the third pair. The bars over the last pair of the variables indicate that the latter form contravariant representation (of course, contravariant representations of $su(2)$ are unitary equivalent). Obvious invariants of the $su(2)$ algebra are: $(s_1t_2 - t_1s_2)$, $(s_i\bar{s}_3 + t_i\bar{t}_3)$, $i = 1, 2$. So one can form the invariant

$$I = (s_1t_2 - t_1s_2)^n(s_1\bar{s}_3 + t_1\bar{t}_3)^{n_2}(s_2\bar{s}_3 + t_2\bar{t}_3)^{n_1} , \qquad (2.4.9)$$

which can also be written as

$$I = \sum_{mm_1m_2} \langle j_1m_1j_2m_2|jm\rangle Q^{j_1}_{m_1} Q^{j_2}_{m_2} \bar{Q}^{j}_{m} , \qquad (2.4.10)$$

$$m = m_1 + m_2 ,$$

and gives CG coefficients $\langle j_1m_1j_2m_2|jm\rangle$ up to a normalization factor depending on j_1, j_2, j.

After q-deformation, the invariants have the form

$$(q^{1/2}s_1t_2 - q^{-1/2}t_1s_2), \qquad (q^{1/2}s_i\bar{s}_3 + q^{-1/2}t_i\bar{t}_3) , \quad i = 1, 2 .$$

As $D^{(q)}_{s,t}$ obey q-deformed Leibniz rule, a usual n-th power of the invariants are no longer invariant of the q-algebra and one has to use q-deformed n-th power defined by one more q-binomial formula [99](cf. (2.3.8),(1.7.5); notice that the latter is used for *noncommutative* variables)

$$(ax + bx)^{[n]}_q := \sum_{i=0}^{n} \binom{n}{i}_q (ax)^{n-i}(by)^i . \qquad (2.4.11)$$

Again taking into account nontrivial comultiplication and Leibniz rule one can write the invariant in six variables

$$I = (q^a s_1t_2 - q^{-a}t_1s_2)^{[n]}_q (q^{-a_2}s_1\bar{s}_3 + q^{a_2}t_1\bar{t}_3)^{[n]_2}_q (q^{a_1}s_2\bar{s}_3 + q^{-a_1}t_2\bar{t}_3)^{[n]_1}_q ,$$

$$\begin{aligned} a &= \tfrac{1}{2}(j+1) , & n &= j_1 + j_2 - j , \\ a_1 &= \tfrac{1}{2}(j_1+1) , & n_1 &= j_1 - j_2 + j , \\ a_2 &= \tfrac{1}{2}(j_2-1) , & n_2 &= -j_1 + j_2 + j , \end{aligned}$$

which after rewriting in the form (2.4.10), gives q-CG coefficients

$$\langle j_1m_1j_2m_2|jm\rangle = f(j_1, j_2, j)q^{\frac{1}{2}nN + (j_1m_2 - j_2m_1)}$$

$$\times \left\{ \prod_{i=1}^{3} [j_i + m_i]_q! [j_i - m_i]_q! \right\}^{1/2} \sum_{\nu \geq 0} (-1)^\nu q^{-\nu N} D^{-1} ,$$

$$N = j_1 + j_2 + j + 1 , \qquad j_3 \equiv j ,$$

$$D = [\nu]_q![j_1 + j_2 - j - \nu]_q![j_1 - m_1 - \nu]_q![j_2 + m_2 - \nu]_q! \\ \times[j - j_2 + m_1 + \nu]_q![j + j_1 - m_2 + \nu]_q! .$$

The normalization factor $f(j_1, j_2, j)$ proves to be the following

$$f(j_1, j_2, j) = \left\{[2j + 1]_q[n_1]_q![n_2]_q![n]_q! \left([N]_q!\right)^{-1}\right\}^{1/2} .$$

Important applications of CG coefficients are connected with Wigner-Eckart theorem [13] which permits to express through them the matrix elements of physical transition operators (up to an invariant scale factor). To generalize this result to q-deformed case, we need the notion of *q-tensor operators*. Following [24, 200], we present:

Definition 2.2 *Let $\mathcal{O}$ denote a vector space of operators mapping a representation space $\mathcal{M}$ (direct sum of irreps) of a QUEA $\mathcal{U}_q$ into itself: $t : \mathcal{M} \longrightarrow \mathcal{M}, t \in \mathcal{O}$. An irreducible q-tensor operator is a set of operators $\{t_{\Lambda,\xi}\} \in \mathcal{O}$ which carries a finite dimensional irrep Λ with vectors labeled by ξ, of the $\mathcal{U}_q$. That is*

$$X_\alpha(t_{\Lambda,\xi}) = \sum_{\xi'} \langle \Lambda, \xi' | X_\alpha | \Lambda, \xi \rangle t_{\Lambda,\xi'} ,$$

where X_α is a generator of $\mathcal{U}_q$, $X_\alpha(t_{\Lambda,\xi})$ denotes an action of X_α on $\mathcal{O}$, and $\langle \ldots \rangle$ denotes the matrix elements of the generators for the irrep Λ.

As always, an action of generators of $\mathcal{U}_q$ on direct product of any irreps, in particular on $t_{\Lambda,\xi} \otimes |\Lambda', \xi'\rangle$, where $|\Lambda', \xi'\rangle$ is some irrep from M, is defined by comultiplication Δ. If $\Delta(X_\alpha) = \sum_i X_{\alpha i} \otimes x^i_\alpha$, then

$$\Delta(X_\alpha)\left(t_{\Lambda,\xi} \otimes |\Lambda', \xi'\rangle\right) = \sum_i X_{\alpha i}(t_{\Lambda,\xi}) \otimes X^i_\alpha |\Lambda', \xi'\rangle .$$

Define the maps c:

$$c : \mathcal{O} \otimes \mathcal{M} \rightarrow \mathcal{M}, \quad c(t_{\Lambda,\xi} \otimes |\Lambda', \xi'\rangle) = t_{\Lambda,\xi}|\Lambda', \xi'\rangle ,$$

$$c : \mathcal{O} \otimes \mathcal{O} \rightarrow \mathcal{O}, \quad c(t_1 \otimes t_2) = t_1 t_2 , \quad t_1, t_2 \in \mathcal{O} .$$

Definition 2.3 *Compatibility of a Hopf algebra $\mathcal{U}_q$ comultiplication Δ and an action of generators X_α on q-tensor operators means commutativity of the diagram on Fig.2.1*

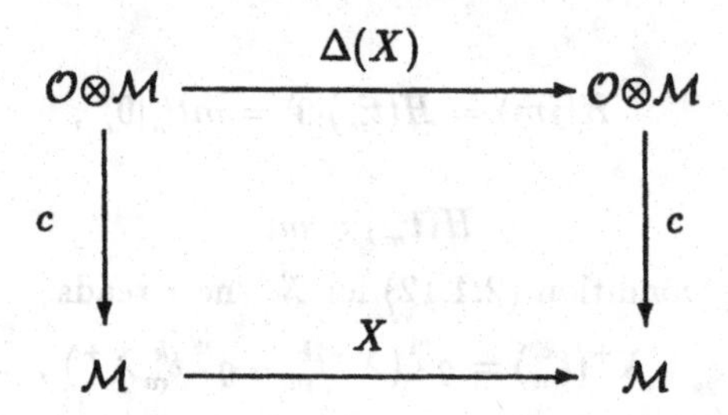

Figure 2.1: **The compatibility of comultiplication and the action of generators**

We now give , without proof, the q-generalization of Wigner-Eckart theorem [24].

- **If $t_{\Lambda,\xi}$ is a q-tensor operator of a QUEA $\mathcal{U}_q$ such that comultiplication in $\mathcal{U}_q$ is compatible with the action $X_\alpha(t_{\Lambda,\xi})$, then the matrix elements of $t_{\Lambda,\xi}$ in $\mathcal{M}$ are proportional to q-CG coefficients of $\mathcal{U}_q$ with invariant constant of proportionality. Conversely, if $t_{\Lambda,\xi}$ is a q-tensor operator and the matrix elements of $t_{\Lambda,\xi}$ are proportional to the q-CG coefficients, then the comultiplication and the action of generators on the q-tensor operators are compatible.**

As usual, we illustrate the general definitions and statements on the example of $su_q(2)$ algebra. Let $t^k_m(q)$ be an irreducible tensor operator of rank k for $su_q(2)$; $|jm\rangle$ is any basis of representation space of $su_q(2)$. Then from the comultiplication (1.3.7) and representation (1.1.40) for $su_q(2)$, we have

$$J_+t^k|jm\rangle = X^+(t^k_m)q^{-\frac{m}{2}}|jm\rangle + q^{\frac{H}{2}}(t^k_m)X^+|jm\rangle\ , \tag{2.4.12}$$

$$Ht^k_m|jm\rangle = H(t^k_m)|jm\rangle + t^k_m H|jm\rangle\ , \tag{2.4.13}$$

and the compatibility gives

$$H(t^k_m) = Ht^k_m - t^k_m H\ .$$

Consider vectors $|jm\rangle$ of the form

$$|jm\rangle = t^j_m|0\rangle\ , \tag{2.4.14}$$

where $|0\rangle$ is the basis vector of identity representation,

$$X^\mp|0\rangle = H|0\rangle = 0\ .$$

Then

$$H|jm\rangle = H(t^j_m)|0\rangle = mt^j_m|0\rangle \ ,$$

i.e.

$$H(t^j_m) = mt^j_m \ .$$

The compatibility condition (2.4.12) for X^+ now reads

$$X^+(t^k_m) = q^{\frac{m}{2}}(X^+t^k_m - q^{\frac{m}{2}}t^k_m X^+) \ .$$

Applying this to the vector (2.4.14), one obtains the CR for tensor operators of $su_q(2)$

$$\begin{aligned} X^\pm t^k_m - q^{\frac{m}{2}} t^k_m X^\pm &= q^{-\frac{m}{2}}\sqrt{[k\mp m]_q[k\pm m+1]_q} t^k_{m\pm 1} \ , \\ [H, t^k_m] &= mt^k_m \ . \end{aligned}$$

Using JS map (2.4.2) for $su(2)_q$, one can explicitly check that the operator pair

$$t^{1/2}_{1/2} := a^+_1 q^{-N_2/2} \ , \qquad t^{1/2}_{-1/2} := a^+_2 q^{N_1/2} \ ,$$

forms a q-tensor operator of $su_q(2)$. We suggest this verification to the reader as a useful exercise.

2.4.3 Covariant systems of q-oscillators

The last topic we consider in this section is, in a sense, the reverse of JS map construction and concerns systems of q-oscillators with CR which are covariant with respect to quantum group transformations.

Sometimes it is necessary for applications, to realize quantum algebras in terms of more than minimal number of the oscillators. For example, in the case of $su(2)$ we can express its generators in terms of two pairs of oscillator operators $(\beta_1, \beta_2), (\beta_3, \beta_4)$

$$\begin{aligned} J_0 &= \beta^+_1\beta_1 - \beta^+_2\beta_2 + \beta^+_3\beta_3 - \beta^+_4\beta_4 \ , \\ J_+ &= \beta^+_1\beta_2 + \beta^+_3\beta_4 \ , \\ J_- &= \beta^+_2\beta_1 + \beta^+_4\beta_3 \ . \end{aligned} \tag{2.4.15}$$

Straightforward substitution of q-oscillators does not lead to $su_q(2)$ CR and the corresponding formulas for the quantum case prove to be more complicated [38]:

$$\begin{aligned} X^+ &= Aa^+b + Cc^+d \ , \\ X^- &= \tilde{A}b^+a + \tilde{C}d^+c \ , \\ H &= N_a - N_b + N_c - N_d \ , \end{aligned} \tag{2.4.16}$$

provided $A\tilde{A} = q^{N_c - N_d}, C\tilde{C} = q^{-(N_a - N_b)}$. One possible solution for $A, \tilde{A}, C, \tilde{C}$ is the following

$$A = \tilde{A} = q^{\frac{N_c - N_d}{2}} \ , \ C = \tilde{C} = q^{-\frac{N_a - N_b}{2}} \ .$$

Thus the expressions for $X_\pm$ in quantum case are non-quadratic contrary to the $q = 1$ case.

Generalization to arbitrary number of modes $\{a_{1i}, a_{2i}\}$, $i = 1, ..., n$ is straightforward. One has for generators of $su_q(2)$

$$\begin{aligned} X^+ &= \sum_{i=1}^{n} C_i a_{1i}^+ a_{2i} \\ X^- &= \sum_{i=1}^{n} C_i a_{2i}^+ a_{1i} \\ C_i &= \prod_{m<i} q^{\frac{1}{2}(N_{1m} - N_{2m})} \prod_{m>i} q^{-\frac{1}{2}(N_{1m} - N_{2m})} \ , \\ H &= \sum_{m=1}^{n} (N_{1m} - N_{2m}) \ . \end{aligned}$$

The system of independent q-oscillators (bosonic as well as fermionic) have been used for the q-oscillator realizations of different quantum Lie algebras [118].

The CR of systems of independent q-oscillators considered above are not invariant with respect to the unitary transformations

$$a_i :\to U_{ij} a_j \ , \ U^\dagger U = 1 \ . \tag{2.4.17}$$

On the other hand, it is well known that in the case $q = 1$ the set of n creation or annihilation operators forms a basis for the $U(n)$ fundamental representations. Using the equivalence between the first order differential calculus and the second quantization procedure (BF representation considered in Section 2.2) the algebra of n q-oscillators covariant under the coaction of $SU_q(n)$ was given first by Pusz and Woronowicz [194]. Interestingly enough, it appeared that $SU_q(n)$-covariance requires a particular coupling of different modes, which determines also the noncommutativity factors in $SU_q(n)$-covariant differential calculus (Section 1.6). As a result, the $SU_q(n)$-covariant generators A_i, A_i^+ $(i = 1, ..., n)$ satisfy the relations

$$\begin{aligned} A_i A_j &= q A_j A_i \ , \quad A_i^+ A_j^+ = q^{-1} A_j^+ A_i^+ \ , \qquad i < j \ , \\ A_i A_j^+ &= q A_j^+ A_i \ , \qquad i \neq j \ , \\ A_i A_i^+ &- q^2 A_i^+ A_i = 1 + (q^2 - 1) \sum_{k<i} A_k^+ A_k \ . \end{aligned} \tag{2.4.18}$$

Their invariance under the transformations

$$A_i' = \sum_k T_{ik} A_k \ , \ A_j^{+'} = \sum_k T_{jk}^\dagger A_k^+ \ ,$$

can be proven provided that the quantities T_{ij} and $T_{jk}^\dagger$ are not c-numbers but are the generators of the $Fun_q(SU(n))$ commuting with A_i, A_j^+.

Generators A_i, A_j^+ can be expressed in terms of the independent q-oscillators a_i, a_j^+ with the CR of the type (2.1.1),(2.1.2)

$$\begin{aligned} a_i a_j^+ - q^{\delta_{ij}} a_j^+ a_i &= q^{-N_i} \delta_{ij} \ , \\ [a_i, a_j] = [a_i^+, a_j^+] &= 0 \ , \end{aligned} \tag{2.4.19}$$

$$a_i^+ a_i = [N_i]_q \ , \quad \text{no summation} \tag{2.4.20}$$

The relations (2.4.20) show that we use restricted q-oscillator algebra or, in other words, Fock representation only for each of the independent oscillators. The expressions for oscillators A_i, A_i^+ in terms of a_i, a_i^+ are

$$A_i = q^{\sum_{k<i} N_k} q^{N_i/2} a_i \ , \ A_i^+ = (A_i)^\dagger \ .$$

The $SU_q(n)$ invariant operator has the form

$$h = \sum_{k=1}^n A_k^+ A_k = q^{-1}[N_1 + \ldots + N_n]_q q^{\sum_j N_j} \ .$$

This expression can be used as a Hamiltonian for the system of $SU_q(n)$ covariant oscillators.

Now we use unitary irreducible representations of n independent q-oscillators for the construction of all unitary irreducible representations of the $SU_q(n)$-covariant system of q-oscillators.

For this aim, it is more convenient to use b, b^+ form of q-oscillators (see (2.1.7),(2.1.8)). In Section 2.1 it was shown that the unitary irreducible representations of the oscillator algebra can be classified according to the sign of the operator $K = [b, b^+]$.

The generalization of this classification to the case of n pairs of independent q-oscillators b_i, b_i^+, $i = 1, ..., n$, satisfying the relations

$$\begin{aligned} b_i b_j^+ - q^{2\delta_{ij}} b_j^+ b_i &= \delta_{ij}, \quad q > 0 \ , \\ [b_i, b_j] = [b_i^+, b_j^+] &= 0 \ , \end{aligned} \tag{2.4.21}$$

is straightforward. Representations of these operators will be given as the tensor product of the representations of each pair of q-oscillators entering in (2.4.21).

First we notice that in any unitary irreducible (infinite-dimensional) representation, there exist operators $N_i = N_i^+$ and real parameters λ_i, $i = 1, ..., n$ (specified below) which satisfy

$$[N_i, N_j] = 0 \ ,$$

$$[N_i, b_j] = [N_i, b_j^+] = 0 \ , \ i \neq j \ , \tag{2.4.22}$$

$$[N_i, b_i] = -b_i \ , \ [N_i, b_i^+] = b_i^+ \ ,$$

and

$$b_i^+ b_i = \frac{1 + \lambda_i q^{2N_i}}{1 - q^2} \ , \ b_i b_i^+ = \frac{1 + \lambda_i q^{2N_i+2}}{1 - q^2} \ . \tag{2.4.23}$$

The admissible λ_i and N_i in (2.4.22), (2.4.23) give two classes of representations:

- **(A)** Fock representations, when $\lambda_i = -1$ and the eigenvalues n_i of N_i: $n_i \in \mathbb{Z}_+$. In this case all the values of $q > 0$ are allowed.

- **(B)** non-Fock representations, when $\lambda_i \geq 0$ and the eigenvalues n_i of N_i: $n_i \in \mathbb{Z}$. In this case only $0 < q < 1$ are allowed.

Let us note that the non-Fock representations of two pairs of q-oscillators, b, b^+ and $\tilde{b}, \tilde{b}^+$ satisfying

$$b^+ b = \frac{1 + \lambda q^{2N}}{1 - q^2} \ , \ \tilde{b}^+ \tilde{b} = \frac{1 + \tilde{\lambda} q^{2\tilde{N}}}{1 - q^2} \ ,$$

are unitarily equivalent, if $\tilde{\lambda} = q^k \lambda$, with k being an integer number. Furthermore, by putting

$$b_0 = (1 + \lambda q^{2N+2})^{-1/2} \, b \ , \ b_0^+ = b^+ (1 + \lambda q^{2N+2})^{-1/2} \ ,$$

we obtain a non-Fock pair of q-oscillators that satisfy the relations

$$b_0 b_0^+ = b_0^+ b_0 = \frac{1}{1 - q^2} \ ,$$

and thus they correspond to the degenerate case $\lambda_0 = 0$. We note, however, that the mapping $b \to b_0$, $b^+ \to b_0^+$ is not unitary.

Below we shall use the unitary irreducible representations of the algebra (2.4.21) to construct all the irreducible representations of the $SU_q(n)$-covariant system of q-oscillators. A general classification of the latter representations was given in [194] by using first-order differential calculus on a q-plane. However, as we shall see below, the construction of such representations in terms of q-oscillators can be performed in the simplest way [49].

Let A_i, A_i^+, $i = 1, ..., n$, be a system of q-oscillators $(0 < q < 1)$ satisfying the $SU_q(n)$-covariant algebra (2.4.18). It is convenient to denote

$$\Lambda_i = 1 - (1 - q^2) \sum_{k \geq i}^{n} A_k^+ A_k \ , \ i = 1, ..., n \ ,$$
$$\Lambda_{n+1} = 1 \ . \tag{2.4.24}$$

From (2.4.18) and (2.4.24) the following basic relations for the operators Λ_j can be deduced:

$$\Lambda_{j+1} - \Lambda_j = (1 - q^2) A_j^+ A_j \ ,$$
$$\Lambda_{j+1} - q^2 \Lambda_j = (1 - q^2) A_j A_j^+ \ , \tag{2.4.25}$$

and

$$\Lambda_j A_k^+ = A_k^+ \Lambda_j \ , \ A_k \Lambda_j = \Lambda_j A_k \ , \ j > k \ ,$$
$$\Lambda_j A_k^+ = q^2 A_k^+ \Lambda_j \ , \ A_k \Lambda_j = q^2 \Lambda_j A_k \ , \ j \leq k \ . \tag{2.4.26}$$

To construct the irreducible representations of (2.4.18), we shall consider in what follows four cases:

(i) Assume that $\Lambda_{i+1} = R_{i+1}^2 > 0$ and let us define

$$A_i = F_i(N_i, ...) b_i \ , \ A_i^+ = b_i^+ F_i(N_i, ...) \ , \tag{2.4.27}$$

where b_i, b_i^+ satisfy the q-oscillator algebra (2.4.21) and the N_i satisfy the relations (2.4.22). The dots in (2.4.27) indicate the dependence of the functions F_i on N_j, $j \neq i$ and hereafter will be omitted. Then according to (2.4.22) and (2.4.23)

$$A_i A_i^+ = \frac{1 + \lambda_i q^{2N_i+2}}{1 - q^2} F_i^2(N_i) \ ,$$
$$A_i^+ A_i = \frac{1 + \lambda_i q^{2N_i}}{1 - q^2} F_i^2(N_i - 1) \ , \tag{2.4.28}$$

where the values of λ_i and N_i are determined in correspondence with the representation (Fock or non-Fock) chosen for the q-oscillators b_i (see statements (A) and (B) after Eq. (2.4.23)).

From relations (2.4.26) and definitions (2.4.27), it is easy to conclude that

$$R_{i+1} = R_{i+1}(N_{i+1}, ..., N_n) \ . \tag{2.4.29}$$

Indeed, since the operator Λ_{i+1} commutes with A_j and A_j^+ for $j = 1, ..., i$ (cf. Eqs. (2.4.26)), then it cannot depend on N_j. Therefore, $\Lambda_{i+1} = \Lambda_{i+1}\,(N_{i+1}, ..., N_n)$. Let us now assume that the q-oscillators b_i are given in the Fock representation, i.e. $\lambda_i = -1$. Then inserting relations (2.4.28) into the last equation in (2.4.18), we obtain the recurrence relation

$$\frac{1-q^{2N_i+2}}{1-q^2}F_i^2(N_i) - q^2\frac{1-q^{2N_i}}{1-q^2}F_i^2(N_i-1) = \Lambda_{i+1} \ , \tag{2.4.30}$$

which has the general solution

$$F_i^2(N_i) = \Lambda_{i+1} + \frac{cq^{2N_i+1}}{1-q^{2N_i+2}} \ , \tag{2.4.31}$$

where c is a constant.

In particular, for $c = 0$ we obtain the solution

$$F_i = R_{i+1}(N_{i+1}, ..., N_n) \ . \tag{2.4.32}$$

Recalling that $\Lambda_i = R_i^2 > 0$, we obtain from (2.4.25), (2.4.28) and (2.4.32) the recurrence relation

$$R_i = q^{N_i} R_{i+1} \ . \tag{2.4.33}$$

Since by definition $R_{n+1} = 1$, finally we have the solution

$$F_i = R_{i+1} = q^{N_{i+1}+...+N_n} \ . \tag{2.4.34}$$

(ii) Let us now consider the case $\Lambda_{i+1} = -R_{i+1}^2 < 0$, and define

$$A_i = b_i^+ G_i(N_i) \ , \ A_i^+ = G_i(N_i) b_i \ , \tag{2.4.35}$$

where b_i, b_i^+ satisfy relations (2.4.21)-(2.4.23) and the dependence of the functions G_i on N_j, $j \neq i$, $j \neq 1$, has been omitted.

Then from (2.4.23), it follows that

$$\begin{aligned} A_i A_i^+ &= \frac{1+\lambda_i q^{2N_i}}{1-q^2} G_i^2(N_i - 1) \ , \\ A_i^+ A_i &= \frac{1+\lambda_i q^{2N_i+2}}{1-q^2} G_i^2(N_i) \ . \end{aligned} \tag{2.4.36}$$

Assuming that the operators b_i, b_i^+ are the Fock oscillators (i.e. $\lambda_i = -1$) and substituting (2.4.36) into (2.4.18), we arrive at the relation

$$\frac{1-q^{2N_i}}{1-q^2}\,G_i^2(N_i-1) - q^2\,\frac{1-q^{2N_i+2}}{1-q^2}\,G_i^2(N_i) = \Lambda_{i+1}\,, \tag{2.4.37}$$

with $\Lambda_{i+1} = \Lambda_{i+1}(N_{i+1}, ..., N_n)$.

The general solution of (2.4.37) is given by

$$G_i^2(N_i) = -q^{-2N_i-2}\Lambda_{i+1} + \frac{cq^{-2N_i}}{1-q^{2N_i+2}}\,, \tag{2.4.38}$$

where c is an arbitrary constant.

Since $\Lambda_{i+1} = -R_{i+1}^2 < 0$, for $c = 0$ we have

$$G_i = q^{-N_i-1}R_{i+1}(N_{i+1}, ..., N_n)\,. \tag{2.4.39}$$

Furthermore, from relations (2.4.18) we obtain the recurrence relation

$$R_i = q^{-N_i-1}R_{i+1}\,. \tag{2.4.40}$$

If $\Lambda_i < 0$, for $i = 1, ..., p < n$, then from (22) we find

$$G_i = q^{-(N_i+1)-...-(N_{p-1}+1)}R_p(N_p, ..., N_n)\,, \tag{2.4.41}$$

valid for $i = 1, ..., p-1$, where R_p is determined by $\Lambda_p = -R_p^2$.

(iii) Let us assume that $\Lambda_{i+1} = R_{i+1}^2 > 0$ and that the q-oscillators b_i are given in the non-Fock representation, i.e.

$$b_i^+ b_i = \frac{1+\varepsilon^2 q^{2N_i}}{1-q^2}\,,\quad b_i b_i^+ = \frac{1+\varepsilon^2 q^{2N_i+2}}{1-q^2}\,, \tag{2.4.42}$$

where we have denoted $\lambda_i \equiv \varepsilon^2 \geq 0$.

Defining A_i, A_i^+ as in (2.4.27) and repeating all the steps from the case (i), we find $F_i = R_{i+1}$. Substituting A_i and A_i^+ into Eq. (2.4.25) and using (2.4.42), we obtain

$$\Lambda_i = -\varepsilon^2 q^{2N_i}\Lambda_{i+1} \leq 0\,. \tag{2.4.43}$$

Thus we conclude that the non-Fock q-oscillators b_i, b_i^+ allow us to switch from the pattern (i) to the pattern (ii), provided that $\varepsilon \neq 0$. If $\varepsilon = 0$, then we arrive at $\Lambda_i = 0$.

(iv) Assume that $\Lambda_{i+1} = 0$. Then we can take b_i, b_i^+ in the non-Fock representation with $\lambda_i = 0$, satisfying

$$b_i^+ b_i = b_i b_i^+ = \frac{1}{1-q^2} \,. \tag{2.4.44}$$

Defining $A_i = F_i(N_i) b_i$, $A_i^+ = b_i^+ F_i(N_i)$, we obtain from (2.4.18) the equation

$$F_i^2(N_i) - q^2 F_i^2(N_{i-1}) = 0 \,, \tag{2.4.45}$$

which has the solution

$$F_i = \xi(N_{i+1}, ..., N_n) q^{N_i} \,. \tag{2.4.46}$$

Using $\Lambda_{i+1} = 0$, finally we find from Eq. (8)

$$\Lambda_i = -q^{-2} F_i^2 = -\xi^2 q^{2N_i - 2} < 0 \,. \tag{2.4.47}$$

Therefore we see that this case leads again to the case (ii).

Thus, from the four cases considered above we conclude that the most general pattern for the classification of all irreducible representations of the $SU_q(n)$-covariant system of q-oscillators defined by (2.4.18) is the following:

- Let b_i, b_i^+, $i = m+1, ..., n$, be the Fock oscillators and define A_i, A_i^+ as in (2.4.27). Then according to case (i), $\Lambda_{i+1} = q^{2(N_{i+1} + \cdots + N_n)} > 0$ for $i = m+1, ..., n-1$, and $\Lambda_{n+1} = 1$, by definition.

- Now if b_m is a degenerate non-Fock oscillator with $\varepsilon = 0$, then from case (iv) we have $\Lambda_{m+1} = 0$ and $\Lambda_m < 0$. In principle one can add some $\Lambda_i = 0$, $i = m, ..., p+1 (m > p)$ and then, using b_p as a degenerate non-Fock q-oscillator, switch to the case $\Lambda_p < 0$. Therefore, the use of a degenerate oscillator b_m allows us to switch from positive to negative values of Λ_i and case (ii) implies that $\Lambda_i < 0$ for $i = m-1, ..., 1$, if b_i are Fock q-oscillators.

- Finally, if b_m is a non-Fock oscillator with $\varepsilon \neq 0$, then according to case (iii) $\Lambda_{m+1} > 0$ and $\Lambda_m < 0$. Therefore we arrive again at case (ii), where for the Fock q-oscillators $b_1, b_2, ..., b_{m-1}$, one has $\Lambda_i < 0$, $i = 1, ..., m-1$.

We notice that the above construction exactly reproduces all the unitary irreducible representations of the $SU_q(n)$-covariant algebra (2.4.18) found in [194].

2.5 q-Deformation of supergroups and conception of braided groups

As we have already mentioned, quantum groups and quantum spaces are not direct generalization of "supermathematics" (superspaces, supertransformations, supergroups, etc.). The main point of distinction is that q-deformation which we described in preceding sections is not strongly correlated with any kind of statistics. In other words, permutations of *different* copies of the same objects (matrix elements, coordinates of different copies of quantum spaces) satisfy usual bosonic (commutative) rule. It seems natural that instead one can consider fermionic anticommutative permutation rule. This is indeed possible [178, 156, 38, 229, 141, 213] and we will consider the corresponding objects briefly in the first subsection.

Another possible line of development uses the idea of deformation of the very permutation rules ("statistics"). This leads to the so-called *braided groups* ([174, 175] and refs. therein).

2.5.1 q-Supergroups, q-superalgebras and q-superoscillators

Detailed discussion of quantum supergroups is outside of the frame of the present introductory book. Instead, we will give the basic definitions, illustrate them with an example and discuss quantum superoscillators and their covariant properties.

Let L_S be a Lie superalgebra [132, 186] with a Cartan matrix $A = \{a_{ij}\}$, $i, j \in \{1, 2, ..., r\} = \mathcal{I}$. The simple roots α_j of L_S are divided into two sets, even and odd ($j \in \tau \subset \mathcal{I}$) roots. As in the case of ordinary Lie algebras (Section 1.3), a quantum universal enveloping algebra $\mathcal{U}_q(L_S)$ of superalgebra corresponding to L_S has the same set of generators but deformed multiplication map which is defined by the relations (in Chevalley basis)

$$\begin{aligned} [H_i,\, H_j\} &= 0\,, \\ [H_i,\, X_j^\pm\} &= \pm a_{ij} X_j^\pm\,, \qquad (2.5.1) \\ [X_i^+,\, X_j^-\} &= \delta_{ij}[H_j]_{q_j}\,, \end{aligned}$$

$$(ad^{(q)}_{X_i^\pm})^{n_{ij}} X_j^\pm = 0\,, \qquad \text{for } i \neq j\,. \qquad (2.5.2)$$

Here $[\cdot,\cdot\}$ is supercommutator

$$[X,Y\} := XY - (-1)^{p(X)p(Y)} YX\,, \qquad X, Y \in \mathcal{U}_q(L_S)\,,$$

$p(X)$ is a parity of the elements of $\mathcal{U}_q(L_S)$

$$\begin{array}{rcll} p(H_i) & = & 0\,, & i \in \mathcal{I}\,,\\ p(X_i^{\pm}) & = & 0\,, & i \notin \tau \subset \mathcal{I}\,,\\ p(X_i^{\pm}) & = & 1\,, & i \in \tau \subset \mathcal{I}\,, \end{array}$$

$ad_X^{(q)}$ denotes, as in Section 1.3, a deformed supercommutator

$$ad^{(q)}_{X_i^{\pm}} X_j^{\pm} = X_i^{\pm} X_j^{\pm} - (-1)^{p(X_i^{\pm})p(X_j^{\pm})} q^{(\alpha_i,\alpha_j)} X_j^{\pm} X_i^{\pm}\ .$$

The matrix n_{ij} in the super-q-Serre relations (2.5.2) is the following

$$n_{ij} = \begin{cases} 1 - a_{ij}\,, & \text{if } (\alpha_i,\alpha_i) \neq 0\,,\\ 2\,, & \text{if } (\alpha_i,\alpha_i) = 0\,, (\alpha_i,\alpha_j) \neq 0\,,\\ 1\,, & \text{if } (\alpha_i,\alpha_i) = (\alpha_i,\alpha_j) = 0\,. \end{cases}$$

The maps of comultiplication, antipode and counit are:

$$\begin{array}{rcl} \Delta(H_i) & = & H_i \otimes 1 + 1 \otimes H_i\,,\\ \Delta(X_i^+) & = & X_i^+ \otimes 1 + q_i^{-H_i} \otimes X_i^+\,,\\ \Delta(X_i^-) & = & X_i^- \otimes q_i^{H_i} + 1 \otimes X_i^-\,,\\ S(\mathbf{1}) & = & 1, \quad S(H_i) = -H_i\,,\\ S(X_i^+) & = & -q_i^{H_i} X_i^+\,, \quad S(X_i^-) = -X_i^- q_i^{-H_i}\,,\\ \varepsilon(\mathbf{1}) & = & 1, \quad \varepsilon(H_i) = \varepsilon(X_i^{\pm}) = 0\,. \end{array}$$

The corresponding quantum matrix supergroups are defined again as dual objects. Consider the simplest example of $GL_q(1|1)$ q-supergroup, its QUEA $\mathcal{U}_q(gl(1|1))$ and related sets of quantum superoscillators.

The quantum matrix supergroup $GL_q(1|1)$ is a graded associative Hopf algebra with two even, (a,d), and two odd, (β,γ), generators satisfying the following algebraic relations

$$GL_q(1|1): \quad \begin{array}{ll} a\beta = q\beta a\,, & d\beta = q\beta d\,,\\ a\gamma = q\gamma a\,, & d\gamma = q\gamma d\,,\\ \beta\gamma = -\gamma\beta\,, & ad - da = (q - q^{-1})\beta\gamma\,. \end{array} \tag{2.5.3}$$

The comultiplication Δ is defined by the matrix corepresentation

$$T^{1|1} := \begin{pmatrix} a & \beta \\ \gamma & d \end{pmatrix}\,,$$

$$\Delta((T^{1|1})^i{}_k) = (T^{1|1})^i{}_l \otimes (T^{1|1})^l{}_k \,.$$

Introducing the q-deformed superdeterminant, one defines

$$SL_q(1|1): \quad sdet_q T^{1|1} := ad^{-1} - \beta d^{-1}\gamma d^{-1} = 1 \,. \tag{2.5.4}$$

Further, the *-involution restricts consistently the algebra (2.5.3) by imposing the relations

$$U_q(1|1): \quad T^{1|1} \in GL_q(1|1);$$

$$\gamma = -\frac{1}{a^*}\beta^*\frac{1}{a^*}\,, \qquad d = (a^*)^{-1}\,, \tag{2.5.5}$$

and for $SU_q(1|1)$ the condition of unimodularity becomes

$$SU_q(1|1): \quad T^{1|1} \in U_q(1|1);$$

$$sdet_q T^{1|1} := a^*a + \beta\beta^* = 1 \,. \tag{2.5.6}$$

Commutation relation for a pair of bosonic q-oscillator operators B^+, B and a pair of fermionic operators F^+, F which are invariant with respect to an action of $SU_q(1|1)$ q-group were found in [43]

A supersymmetric pair of q-oscillators B, F is defined by the CR

$$BF = qFB\,, \qquad B^+F^+ = q^{-1}F^+B^+\,, \tag{2.5.7}$$

$$BF^+ = q^{-1}F^+B\,, \qquad B^+F = qFB\,, \tag{2.5.8}$$

$$F^2 = (F^+)^2 = 0\,, \tag{2.5.9}$$

$$BB^+ - q^{-2}B^+B = 1 + (q^{-2} - 1)F^+F\,, \tag{2.5.10}$$

$$FF^+ + F^+F = 1\,. \tag{2.5.11}$$

The coaction of $GL_q(1|1)$,

$$\begin{aligned} B' &= aB + \beta F\,, & F' &= \gamma B + dF\,, \\ B^{+'} &= a^*B^+ - \beta^*F^+\,, & F^{+'} &= \gamma^*B^+ + d^*F^+\,, \end{aligned}$$

with the generators of the quantum group $GL_q(1|1)$ *graded* (*super-*) commuting with the q-oscillators (B, F, B^+, F^+) leave invariant

- (2.5.7),(2.5.9), for arbitrary $T^{1|1} \in GL_q(1|1)$;
- (2.5.7)-(2.5.10), if $T^{1|1} \in U_q(1|1)$);

- (2.5.7)-(2.5.11), if $T^{1|1} \in SU_q(1|1)$.

As in pure bosonic case, the invariant CR (2.5.7)-(2.5.10) are closely related to the quantum supergroup invariant differential calculus [146]. Thus analogously to bosonic case there is another "permuted" set of covariant q-superoscillators

$$
\begin{aligned}
&BF = qFB\,, \qquad BF^+ = qF^+B\,, \\
&BB^+ - q^2 B^+B = 1\,, \\
&FF^+ + F^+F = 1 + (q^2-1)B^+B\,.
\end{aligned}
$$

The $U_q(1|1)$ invariant Hamiltonian is described by the following bilinear form

$$H^{1|1} = B^+B + F^+F\,. \tag{2.5.12}$$

Superanalog of JS map in this case, i.e. the construction of the generators of $\mathcal{U}_q(U(1|1))$ out of the q-superoscillator operators, has the form [43]

$$
\begin{aligned}
Q &= B^+F\,, \qquad & Q^+ &= F^+B\,, \\
Y &= B^+B - F^+F\,, \qquad & Z &= H^{1|1}\,.
\end{aligned}
\tag{2.5.13}
$$

The CR (2.5.7)-(2.5.11) give the defining relations of $\mathcal{U}_q(U(1|1))$

$$Q^2 = Q^{+^2} = 0\,, \tag{2.5.14}$$

$$
\begin{aligned}
q^2QQ^+ + q^{-2}Q^+Q = q^{-2}Z \;&+\; \frac{1}{2}(1-q^{-2})(Y+Z) \\
&+\; \frac{1}{4}(q^2-1)(Z+Y)^2\,,
\end{aligned}
\tag{2.5.15}
$$

$$Q^+Y - q^2YQ^+ = -2Q^+ + (1-q^2)ZQ^+\,, \tag{2.5.16}$$

$$QY - q^2YQ^+ = 2Q - (1-q^2)ZQ\,, \tag{2.5.17}$$

$$[Q,Z] = \left[Q^+,Z\right] = [Y,Z] = 0\,. \tag{2.5.18}$$

Due to the relation (2.5.18), we see that the Hamiltonian (2.5.12) describes the central extension of the quantum algebra with basic generators Q, Q^+, Y, obtained by putting $Z = 0$ in the relations (2.5.15)-(2.5.17). Together with (2.5.18), these CR describe the q-extension of the $N = 2$ supersymmetry in quantum mechanics

$$\{Q, Q^+\} = H^{1|1},\; Q^2 = Q^{+^2} = 0\,,$$

with the supersymmetric doublets

$$|E\rangle\ , \quad Q|E\rangle\ ,$$

having the same energy values (see (2.5.18)). This CR and the corresponding comultiplication can be written in the equivalent form in terms of $L^{\pm}$-matrices (cf. Section 1.6.1, eqs. (1.6.1),(1.6.4),(1.6.5)) using the R-matrix

$$R = \begin{pmatrix} q & 0 & 0 & 0 \\ 0 & 1 & q-q^{-1} & 0 \\ 0 & 0 & 1 & 0 \\ 0 & 0 & 0 & q^{-1} \end{pmatrix} .$$

The CR for a set of n bosonic and m fermionic q-oscillators which are invariant with respect to the coaction of $SU_q(n|m)$ has been presented in [43]

$$\begin{aligned} B_iB_j = qB_jB_i\ ,\ B_i^+B_j^+ = q^{-1}B_j^+B_i^+\ ,\ i<j\ , \\ B_iB_j^+ = qB_j^+B_i\ , \quad i \neq j\ , \\ B_iB_i^+ - q^2B_i^+B_i = 1 + (q^2-1)\sum_{k<i} B_k^+B_k\ , \\ B_iF_r = qF_rB_i\ ,\ B_i^+F_r^+ = q^{-1}F_r^+B_i^+\ , \\ B_iB_j^+ = qB_j^+B_i\ , \quad i \neq j\ , \\ B_i^+F_r = qF_r^+B_i\ , \quad 1 \leq l \leq n,\ \ 1 \leq r \leq m\ , \\ F_rF_s = -qF_sF_r\ ,\ F_r^+F_s^+ = -q^{-1}F_s^+F_r^+\ , \quad r<s\ , \\ F_rF_s^+ = qF_s^+F_r\ , \quad r \neq s\ , \\ F_rF_r^+ + F_r^+F_r = 1 + (q^2-1)\sum_{k<r} F_k^+F_k\ . \end{aligned} \tag{2.5.19}$$

The invariant Hamiltonian is given by the formula

$$H^{n|m} = \sum_{i=1}^{n} B_i^+B_i + \sum_{r=1}^{m} F_r^+F_r\ , \tag{2.5.20}$$

which is commuting with the $\mathcal{U}_q(U(n|m))$ quantum algebra with the q-deformed Cartan-Weyl basis obtained by the bilinear hermitian products of the superoscillators (2.5.19).

In order to obtain the formulae (2.5.19), one can start from a set of n bosonic and m fermionic *independent* q-oscillators

$$\begin{aligned} a_ia_j^+ - q^{\delta_{ij}}a_j^+a_i = q^{-N_i}\delta_{ij}\ , \\ [a_i, a_j] = [a_i^+, a_j^+] = 0\ , \end{aligned} \tag{2.5.21}$$

$$a_i^+ a_i = [N_i]_q \ , \tag{2.5.22}$$

$$\begin{aligned} f_r f_s^+ + q^{-\delta_{rs}} f_s^+ f_r &= q^{-M_r}\delta_{rs} \ , \\ \{f_r, f_s\} = \{f_r^+, f_s^+\} &= 0 \ , \end{aligned} \tag{2.5.23}$$

$$f_r^+ f_r = [M_r]_{q^{-1}} \ , \quad \text{no summation.} \tag{2.5.24}$$

Note that (2.5.22),(2.5.24) show that we use Fock representation for independent q-superoscillators. In order to obtain (2.5.19), one should introduce

$$B_i = q^{L_i} a_i \ , \quad L_i := \sum_{k<i} N_k + \frac{N_i}{2} \ ,$$

$$B_i^+ = (B_i)^\dagger \ ,$$

$$F_r = q^{K_r} f_r \ , \quad K_r := \sum_{i=1}^{n} N_k + \sum_{s>r} M_s + \frac{M_r}{2} \ ,$$

$$F_i^+ = (F_i)^\dagger \ .$$

Fock representations of (2.5.21)-(2.5.24) can be constructed in a space spanned by the normalized eigenstates $|n_{a_i}, m_{f_r}>$ of the number operators N_i, M_r ($i = 1, ..., n$; $r = 1, ..., m$). In particular the spectrum of the Hamiltonian (2.5.20) is given by the formula

$$H^{n|m} |n_{a_i}, m_{f_r}> = [\bar{n} + \bar{m}]_q |n_{a_i}, m_{f_r}> \ ,$$

where

$$\bar{n} + \bar{m} = \sum_{l=1}^{n} n_{a_i} + \sum_{r=1}^{m} m_{f_r} \ .$$

2.5.2 Braided groups and spaces

The basis of this generalization [174, 175] is to replace the ± 1 phases encountered for superspace variables by a more general phase or more generally by an R-matrix. These are the generalized braid statistics which play the role of usual transposition or super-transposition. The essentially new feature of braided transposition Ψ compared with the usual bosonic or super-transpositions σ, where $\sigma^2 = \text{id}$, is that in this case $\Psi \neq \Psi^{-1}$. Yang-Baxter equation again plays a crucial role because the same phase factors or R-matrices can be used from one hand to define (as discussed in the Chapter 1) associative non-commutative algebras and on the other hand the generalizations of super-geometry called braided geometry.

Although quantum and braided groups have different physical and mathematical meaning, the fact that they appear from the same data (R-matrix, solution of the quantum Yang-Baxter equation) provide interesting relations between them. In fact, every quantum group equipped with a universal R-matrix has a braided-group analogue. This process of conversion of a strict quantum group into a braided group, is a process called *transmutation*. There is also an 'adjoint' process that converts any braided-group of a certain type into an ordinary (bosonic) quantum group called *bosonization*.

The starting point for the development of the theory of braided matrices is the fact that in the case of supermatrices [17, 111] not only entries of the same matrix do not commute (but supercommute according to their grading)

$$t^i{}_j t^k{}_l = (-1)^{p(t^i{}_j)p(t^k{}_l)} t^k{}_l t^i{}_j \,, \tag{2.5.25}$$

where grading is $p(t^i{}_j) = i + j$, but matrix elements $t^i{}_j$ and $t'{}_l{}^k$ of different copies T and T' of supermatrices also do not commute (*in contrast to the case of quantum matrix group*)

$$t'^i{}_j t^k{}_l = (-1)^{p(t'^i{}_j)p(t^k{}_l)} t^k{}_l t'^i{}_j \,. \tag{2.5.26}$$

The product of the supermatrices has the same CR

$$t''^i{}_j = t^i{}_k t'^k{}_j \;\Rightarrow\; t''^i{}_j t''^k{}_l = (-1)^{p(t''^i{}_j)p(t''^k{}_l)} t''^k{}_l t''^i{}_j \,. \tag{2.5.27}$$

It turns out that, if one requires appropriate generalization of this situation to more complicated case in which CR between different copies are defined by R-matrix, one has to modify the CR for entries $u^a{}_b$ of the same matrix also compared with the quantum matrix CR (cf. (1.1.22) or (1.4.10))

$$R^k{}_a{}^i{}_b u^b{}_c R^c{}_j{}^a{}_d u^d{}_l = u^k{}_a R^a{}_b{}^i{}_c u^c{}_d R^d{}_j{}^b{}_l \,,$$

that is

$$R_{21} u_1 R_{12} u_2 = u_2 R_{21} u_1 R_{12} \,. \tag{2.5.28}$$

These CR have the form of the so-called *reflection equation* [152]. If u' is another independent braided matrix obeying the same relations, then appropriate generalization of the super-CR (2.5.26) (which can be called superstatiscs relations) has the form

$$R_{12}^{-1} u'_1 R_{12} u_2 = u_2 R_{12}^{-1} u'_1 R_{12} \,. \tag{2.5.29}$$

The relations between u, u' are the *braid-statistics relations* between two independent identical braided matrices. These relations provide that the product uu' is also a braided matrix

$$u''^i{}_k = u^i{}_k u'^k{}_j \;\Rightarrow\; R_{21} u''_1 R_{12} u''_2 = u''_2 R_{21} u''_1 R_{12} \,. \tag{2.5.30}$$

Indeed,

$$\begin{aligned} R_{21}u_1''R_{12}u_2'' &= R_{21}u_1u_1'R_{12}u_2u_2' = R_{21}u_1R_{12}(R_{12}^{-1}u_1'R_{12}u_2)u_2' \\ &= (R_{21}u_1R_{12}u_2)R_{12}^{-1}R_{21}^{-1}(R_{21}u_1'R_{12}u_2') \\ &= u_2R_{21}(u_1R_{21}^{-1}u_2'R_{21})u_1'R_{12} = u_2R_{21}R_{21}^{-1}u_2'R_{21}u_1u_1'R_{12} \\ &= u_2''R_{21}u_1''R_{12} \,. \end{aligned}$$

For a triangular R-matrix (so that $R_{21} = R_{12}^{-1}$) the braid-statistics relations (2.5.29) and the braided-commutativity relations (2.5.28) coincide. This is the case for super-matrices (which fit into this framework for suitable R) but in the general braided case, the notion of braided-commutativity and braid-statistics are slightly different.

Explicit formulae in the simplest nontrivial case of 2×2 braided matrices $BGL_q(2)$

$$u = \begin{pmatrix} a & b \\ c & d \end{pmatrix},$$

with R being the same (1.1.25) as for $GL_q(2)$ quantum group, are the following:

- the braided-commutativity relations

$$\begin{aligned} &ba = q^2ab\,, \qquad ca = q^{-2}ac\,, \\ &da = ad\,, \qquad bc = cb + (1-q^{-2})a(d-a)\,, \\ &db = bd + (1-q^{-2})ab\,, \qquad cd = dc + (1-q^{-2})ca\,; \end{aligned}$$

- the braid-statistics relations

$$\begin{aligned} &a'a = aa' + (1-q^2)bc'\,, \qquad a'b = ba'\,, \\ &a'c = ca' + (1-q^2)(d-a)c'\,, \qquad a'd = da' + (1-q^{-2})bc'\,, \\ &b'a = ab' + (1-q^2)b(d'-a')\,, \qquad b'b = q^2bb'\,, \\ &b'c = q^{-2}cb' + (1+q^2)(1-q^{-2})^2bc' - (1-q^{-2})(d-a)(d'-a')\,, \\ &b'd = db' + (1-q^{-2})b(d'-a')\,, \\ &c'a = ac'\,, \qquad c'b = q^{-2}bc'\,, \\ &c'c = q^2cc'\,, \qquad c'd = dc'\,, \\ &d'a = ad' + (1-q^{-2})bc'\,, \qquad d'b = bd'\,, \\ &d'c = cd' + (1-q^{-2})(d-a)c' \qquad d'd = dd' - q^{-2}(1-q^{-2})bc'\,. \end{aligned}$$

In this example, as $q \to 1$ the algebra becomes commutative and the statistics also becomes commutative, so that we return to the case of usual matrix with commutative entries.

There are also deformed trace and determinant for braided algebras. Let $\widetilde{R} = ((R^{\mathsf{T}_2})^{-1})^{\mathsf{T}_2}$, where T_2 is transposition in the second matrix factor of $M_n \otimes M_n$ and we assume here that the relevant inverse of the R-matrix exists. Let $\vartheta^i{}_j = \widetilde{R}^i{}_k{}^k{}_j$. Then

$$c_k = \mathrm{Tr}\,\vartheta u^k \quad \Rightarrow \quad u' c_k = c_k u', \qquad u c_k = c_k u \,, \tag{2.5.31}$$

so that the c_k are bosonic and central elements of the braided matrix algebra $B(R)$. The element c_1 is the braided-trace while products of the c_k can be used to define the braided-determinant. The braided-determinant for the $SL_q(2)$ R-matrix is

$$\det_B(u) = ad - q^2 cb. \tag{2.5.32}$$

Setting this to 1 in $BGL_q(2)$, gives the braided group $BSL_q(2)$.

The analogy of the approach based on braided permutation rules with supermathematics is not restricted to the matrix algebra only. Recall that in superspace there are supervectors and supercovectors with a linear addition law. If θ_i are supervariables with parity $p(\theta_i) = i - 1 (\mathrm{mod}\ 2)$, we have supercommutativity

$$\theta_i \theta_j = (-1)^{p(\theta_i)p(\theta_j)} \theta_j \theta_i \,, \tag{2.5.33}$$

and if θ' is another copy of the super-plane with super-statistics relative to θ, we can add them. Thus

$$\theta'_i \theta_j = (-1)^{p(\theta'_i)p(\theta_j)} \theta_j \theta'_i \,, \quad \theta''_i = \theta_i + \theta'_i \;\Rightarrow\; \theta''_i \theta''_j = (-1)^{p(\theta''_i)p(\theta''_j)} \theta''_j \theta''_i \,. \tag{2.5.34}$$

Generalization of these relations leads to the notion of braided vectors and braided covectors. In fact, we already encountered analogous objects when we considered (quasi)classical limit for q-oscillator in Section 2.3. Indeed, the relations (2.3.18) define nontrivial CR for *different* copies of "classical" phase space coordinates. Those relations were physically motivated and relatively simple. So we managed to avoid special mathematical tools to derive and use them. Note, however, that different copies of the same coordinate on our q-oscillator phase space still commute. In more general braided spaces, this is not the case and to construct more complicated relations, R-matrix formalism again proves to be very useful.

Let the covector generators x_i generate a braided-commutative algebra $\mathbb{C}^B_{R'}$ of 'co-ordinate' functions with braided-commutation relations

$$x_j x_l = x_n x_m R'^m{}_j{}^n{}_l \,,$$

that is

$$x_1 x_2 = x_2 x_1 R'_{12} . \tag{2.5.35}$$

Here the matrix R' is characterized by the equation

$$(PR+1)(PR'-1) = 0. \tag{2.5.36}$$

The linear addition of braided-covectors means that if x' is another copy with braid-statistics relative to x, then $x+x'$ is also a realization of the same algebra,

$$x'_1 x_2 = x_2 x'_1 R_{12}, \quad x'' = x + x' \;\Rightarrow\; x''_1 x''_2 = x''_2 x''_1 R'_{12}. \tag{2.5.37}$$

Indeed,

$$\begin{aligned} (x_1 + x'_1)(x_2 + x'_2) &= x_1 x_2 + x'_1 x_2 + x_1 x'_2 + x'_1 x'_2 \\ &= x_2 x_1 R'_{12} + x_1 x'_2 (PR_{12} + 1) + x'_2 x'_1 R'_{12} \end{aligned} \tag{2.5.38}$$

On the other hand, the use of equality

$$PR + 1 = PRPR' + PR' ,$$

which is consequence of (2.5.36), and hence

$$x'_2 x_1 R'_{12} + x_2 x'_1 R'_{12} = x_1 x'_2 (PRP + P) R' ,$$

allows to prove equality of $(x_2 + x'_2)(x_1 + x'_1) R'_{12}$ to (2.5.38).

If R satisfies Hecke condition, i.e. $(PR+1)(PR-q^2) = 0$ (as for all the $SL_q(n)$ R-matrices; see Section 1.5), we can take $R' = q^{-2} R$. For example, the $SL_q(2)$ R-matrix gives for $\mathbb{C}^B_R$ the usual quantum planes with generators x, y and braided-commutativity relations $xy = q^{-1} yx$. The corresponding braid-statistics relations are

$$\begin{aligned} x'x &= q^2 x x' , & x'y &= q y x' , \\ y'y &= q^2 y y' , & y'x &= q x y' + (q^2 - 1) y x' . \end{aligned} \tag{2.5.39}$$

Corresponding differentials have generators θ, η with braided-commutativity relations $\theta^2 = 0, \eta^2 = 0, \theta\eta = -q\eta\theta$, and braid-statistics relations

$$\begin{aligned} \theta'\theta &= -\theta\theta' , & \theta'\eta &= -q^{-1}\eta\theta' , \\ \eta'\eta &= -\eta\eta' , & \eta'\theta &= -q^{-1}\theta\eta' + (q^{-2} - 1)\eta\theta' . \end{aligned} \tag{2.5.40}$$

We have already mentioned the close relation and possible transition from quantum to braided groups and *vice versa* (the so-called *transmutation* and *bosonization* processes). We do not give an exposition of this rather involved theory (see [174]). Here we indicate only how quantum groups with universal R-matrix $\mathcal{R}$ can generate nontrivial statistics.

Let V and W be two (co)representations of quasi-triangular Hopf algebra A. The tensor product representation $V \otimes W$ is given by the action of $\Delta(A) \subset A \otimes A$, the first factor acting on V and the second factor on W. Then the braiding is given by

$$\Psi_{V,W}(v \otimes w) = P(\mathcal{R} \triangleright (v \otimes w)) \,, \tag{2.5.41}$$

where $\triangleright$ is the action of $\mathcal{R} \in A \otimes A$, with its first factor on V and its second factor on W, and this is then followed by the usual vector-space transposition P. It is important that this braiding is consistent with the q-group action, i.e., $\Psi_{V,W}$ is intertwiner of the representations. Indeed, let $h \in A$, where A is some quasi-triangular Hopf algebra (i.e. QUEA of some Lie algebra). Then for its action we have (it is understood that h is in an appropriate representation)

$$\begin{aligned} h \bullet \Psi(v \otimes w) &:= (\Delta h) \triangleright P(\mathcal{R} \triangleright (v \otimes w)) = P((\Delta' h)\mathcal{R} \triangleright (v \otimes w)) \\ &= P(\mathcal{R}(\Delta h) \triangleright (v \otimes w)) = \Psi(h \bullet (v \otimes w)) \,. \end{aligned}$$

If $\mathcal{R}_{21} = \mathcal{R}^{-1}$ (the triangular case), Ψ is symmetric rather than braided.

In particular, $\mathbb{Z}_2$ algebra with a generator g (such that $g^2 = 1$) and a nonstandard triangular structure,

$$\Delta(g) = g \otimes g \,, \quad \varepsilon(g) = 1 \,, \quad S(g) = g \,, \quad \mathcal{R} = 1 - 2t \otimes t \,, \; t := \frac{1-g}{2} \,, \tag{2.5.42}$$

generates usual superstatistics

$$\begin{aligned} \Psi(v \otimes w) &= P(\mathcal{R} \triangleright (v \otimes w)) = (1 - 2t \otimes t)(w \otimes v) \\ &= (1 - 2p(v)p(w)) w \otimes v = (-1)^{p(v)p(w)} w \otimes v \,. \end{aligned}$$

This construction can be generalized to the so-called *anyonic* statistics [160]. One has to consider a group $\mathbb{Z}_n$, generated by g with the relation $g^n = 1$ and

$$\Delta(g) = g \otimes g \,, \quad \varepsilon(g) = 1 \,, \quad S(g) = g^{-1} \,,$$

$$\mathcal{R} = n^{-1} \sum_{a,b=0}^{n-1} e^{-\frac{2\pi i ab}{n}} g^a \otimes g^b \,. \tag{2.5.43}$$

Vector spaces of $\mathbb{Z}_n$-representations split as $V = \oplus_{a=0}^{n-1} V_a$ with the degree of an element defined by the action $g \bullet v = e^{\frac{2\pi i p(v)}{n}} v$. From (2.5.41) and (2.5.43) one finds

$$\Psi_{V,W}(v \otimes w) = e^{\frac{2\pi i p(v)p(w)}{n}} w \otimes v. \tag{2.5.44}$$

The case $n = 2$ is that of superspaces. For $n > 2$ the object is strictly braided in the sense that $\Psi \neq \Psi^{-1}$.

Note that these quantum groups are discrete and differ from those we considered in the Chapter 1.

2.6 Quantum symmetries and q-deformed algebras in physical systems

To illustrate physical meaning of the different aspects of q-deformation theory described in the previous sections, we present here examples of physical models exhibiting q-symmetry, constructed using the q-deformed algebras. Nowadays there are many models related to q-deformed objects. We selected some of them from different branches of physics: integrable systems, solid state physics, thermodynamics, quantum optics, quantum field and string theory. It is necessary to stress again that the very theory of quantum groups appeared as an extraction from constructions of quantum inverse scattering method (QISM) and so the latter and corresponding integrable systems play special role in the development of quantum group theory and are the most important and obvious area of its applications. QISM is an extensive part of mathematical physics and our brief description (we follow [41]) of the particular systems by no means can serve even as introduction to the subject. For reviews we refer to [100, 91, 152, 101, 105]. Other examples are from the areas of physics where applications of q-deformed objects are rather tempting and promising but not so obvious and well developed. The examples in Subsections 2.6.4 and 2.6.5 can be considered as introductory ones for more systematic presentation of q-deformations of space-time symmetries and the possibility of field theory regularization given in the next Chapter.

2.6.1 Integrable one-dimensional spin-chain model

One example of a system with the symmetry algebra $su(2)$ or $so(3)$ which can be generalized to the quantum algebra case is the spin-1/2 XXX-model with the Hamiltonian

$$H = g \sum_{n=1}^{N} (\sigma_n^X \sigma_{n+1}^X + \sigma_n^Y \sigma_{n+1}^Y + \sigma_n^Z \sigma_{n+1}^Z) . \qquad (2.6.1)$$

The commutativity of H with the total spin operators, $[S^j, H] = 0$,

$$S^j = \frac{1}{2}\sum_{k=1}^{N} \sigma_k^j \,, \tag{2.6.2}$$

is evident. The components S^j are generators of the $su(2)$ Lie algebra. The symmetry means that in the subspace of the Hamiltonian eigenvectors corresponding to a given eigenvalue, the representation of $su(2)$ is defined. More information about the XXX-model follows from the property of its exact solution using the Bethe Ansatz or the quantum inverse scattering method (QISM). The starting relation of the QISM is the Yang-Baxter equation

$$R(\lambda - \nu)L_n(\lambda) \otimes L_n(\nu) = (1 \otimes L_n(\nu))(L_n(\lambda) \otimes 1)R(\lambda - \nu) \,, \tag{2.6.3}$$

with a spectral parameter λ-dependent R-matrix and an auxiliary matrix $L_n(\lambda)$. For the XXX-model they are

$$R(\lambda) = \begin{pmatrix} a & 0 & 0 & 0 \\ 0 & b & c & 0 \\ 0 & c & b & 0 \\ 0 & 0 & 0 & a \end{pmatrix} , \quad \begin{array}{l} a = \lambda + \eta, \\ b = \lambda, \\ c = \eta, \end{array} \tag{2.6.4}$$

$$L_n(\lambda) = \begin{pmatrix} \lambda + \eta\sigma_n^z/2 & \eta\sigma_n^- \\ \eta\sigma_n^+ & \lambda - \eta\sigma_{n/2}^z \end{pmatrix} . \tag{2.6.5}$$

The transformation from local variables σ_n^j to the new ones, $A(\lambda)$, $B(\lambda)$, $C(\lambda)$, $D(\lambda)$, which are called quantum scattering data uses the monodromy matrix

$$T_N(\lambda) = L_N(\lambda)L_{N-1}(\lambda) \ldots L_1(\lambda) = \begin{pmatrix} A(\lambda) & B(\lambda) \\ C(\lambda) & D(\lambda) \end{pmatrix} \tag{2.6.6}$$

(roughly speaking, *monodromy* describes transformations of discrete fibre over some space when one goes around a closed path in this space; in more mathematical terms this means that the monodromy is a representation of the first homotopy group of the base space of some covering in the permutation group of elements of the discrete fibre). The matrix $T_N(\lambda)$ satisfies the same relations (2.6.3) as the L-operator. It is not difficult to see that

$$\eta S^- = \lim_{\lambda\to\infty} \lambda^{-N+1}B(\lambda) \,, \quad \eta S^+ = \lim_{\lambda\to\infty} \lambda^{-N+1}C(\lambda) \,, \tag{2.6.7}$$

and as a result of (2.6.3) and (2.6.6),

$$[S^j, A(\lambda) + D(\lambda)] = 0 \ . \tag{2.6.8}$$

It is possible to prove that the eigenvectors ψ of H constructed by the algebraic Bethe Ansatz are the highest-weight vectors for the total spin: $S^+\psi = (\sum_{k=1}^{N} \sigma_k^+)\psi = 0$.

More complicated analysis of this symmetry can be extended straightforwardly to many other algebras. The Hamiltonian density h of the XXX-model can be represented

$$h_n = \vec{\sigma}_n \vec{\sigma}_{n+1} = 2(P_1(n) - P_0(n)) - 1 \ , \tag{2.6.9}$$

as a linear combination of the orthogonal projectors into irreducible representations of the Clebsch-Gordan decomposition

$$V_n \otimes V_{n+1} = \mathbb{C}^3 \oplus \mathbb{C}^1 \ , \tag{2.6.10}$$

where $V_n \simeq \mathbb{C}^2$ is the spin 1/2 irreducible representation space in the n-th site. The sum of such densities with the nearest-neighbour interactions $\sum h_k$ and the periodic boundary conditions $h_N(\sigma_N, \sigma_{N+1}) \equiv h_N(\sigma_N, \sigma_1)$ will commute with the generators of total spin.

In this construction we can take:

1. any Lie algebra L with corresponding generators X_n in the n-th site;
2. any representations V_n of L with Clebsch-Gordan decomposition of $V_n \otimes V_{n+1}$ and the corresponding orthogonal projectors $P_j(n)$.

Then

$$H = \sum_{k=1}^{N} h_k(P_j(k)) \ , \tag{2.6.11}$$

is invariant with respect to global Lie algebra L with generators

$$\hat{X} = \sum_{k=1}^{N} X_k = \sum_{k=1}^{N} 1 \otimes ... \otimes 1 \otimes \underbrace{X}_{k-\text{th site}} \otimes ... \otimes 1 = \Delta^{(N)}(X) \ . \tag{2.6.12}$$

The essential part of this construction is the existence, for the considered algebra, of its action in tensor product of representations or comultiplication

(2.6.12). The peculiar property of a given density (2.6.9) is that the corresponding Hamiltonian (2.6.1) is one in the sequence of mutually commuting integrals of motion.

There exists, due to the QISM, direct generalization of the XXX-model to arbitrary spin s by substitution in the L-operator (2.6.5) the Pauli matrices by the $su(2)$ generators S_n^Z, $S_n^\pm$ in the irreducible representation $V_n \simeq \mathbb{C}^{2s+1}$.

The XXZ-model with the Hamiltonian

$$H = g\sum_{k=1}^{N}(\sigma_k^X\sigma_{k+1}^X + \sigma_k^Y\sigma_{k+1}^Y + \frac{q+q^{-1}}{2}\sigma_k^Z\sigma_{k+1}^Z) + g\frac{q-q^{-1}}{2}(\sigma_1^Z - \sigma_N^Z)\,, \tag{2.6.13}$$

is solved by the QISM using the same relation (2.6.3) with the R-matrix

$$R(u) = \begin{pmatrix} a & 0 & 0 & 0 \\ 0 & b & c & 0 \\ 0 & c & b & 0 \\ 0 & 0 & 0 & a \end{pmatrix}, \quad \begin{array}{l} a = \sinh(u+\eta), \\ b = \sinh u, \\ c = \sinh \eta, \end{array} \tag{2.6.14}$$

and the L-operator which we will write down for arbitrary value of spin S as

$$L_n(u) = \begin{pmatrix} \sinh(u+\eta S_n^Z) & \sinh\eta\ S_n^- \\ \sinh\eta\ S_n^+ & \sinh(u-\eta S_n^Z) \end{pmatrix}. \tag{2.6.15}$$

For the operators $S_n^\pm$, S_n^Z in the matrix L_n to satisfy (2.6.3), they must be now the generators of the quantum algebra $su_q(2)$ [151]:

$$[S_n^Z, S_m^\pm] = \pm\delta_{mn}S_n^\pm, \quad [S_n^+, S_m^-] = \delta_{nm}\frac{\sinh(2\eta S_n^Z)}{\sinh\eta}\,. \tag{2.6.16}$$

Let us remind once more that the comultiplication of $S^\pm$ or the action of the algebra in the tensor product of representations is different from the usual ones:

$$\Delta^{(N)}(S^\pm) = \sum_{k=1}^{N} q^{-S_1^Z}\ldots q^{-S_{k-1}^Z}S_k^\pm q^{S_{k+1}^Z}\ldots q^{S_N^Z}\,, \tag{2.6.17}$$

$$\Delta^{(N)}(S^Z) = \sum_{k=1}^{N} S_k^Z\,.$$

These operators which we can name as the global q-spin components can be extracted from the limit of the monodromy matrix $T(\lambda)$, $\lambda = \exp u$, $q = \exp \eta$, after the symmetry breaking transformation

$$T_{+} = \lim_{\lambda\to\infty} \lambda^{-N} U_1(\lambda) T(\lambda) U_1^{-1}(\lambda) = \begin{pmatrix} A_{+} & B_{+} \\ O & D_{+} \end{pmatrix} , \qquad (2.6.18)$$

$$R_{+}(\nu) = \lim_{\lambda\to\infty} \lambda^{-1} U_1(\lambda) R(\lambda/\nu) U_1^{-1}(\lambda) = \nu^{-1} \begin{pmatrix} q & 0 & 0 & 0 \\ 0 & 1 & \nu(q-q^{-1}) & 0 \\ 0 & 0 & 1 & 0 \\ 0 & 0 & 0 & q \end{pmatrix} , \qquad (2.6.19)$$

$$R_{++} = \lim_{\nu\to\infty} \nu U_2(\nu) R_{+}(\nu) U_2^{-1}(\nu) = \begin{pmatrix} q & 0 & 0 & 0 \\ 0 & 1 & q-q^{-1} & 0 \\ 0 & 0 & 1 & 0 \\ 0 & 0 & 0 & q \end{pmatrix} , \qquad (2.6.20)$$

$$U_1 = U \otimes 1, \quad U_2 = 1 \otimes U ,$$

$$U(\lambda) = \exp(\frac{u}{2}\sigma^z) = \begin{pmatrix} \lambda^{1/2} & 0 \\ 0 & \lambda^{-1/2} \end{pmatrix} . \qquad (2.6.21)$$

For these matrices the following relations are valid:

$$R_{+}(\nu) T_{+} \otimes T(\nu) = (1 \otimes T(\nu))(T_{+} \otimes 1) R_{+}(\nu) , \qquad (2.6.22)$$

$$R_{++} T_{+} \otimes T_{+} = (1 \otimes T_{+})(T_{+} \otimes 1) R_{++} . \qquad (2.6.23)$$

There are also analogous relations for T_{-}:

$$T_{-} = \lim_{\lambda\to 0} \lambda^{N} U(\lambda) T(\lambda) U^{-1}(\lambda) = \begin{pmatrix} A_{-} & 0 \\ C_{-} & D_{-} \end{pmatrix} , \qquad (2.6.24)$$

and for $R_{-}(\nu)$, R_{+-}, ..., etc.

It is easy to see that $T_{\pm}$ and relations between them are nothing but $L^{\pm}$ operators in the Faddeev-Reshetikhin-Takhtajan approach to the quantum Lie algebras (Section 1.6.1). The entries of $T_{\pm}$ coincide with $\Delta^{(N)}(S^j)$ for the quantum algebra $su_q(2)$:

$$A_{+} = \prod_{k=1}^{N} q^{S_k^z} , \quad D_{+} = \Delta^{(N)}(q^{-S^z}) = \prod_{k=1}^{N} q^{-S_k^z} , \qquad (2.6.25)$$

$$B_+/\sinh\eta = \Delta^{(N)}(S^-) = \sum_{k=1}^{N} q^{-S_1^z} \dots q^{-S_{k-1}^z} S_k^- q^{S_{k+1}^z} \dots q^{S_N^z} .$$

The Hamiltonian of the XXZ-model with periodic boundary conditions can be extracted from the transfer matrix $t(u) = tr\ T(u)$. As it follows from (2.6.22), it does not commute with the entries of $T_\pm$ which are generators of $su_q(2)$.

2.6.2 A model in quantum optics

We now turn to an application of the q-oscillators and consider following [39, 41] a fundamental quantum optical system, the Jaynes-Cummings Model (JCM), which describes the coupling of a single bosonic mode (laser field) with a two-level atom during the passing of a beam of such atoms through a cavity. For the coherent excitation of the atoms we consider the JCM Hamiltonian with intensity-dependent coupling, which physically signifies the fact that the strength of the laser-atom interaction depends on the number of photons in the cavity:

$$H_{int} = \lambda(\sqrt{N} b_0^+ \sigma^- + \beta\sqrt{N}\sigma^+) , \tag{2.6.26}$$

where N is the bosonic number operator ($[b_0, b_0^+] = 1$) and $\sigma^\pm$ are the Pauli matrices. Utilizing the Holstein-Primakoff (HP) realization of the $su(1,1)$ algebra, one can write the Hamiltonian as

$$H_{int} = \lambda(L_+\sigma^- + L_-\sigma^+) , \tag{2.6.27}$$

where $L_+ = \sqrt{N}a^+$, $L_- = a\sqrt{N}$, $L_0 = N + \frac{1}{2}$ are the $su(1,1)$ generator. The deformation of this model formally consists of replacing the usual bosons, b_0, b_0^+, by deformed bosons a, a^+.

Using the q-analogue of HP realization for the $su(1,1)_q$ algebra

$$K_+ = \sqrt{[N]_q}a^+ = L_+\frac{[N+1]_q}{N+1} ,\ K_- = a\sqrt{[N]_q} = \frac{[N+1]_q}{N+1}L_- ,\ K_0 = L_0 , \tag{2.6.28}$$

the deformed interaction Hamiltonian [39]

$$H_{int}^{(q)} = \lambda(\sqrt{[N]_q}a^+\sigma^- + a\sqrt{[N]_q}\sigma^+) , \tag{2.6.29}$$

can be written as

$$H_{int}^{(q)} = \lambda(K_+\sigma^- + K_-\sigma^+) = \lambda(L_+\sigma^-\frac{[N+1]_q}{N+1} + \frac{[N+1]_q}{N+1}L_-\sigma^+) . \tag{2.6.30}$$

The last form of the Hamiltonian shows that the deformation of bosons introduces a q-dependence in the coupling constant in addition to its dependence on the field intensity. Also it is clear that the dynamical algebra of the model from $su(1,1) \oplus su(2)$ becomes, after the deformation, the quantum algebra $su(1,1)_q \oplus su(2)$.

The unitary evolution operator is given by exponentiation of the Hamiltonian, $U(t) = \exp\left(-itH_{int}^{(q)}\right)$, which gives

$$U(t) = \begin{pmatrix} \cos(\lambda t\sqrt{K_-K_+}) & -i\dfrac{\sin(\lambda t\sqrt{K_-K_+})}{\sqrt{K_-K_+}}K_- \\ -iK_+\dfrac{\sin(\lambda t\sqrt{K_-K_+})}{\sqrt{K_-K_+}} & \cos(\lambda t\sqrt{K_+K_-}) \end{pmatrix}, \qquad (2.6.31)$$

with $K_+K_- = [N]_q^2$ and $K_-K_+ = [N+1]_q^2$.

Calculations of the evolution of the population inversion with the initial condition $|\psi(0)\rangle = |+\rangle \otimes |\psi\rangle$, give

$$< \sigma_3(t) >= \sum_{n=0}^{\infty} \cos(2\lambda t[n+1]_q)\, |\langle\psi|n\rangle|^2 \,. \qquad (2.6.32)$$

If one specifies the initial state of the field $|\psi\rangle$ as a q-deformed coherent state [23],

$$|\psi\rangle \equiv |\alpha\rangle = (\exp_q(-|\alpha|^2))^{-\frac{1}{2}} \sum_{n=0}^{\infty} \frac{\alpha^n}{\sqrt{[n]_q!}}|n\rangle \,,$$

then one obtains

$$< \sigma_3(t) >= (\exp_q |\alpha|^2)^{-\frac{1}{2}} \sum_{n=0}^{\infty} \cos(2\lambda t[n+1]_q)\, \frac{|\alpha|^{2n}}{[n]_q!} \,. \qquad (2.6.33)$$

The deformed quantities in the sum forbid the exact summation and a numerical calculation shows that periodicity for $q = 1$ is destroyed progressively in time for $q > 1$ values. For details see [39] where the q-analogs of the Barut-Girardello and the Perelomov coherent states for the quantum algebras were introduced.

2.6.3 Magnetic translations and the algebra $sl_q(2)$

Consider a spinless particle which moves in a plane and experiences a uniform external magnetic field along the orthogonal z-direction, $\vec{B} = Be_z$. The Hamiltonian of the system can be written as

$$h = \frac{1}{2m}(\vec{p} + e\vec{A})^2 \,, \qquad (2.6.34)$$

where m, e are mass and charge of the particle, respectively; $\vec{A}$ is the vector potential which satisfies

$$\nabla \times \vec{A} = B\vec{e}_z \ .$$

The above problem can be easily solved in a proper gauge [159].

The system is not invariant with respect to usual translations but exhibits the so-called magnetic translation invariance generated by the operator

$$\vec{\kappa} = \vec{p} + e\vec{A} + e\vec{r} \times \vec{B} \ ,$$

$\vec{\kappa}$ being the two dimensional vector in the plane: $\vec{\kappa} = \kappa_x \vec{e}_x + \kappa_y \vec{e}_y$. It is easy to see that

$$[\vec{\kappa}, h] = 0 \ . \tag{2.6.35}$$

The magnetic translation operator [117]

$$t(\vec{a}) = \exp\{\frac{i}{\hbar}\vec{a}\vec{\kappa}\} \ ,$$

($\vec{a}$ is an arbitrary two-dimensional vector) satisfies the CR for coordinates on a q-plane

$$t(\vec{a})t(\vec{b}) = \exp\{-i\frac{\vec{e}_z(\vec{a} \times \vec{b})}{a_0^2}\}t(\vec{b})t(\vec{a}) \ , \tag{2.6.36}$$

where $a_0 = \sqrt{\hbar/eB}$ is the magnetic length.

Using the magnetic translation operator one can construct the following operators ([240, 62] and refs. therein)

$$\begin{aligned} X^+ &= \frac{t(\vec{a}) + t(\vec{b})}{q + q^{-1}} \ , \\ X^- &= -\frac{t(-\vec{a}) + t(-\vec{b})}{q + q^{-1}} \ , \\ q^H &= t(\vec{b} - \vec{a}) \ , \end{aligned} \tag{2.6.37}$$

with

$$q = \exp\left\{i2\pi\frac{\Phi}{\Phi_0}\right\} \ ,$$

where $\Phi = \frac{1}{2}\vec{B}(\vec{a} \times \vec{b})$ is the magnetic flux through the triangle enclosed by vectors $\vec{a}$ and $\vec{b}$, $\Phi_0 = \frac{\hbar}{e}$ is the magnetic flux quanta. A straightforward calculations shows that the operators $X^{\pm}$ and H satisfy the CR (1.1.36) of the quantum algebra $s\ell_q(2)$. From (2.6.35) it follows that

$$[X^{\pm}, h] = 0 \ , \qquad [q^{\pm H}, h] = 0 \ ,$$

which indicates that $X^{\pm}, H$ are the conserved quantities of the system.

The periodic boundary conditions

$$t(\vec{L}_1)\Psi = \Psi \,, \qquad t(\vec{L}_2)\Psi = \Psi \,,$$

where $\vec{L}_1 = L_1\vec{e}_x$ and $\vec{L}_2 = L_2\vec{e}_y$, together with (2.6.36) yields that

$$\exp\left\{i2\pi\frac{\Phi}{\phi_0}\right\} = 1 \,, \qquad \Phi = \frac{1}{2}BL_1L_2 \,,$$

i.e. the magnetic flux through the triangle enclosed by $\vec{L}_1$ and $\vec{L}_2$ is quantized

$$\Phi = N_s\Phi_0 \,, \qquad N_s \in \mathbb{Z}_+ \,.$$

The boundary conditions reduce the magnetic translation invariance to transformations generated by the operators

$$T_x = t\left(\frac{\vec{L}_1}{N_s}\right) \,, \qquad T_y = t\left(\frac{\vec{L}_2}{N_s}\right) \,,$$

which also satisfy $s\ell_q(2)$ commutation relations, but with *the root of unity* deformation parameter

$$q = \exp\left\{i\frac{\pi}{N_s}\right\} \quad \Rightarrow \quad q^{2N_s} = 1 \,.$$

As $X^{\pm}, H$ are conserved quantities, the degeneracy of the Hamiltonian h is described by irreducible representations of $s\ell_q(2)$. In particular, as we discussed in Section 1.7.2, there exist cyclic representations of dimension $2N_s$. This dimension coincides with the degree of levels (called *the Landau levels*) of a particle in an external magnetic field, obtained by direct computation in a particular gauge (see, e.g. [159]).

2.6.4 Pseudoeuclidian quantum algebra as symmetry of phonons

Let us so consider following [27] the linear chain of equal masses lying at a distance a from one another, with nearest neighbor harmonic interaction. The equations of motion are:

$$\ddot{z}_j(t) = \omega^2\left(z_{j-1}(t) + z_{j+1}(t) - 2z_j(t)\right) \,,$$

where $z_j(t)$ is the displacement of the j-th mass ($j = 0, 1, \ldots, N$). Periodic boundary conditions are assumed and initial conditions $z_j(0)$, $\dot{z}_j(0)$ must be specified.

The ordinary system for displacements can be embedded into the partial differential equation

$$\left(\partial_t^2 + (2v/a)^2 \sin^2(-ia\partial_x/2)\right) z(x,t) = 0 , \tag{2.6.38}$$

where $v = \omega a$. The periodic conditions are $z(0,t) = z(Na,t)$, while the Cauchy data consist in the assignment of smooth functions $z(x,0)$ and $\partial_t z(x,0)$. When $z(ja,0) = z_j(0)$, $\partial_t z(ja,0) = \dot{z}_j(0)$ for all j, it is easy to see that the solutions of the ordinary system are directly obtained as $z_j(t) = z(ja,t)$, irrespectively of the behaviour of the solutions in the points $x \neq ja$.

This partial differential equation gives a realization with Casimir operator $C_q = 0$ of the pseudoeuclidean quantum algebra $E_q(1,1)$ [35] with $q = \exp\{ia\}$. This algebra had been obtained via q-contaction analogous to that described in Section 2.1 for q-Heisenberg group. In the next Chapter the reader will find more details about the contraction procedure.

$E_q(1,1)$ is generated by three generators E_q, P_q, B_q with the defining relations

$$[B_q, P_q] = iE_q , \quad [B_q, E_q] = (i/\chi)\sinh(\chi P_q) , \quad [E_q, P_q] = 0 ,$$

comultiplication

$$\begin{aligned} \Delta(P_q) &= P_q \otimes \mathbf{1} + \mathbf{1} \otimes P_q , \\ \Delta(E_q) &= E_q \otimes e^{\chi P_q/2} + e^{-\chi P_q/2} \otimes E_q , \\ \Delta(B_q) &= B_q \otimes e^{\chi P_q/2} + e^{-\chi P_q/2} \otimes B_q , \end{aligned}$$

and the antipodes

$$S(B_q) = -B_q + (i/2)\,\chi E_q , \quad S(P_q) = -P_q , \quad S(E_q) = -E_q .$$

The Casimir of $E_q(1,1)$ is

$$C_q = E_q{}^2 - 4/\chi^2 \sinh^2(\chi P_q/2) .$$

It is straightforward to verify that this quantum algebra satisfies the Hopf algebra axioms.

The realization of the $E_q(1,1)$-algebra reads

$$E_q = (i/v)\,\partial_t , \quad P_q = -i\partial_x , \quad B_q = i(x/v)\partial_t - (vt/a)\sin(-ia\partial_x) .$$

On the mass-shell $C = 0$ (i.e. on a solution of the equation of motion (2.6.38)) and in the momentum representation, a realization of the $E_q(1,1)$ in terms of the diagonal P_q and the position operator $X = i\partial/\partial p$ is given by

$$\begin{aligned} E_q &= \frac{2}{a}\sin(ap/2), \\ B_q &= \frac{1}{a}(\sin(ap/2)X + X\sin(ap/2)), \\ P_q &= p. \end{aligned}$$

As in the case of semisimple algebras (cf. (1.3.19)), all the defining relations can be rewritten in terms of the generator $k = e^{iaP_q}$ instead of P_q. If we rewrite everything in terms of k the real topology appears and we see that P_q is determined up to an integer multiple of $2iP_q/a$. In such a way, the topology of the q-algebra with $|q| = 1$ implies what in solid state physics is called the reduction to the first Brillouin zone, i.e. the limitation of the values of P_q and E_q: $0 \leq p < 2\pi/a$, $E_q > 0$.

The time derivative of X is given by $\dot{X} = iv\,[E_q, X]$ and the commutator, evaluated from the q-algebra, gives the well known group velocity of the phonons

$$\dot{X} = v_g = v\,\cos(aP_q/2)\,.$$

The fusion of phonons is governed by the comultiplication in $E_q(1,1)$, but one must care about the correct statistics. Phonons are bosons: to save their statistics and generate a correct composite system, we need symmetrical operators in tensor product of two representation spaces $V \otimes V$. So the operators of the coalgebra cannot describe the observables of the two phonons system: they are not symmetric (and not even hermitian). At the same time q-algebras have a symmetry: $q \leftrightarrow q^{-1}$ corresponding to the exchange of the two spaces in $V \hat{\otimes} V$. So we can attempt to impose this symmetry by hand, substituting all the coalgebra with its symmetrized form: $\Delta + \Delta'$ (cf. Section 1.4)

$$\begin{aligned} P_q^s &= P_q^{(1)} + P_q^{(2)}, \\ E_q^s &= \cos\left(aP_q^{(1)}/2\right) E_q^{(2)} + \cos\left(aP_q^{(2)}/2\right) E_q^{(1)}, \\ B_q^s &= \cos\left(aP_q^{(1)}/2\right) B_q^{(2)} + \cos\left(aP_q^{(2)}/2\right) B_q^{(1)}. \end{aligned}$$

Now P_q^s, E_q^s and B_q^s are all symmetric and hermitian and an explicit check shows that they still generate the $E_q(1,1)$ algebra. They are, in such a way, good candidates to describe the observables for the system of two phonons. However, it must be stressed, that they are not a coalgebra of our $E_q(1,1)$

because they do not satisfy all the requirements of a Hopf algebra; in particular, they do not allow for building, by iteration, the operators for three and more phonons: the global operators must be calculated from the original coproduct up to the required number of phonons and then should be completely symmetrized. It is easy to show that for each n this procedure again gives the generators of the $E_q(1,1)$ algebra.

Because P_q is defined up to $2\pi/a$, the composition of the momenta also has the same property: $P_q{}^s = P_q{}^{(1)} + P_q{}^{(2)} + 2\pi n/a$, showing that the Umklapp process is implied by the quantum group symmetry.

From the same definition of the position operator in terms of the algebra, one obtains the position operation of the two phonon system:

$$X^s = \frac{1}{2}(X^{(1)} + X^{(2)}) + \frac{1}{2}\left\{\frac{\sin(a(P_q{}^{(1)} - P_q{}^{(2)})/2)}{\sin(a(P_q{}^{(1)} + P_q{}^{(2)})/2)}, \frac{1}{2}(X^{(1)} - X^{(2)})\right\}_+ ,$$

which reproduces the Heisenberg algebra $[X^s, P_q{}^s] = i$ for the global variables. Finally, the group velocity of the composite system $\dot{X}^s = i\ [\Omega, X^s] = v\ \cos(aP_q{}^s/2)$, appears formally identical to that of the elementary system, having performed the Umklapp process.

2.6.5 q-Oscillators and regularization of quantum field theory

Now we shall discuss, following [139], the Heisenberg algebras of positions and momenta that can be represented on the q-deformed Bargmann-Fock spaces (Section 2.2). Let us define q-Heisenberg algebra that is generated by operators x_i, p_i, $(i = 1, ..., n)$ which are represented, just like in usual quantum mechanics, as

$$\begin{aligned} x_j &= L_j(\bar{z}_j + \bar{\partial}_j) , \\ p_j &= iK_j(\bar{z}_j - \bar{\partial}_j) , \end{aligned} \tag{2.6.39}$$

on the Bargmann-Fock space. Here $\bar{z}_j$, $\bar{\partial}_j$ are coordinates and derivatives of different copies of quantum planes related to the q-deformed Bargmann-Fock representation. As we discussed in Section 2.3, commutation relations between coordinates on different copies of q-planes depend on their concrete meaning. For simplicity, in this subsection we will choose them commutative (there exists another possibility, which leads to covariance with respect to $SU_q(n)$). Then CR for operators x_j, p_j have the form

$$[x_i, p_j] = i\hbar\delta_{ij} + i\hbar\delta_{ij}(q^2 - 1)\left(\frac{1}{4L_i^2}x_i^2 + \frac{1}{4K_i^2}p_i^2\right) , \tag{2.6.40}$$

$$[x_i, x_j] = 0 \ , \tag{2.6.41}$$

$$[p_i, p_j] = 0 \ , \tag{2.6.42}$$

where

$$L_i K_i = \hbar(q^2 + 1)/4 \ . \tag{2.6.43}$$

Note that if $q = 1$, the dimensionful constants K_i and L_i drop out of the commutation relations. This reflects the fact that in ordinary quantum mechanics a length or a momentum scale can only be set by the Hamiltonian, i.e. by choosing a particular system. Here, for $q > 1$, the K_i and L_i do appear in the commutation relations.

The functional analysis for the position and momentum operators is as follows ([139] and refs. therein). Their domain D is the set of Bargmann-Fock functions which are polynomials in $\bar{z}_i$. This is a dense set in the Hilbert space and on it the x_i and p_i are symmetric. But the x_i and p_i are not essentially self-adjoint. Their adjoints x_i^* and p_i^* are closed but nonsymmetrical. The x_i^{**} and p_i^{**} are closed and symmetric, but their deficiency indices (see, e.g., [13]) do not vanish. The deficiency subspaces are still of the same size, so that there are continuous families of self-adjoint extensions. One might be tempted to try to fix the choice of the self-adjoint extensions by the requirement that the domains coincide. A diagonalization of the x_i or the p_i, would then lead as usual to momentum space or position space representations. However, one can show that there are no more eigenstates in the representations of the q-Heisenberg algebras, and there is thus no position or momentum eigenbasis.

It can already be seen in the one-dimensional case, that neither x nor p is diagonalizable on a representation of the Heisenberg algebra and that they have not any eigenvectors. The commutation relation

$$[x, p] = i\hbar + i\hbar(q^2 - 1)\left(\frac{1}{4L^2}x^2 + \frac{1}{4K^2}p^2\right) , \tag{2.6.44}$$

leads to the uncertainty relation:

$$\Delta x \Delta p \geq \frac{\hbar}{2}\left(1 + (q^2 - 1)\left(\frac{(\Delta x)^2 + \langle x\rangle^2}{4L^2} + \frac{(\Delta p)^2 + \langle p\rangle^2}{4K^2}\right)\right) , \tag{2.6.45}$$

(do not confuse uncertainties $\Delta x, \Delta p$ with comultiplication map $\Delta(a)$ in a Hopf algebra; both notations are too traditional to be changed). It implies nonzero minimal uncertainties both in x as well as p measurements. Indeed, as e.g. Δx gets smaller, Δp must increase so that the product $\Delta x \Delta p$ of the lhs remains larger than the rhs. In usual quantum mechanics this is always possible, i.e. Δx can be made arbitrarily small. However, in the case $q > 1$

there is a $(\Delta p)^2$ term on the rhs of (2.6.45) which eventually grows faster with Δp than the lhs. Thus Δx can no longer become arbitrarily small. The minimal uncertainty in position measurements comes out as

$$\Delta x_0 = L\sqrt{1-q^{-2}}\,. \tag{2.6.46}$$

Analogously, one obtains the smallest uncertainty in the momentum

$$\Delta p_0 = K\sqrt{1-q^{-2}} = \frac{(q^2+1)\hbar}{4L}\sqrt{1-q^{-2}}\,, \tag{2.6.47}$$

(the latter equality follows from (2.6.43)).

So one can show that the q-Heisenberg algebra describes ordinary quantum mechanical behaviour in the physical region of not too large and not too small positions and momenta, explicitly where

$$(\Delta x_0)^2 \ll \langle x^2 \rangle \ll \frac{\hbar^2}{4(\Delta p_0)^2}\,,$$

$$(\Delta p_0)^2 \ll \langle p^2 \rangle \ll \frac{\hbar^2}{4(\Delta x_0)^2}\,.$$

Now consider consequences of the q-deformation of canonical commutation relations for a quantum field theory, namely, for simple example of the ϕ^4 model.

The original generating functional, with the fields defined on position space, reads

$$Z[J,J^*] = N\int D\phi\, D\phi^*\, e^{\int d^4x\, \phi^*(-\frac{L^2}{\hbar^2}p_ip_i - \frac{L^2M^2c^2}{\hbar^2})\phi - \frac{\lambda L^4}{4!}(\phi\phi)^*\phi\phi + \phi^*J + J^*\phi}\,. \tag{2.6.48}$$

Generalized ϕ^4 model suggested in [139] is defined by the generating functional

$$\begin{aligned} Z[J,J^*] \;=\; & N\int D\phi\, D\phi^*\, e^{\overline{\phi(\bar z)}\, e_{1/q}^{\partial_i \bar\partial_i}\left(-\bar z_i\bar\partial_i - 2 - \frac{L^2M^2c^2}{\hbar^2}\right)\phi(\bar z)\,|_0} \\ & \times e^{-\frac{\lambda}{4\pi^2 4!}\overline{\phi(\bar z+\partial_{z'})\phi(\bar z+\bar z')}\, e_{1/q}^{\partial_i\bar\partial_i}\, \phi(\bar z+\partial_{z''})\phi(\bar z+\bar z'')\,|_0} \\ & \times e^{\overline{\phi(\bar z)}\, e_{1/q}^{\partial_i\bar\partial_i}\, J(\bar z)\,|_0 + \overline{J(\bar z)}\, e_{1/q}^{\partial_i\bar\partial_i}\, \phi(\bar z)\,|_0}\,. \end{aligned} \tag{2.6.49}$$

In the limit $q \to 1$ the Lagrangian in (2.6.49) reduces to

$$\mathcal{L}' = \phi^*(-\frac{L^2}{\hbar^2}p_ip_i - \frac{x_ix_i}{4L^2} - \frac{L^2M^2c^2}{\hbar^2})\phi - \frac{\lambda L^4}{4!}(\phi\phi)^*\phi\phi\, e^{\frac{x_ix_i}{2L^2}} + \phi^*J + J^*\phi\,. \tag{2.6.50}$$

Thus the model corresponds to q-deformation of ϕ^4 theory with modified long distance behaviour as compared to (2.6.48). Note that in this model one does not care about any (in particular, translational) space-time symmetries. For the derivation of the Feynman rules it is more convenient to work with the coefficients of the fields in the orthonormal basis of ordered polynomials

$$\phi(\bar{z}) = \sum_{s_1,s_2,s_3,s_4=0}^{\infty} \phi_{s_1 s_2 s_3 s_4} \frac{\bar{z}_1^{s_1} \bar{z}_2^{s_2} \bar{z}_3^{s_3} \bar{z}_4^{s_4}}{\sqrt{[s_1]_q![s_2]_q![s_3]_q![s_4]_q!}} . \tag{2.6.51}$$

In this basis the generating functional has the form:

$$Z[J, J^*] = N \int D\phi \, D\phi^* \, e^{-\phi^*_{\vec{r}} M_{\vec{r}\vec{s}} \phi_{\vec{s}} + \phi^*_{\vec{r}} J_{\vec{r}} + J^*_{\vec{r}} \phi_{\vec{r}} - \frac{\lambda}{4\pi^2 4!} V_{\vec{t}\vec{u}\vec{v}\vec{w}} \phi^*_{\vec{t}} \phi^*_{\vec{u}} \phi_{\vec{v}} \phi_{\vec{w}}} . \tag{2.6.52}$$

The term quadratic in ϕ, i.e. the matrix M is diagonal

$$M_{\vec{r}\vec{s}} = \left([r_1]_q + [r_2]_q + [r_3]_q + [r_4]_q + 2 + \frac{L^2 M^2 c^2}{\hbar^2} \right) \delta_{\vec{r},\vec{s}} . \tag{2.6.53}$$

The usual derivation of the Feynman rules (see, e.g., [36]) gives the free propagator [139]

$$\Delta_0(\vec{r}, \vec{s}) = \frac{1}{[r_1]_q + [r_2]_q + [r_3]_q + [r_4]_q + 2 + \frac{L^2 M^2 c^2}{\hbar^2}} \delta_{\vec{r},\vec{s}} . \tag{2.6.54}$$

In the usual, unregularized theory the first order correction to the propagator, namely the truncated tadpole graph, is quadratically divergent. Calculations of the tadpole for the q-deformed model lead to the estimation

$$\Sigma(\vec{0}, \vec{0}) \leq \sum_{r_1,r_2,r_3,r_4=0}^{\infty} \frac{1}{[r_1]_q + [r_2]_q + [r_3]_q + [r_4]_q + 2 + \frac{L^2 M^2 c^2}{\hbar^2}} .$$

One can use the "rotational" symmetry in the discrete summation space $\mathbb{Z}_+^4$ of the r_i to majorize this multiple sum by a simpler sum of the form

$$\sum_{r=0}^{\infty} V(r) \frac{1}{q^r + const} . \tag{2.6.55}$$

Here $V(r)$ is the number of terms which are to be summed over in the "layer", determined by $r^2 \leq \sum_{i=1}^4 r_i^2 \leq (r+1)^2$, which is essentially proportional to r^3. Thus, since the denominator in the sum (2.6.55) grows exponentially, a simple ratio test proves the convergence of the sum. This can be interpreted as the ultraviolet regularization of the tadpole graph in the model based on q-deformed oscillators.

2.6.6 q-Deformed statistics and the ideal q-gas

In order to study more the physical properties of q-oscillators, let us consider now the statistical averages and calculate the deformed "distributions" [50] which follow from the q-algebras defined in Section 2.1.

As is well known the thermodynamical properties are determined by the partition function Z, which in the canonical ensemble is defined by

$$Z = Tr(e^{-\beta H}) \,, \tag{2.6.56}$$

where $\beta = 1/(kT)$, H is the Hamiltonian and the trace must be taken over a complete set of states. For any operator O, the ensemble average is then obtained by the prescription

$$< O >= \frac{1}{Z} Tr(e^{-\beta H} O) \,,$$

and it is crucial that the cyclicity of the trace be consistent with the algebraic structure.

There are two natural Hamiltonians. The first one is identified with the number operator and reads as

$$H = \omega N \,, \tag{2.6.57}$$

(hereafter $\hbar = 1$), while the second one is defined in terms of the creation and annihilation operators, following the classical realization of the harmonic oscillator

$$H = \frac{\omega}{2}(a^+ a + a a^+) \,. \tag{2.6.58}$$

The partition function, and hence the thermodynamics of the system, is determined uniquely by the spectrum of the Hamiltonian. With the choice (2.6.57), the partition function will be that of the harmonic oscillator,

$$Z = \sum_{n=0}^{\infty} e^{-\beta \omega n} = \frac{e^{\beta\omega}}{e^{\beta\omega} - 1} \,, \tag{2.6.59}$$

irrespective of the deformation. Of course, for root of unity values of q for which the Fock space separates into disjoint subspaces each carrying a finite-dimensional representation, it is perhaps physically more meaningful to consider the truncation to one of these subspaces only; i.e. picking only those terms in the sum (2.6.59) corresponding to the subspace chosen. As an example, when $q = e^{i\pi/3}$, the Fock space states $|3>$, $|4>$, $|5>$ span a representation R of (2.1.28) in which

$$a_R = \begin{pmatrix} 0 & i & 0 \\ 0 & 0 & i \\ 0 & 0 & 0 \end{pmatrix} , \; a_R^+ = \begin{pmatrix} 0 & 0 & 0 \\ i & 0 & 0 \\ 0 & i & 0 \end{pmatrix} , \; N_R = \begin{pmatrix} 3 & 0 & 0 \\ 0 & 4 & 0 \\ 0 & 0 & 5 \end{pmatrix} , \tag{2.6.60}$$

(note that a^+ is not the conjugate of a in this representation). The corresponding truncated partition function is

$$Z_R = e^{-3\beta\omega} + e^{-4\beta\omega} + e^{-5\beta\omega} \,. \tag{2.6.61}$$

Summing over all subspaces reproduces the full partition function (2.6.59).

While the thermodynamics for a system with Hamiltonian (2.6.57) is independent of the deformation, Green functions like $< a^+a >$ will depend on the deformation. For the q-bosonic algebra (2.1.1),(2.1.2) with zero central element (2.1.4), we obtain

$$< a^+a > = < [N]_q > = \frac{1}{Z}\sum_n [n]_q e^{-\beta\omega n}$$

$$= \frac{e^{\beta\omega} - 1}{e^{2\beta\omega} - (q+q^{-1})e^{\beta\omega} + 1} \,, \tag{2.6.62}$$

while for the operators b, b^+ obeying (2.1.8), we have the simple expression

$$< b^+b > = < [N;q^2] > = \frac{1}{e^{\beta\omega} - q^2} \,. \tag{2.6.63}$$

The average of N is, of course, always given by the usual Bose-Einstein formula

$$n = < N > = \frac{1}{e^{\beta\omega} - 1} \,, \tag{2.6.64}$$

(when the average is taken over the complete Fock space), to which (2.6.62) and (2.6.63) reduce in the case $q = 1$, as they should.

When $q \neq 1$, the temperature Green functions $< a^+a >$, etc. do not have a direct relation to the thermodynamical quantities of the system. In a multimode context, such as the one which will be considered below, this can be understood by noting that the average $< a_k^+ a_k >$ does not describe the average occupancy of the kth level, $< N_k >$.

When the Hamiltonian is chosen as in (2.6.58), the partition function cannot any longer be computed in a closed form for the generic q.

By an "ideal q-gas" we understand a system defined by the Hamiltonian

$$H = \sum_i \epsilon_i \left(\alpha\, a_i^+ a_i + (1-\alpha)\, a_i a_i^+\right) \,, \tag{2.6.65}$$

where the operators a_i, a_i^+ satisfy the q-oscillator algebra (2.4.19),(2.4.20) and α is a real parameter between 0 and 1. An equivalent form of the Hamiltonian can be written using (2.4.20)

$$H = \sum_i \epsilon_i \left(\alpha[N_i]_q \,+\, (1-\alpha)[N_i+1]_q\right) \,. \tag{2.6.66}$$

We shall interpret a_i, a_i^+, N_i as annihilation, creation and occupation number operators, respectively, of particles in the state (level) i, although the mathematical results are, of course, independent of this interpretation. We shall call ϵ_i the energy of the level i. The ideal q-gas is a special case of the general class of systems, which could be called "general ideal gases". The energy eigenvalues of such a system are uniquely determined by a set of occupation numbers $\{n_i\}$ and a set of functions E_i:

$$E = \sum_i E_i(n_i) . \tag{2.6.67}$$

In a usual ideal gas, the functions E_i are proportional to the occupation number:

$$E_i(n) = \epsilon_i^{\circ} n , \tag{2.6.68}$$

corresponding to the physical interpretation of the total energy as a sum of single-particle energies ϵ_i°, with vanishing interaction energy between the particles. In the general case (2.6.67), the interaction between the particles is such that it causes the energy of the ith level to depend on the occupancy of that level. For the ideal q-gas (2.6.66), this dependence is (for real $q = \exp\chi$)

$$E_i(n) = \frac{\epsilon_i}{\sinh\chi}(\alpha\ \sinh(n\chi) + (1-\alpha)\sinh((n+1)\chi)), \tag{2.6.69}$$

and for $q = \exp(i\theta)$

$$E_i(n) = \frac{\epsilon_i}{\sin\theta}(\alpha\sin(n\theta) + (1-\alpha)\sin((n+1)\theta)) . \tag{2.6.70}$$

The grand canonical partition function

$$Z = Tr\exp(-\beta(H - \mu N)) = \exp(-\beta\Omega) , \tag{2.6.71}$$

where N is the total number operator

$$N = \sum_i N_i , \tag{2.6.72}$$

μ the corresponding chemical potential and Ω the grand canonical potential, factorizes for a general ideal gas into a product of single level partition functions:

$$Z = \prod_i Z_1(i,\beta,\mu) , \tag{2.6.73}$$

$$Z_1(i,\beta,\mu) = \sum_{n=0}^{\infty} e^{-\beta(E_i(n)-\mu n)} . \tag{2.6.74}$$

Correspondingly, the grand canonical potential is given as a sum over the levels i:

$$\Omega = -\frac{1}{\beta}\sum_i \log Z_1(i,\beta,\mu) \,. \tag{2.6.75}$$

The allowed values of the chemical potential μ are determined by the requirement that the sum in (2.6.74) converges. For the ideal q-gas with real q, the absolute value of E_i grows exponentially with n, eq. (2.6.69). In order to have the total energy bounded from below, we require $\epsilon_i \geq 0$ for all i in (2.6.69). Then μ can take any real value if all $\epsilon_i > 0$; if at least one $\epsilon_i = 0$, we must have $\mu < 0$. When q is a pure phase, $|E_i(n)|$ stays bounded for all values of i and n (eq. (2.6.70)), in which case we require $\mu < 0$.

The sum over n in (2.6.74) cannot be explicitly performed in general. However, when $q = \exp\chi$ is close to 1 ($\chi \to 0$), we can calculate the leading correction to the usual Bose gas. To order χ^2, eq. (2.6.69) reads

$$E_i = \epsilon_i(n+1-\alpha+\frac{1}{6}n(n+1)(n+2-3\alpha)\chi^2+\cdots) \,. \tag{2.6.76}$$

Substituting this into (2.6.74) gives the approximate expression

$$Z_1(i,\beta,\mu) \simeq e^{\beta(\alpha-1)\epsilon_i}\sum_{n=0}^{\infty} e^{\beta(\mu-\epsilon_i)n}(1-\frac{\chi^2}{6}\beta\epsilon_i n(n+1)(n+2-3\alpha)+\cdots) \,. \tag{2.6.77}$$

Introducing the variable $x_i = \beta\epsilon_i$ and the fugacity $z = \exp(\beta\mu)$ and performing the sum over n, we obtain

$$Z_1(x_i, Z) \simeq e^{(\alpha-1)x_i} Z_0(z,x_i)(1-\chi^2 x_i z e^{-x_i} Z_0^3(z,x_i)(\alpha z e^{-x_i}+1-\alpha)) \,, \tag{2.6.78}$$

where

$$Z_0(z,x) = \frac{1}{1-ze^{-x}}$$

corresponds to the ideal Bose gas. The grand canonical potential is then to second order in χ

$$\beta\Omega \simeq \sum_i(1-\alpha)x_i - \sum_i \log Z_0(z,x_i) \tag{2.6.79}$$

$$+\chi^2 z\sum_i x_i e^{-x_i} Z_0^3(z,x_i)(\alpha z e^{-x_i}+1-\alpha) \,.$$

In order to study the thermodynamics of the system, we have to specify the energies ϵ_i and the density of states. We shall consider a system where the states i are specified by a d-dimensional momentum vector $\vec{k}$ (the momentum of the q-boson). The energy of the q-boson is given by the dispersion law

$$\epsilon_i = \epsilon(\vec{k}) = \gamma|\vec{k}|^p \,, \tag{2.6.80}$$

covering the cases of nonrelativistic ($\gamma = 1/2m$, $p = 2$) and ultrarelativistic ($\gamma = 1$, $p = 1$) q-bosons. We enclose the system in a large d-dimensional volume V and replace in the usual way the sum over levels by an integral over $\vec{k}$-space:

$$\sum_i \rightarrow V \int \frac{d^d k}{(2\pi)^d} \,.$$

The first term on the right hand side of (2.6.79) is now divergent, giving rise to an infinite (negative) vacuum pressure, and we renormalize it away by simply dropping it. This means in effect that we subtract away the vacuum energy, i.e. replace the functions $E_i(n)$ by $E_i(n) - E_i(0)$.

Performing the $\vec{k}$-space integrals using the general formula

$$\int_0^\infty dx \, \frac{x^{\alpha-1} e^{-px}}{(e^{qx} - z)^n} = \frac{\Gamma(\alpha)}{(n-1)!} \sum_{k=0}^{\infty} \frac{(k+n-1)!}{k!} \frac{z^k}{(qk+qn+p)^\alpha} \tag{2.6.81}$$

and introducing the functions $g_\ell(z)$

$$g_\ell(z) = \sum_{k=1}^{\infty} \frac{z^k}{k^\ell} \,, \tag{2.6.82}$$

we get for the pressure $p = -\Omega/V$

$$\beta p = a \left\{ g_{\frac{d}{p}+1}(z) - \frac{d}{2p} \, \chi^2 \left[g_{\frac{d}{p}-1}(z) + (1-2\alpha) g_{\frac{d}{p}}(z) \right] \right\} \,. \tag{2.6.83}$$

Here

$$a = \frac{A_{d-1}}{(2\pi)^d} \, \frac{\Gamma(\frac{d}{p}+1)}{d} \left(\frac{1}{\gamma\beta} \right)^{d/p} \,, \tag{2.6.84}$$

where A_{d-1} is the area of the unit sphere S^{d-1}:

$$A_{d-1} = \frac{2\pi^{d/2}}{\Gamma(\frac{d}{2})} \,.$$

The q-boson density

$$n = \frac{N}{V} = \left(\frac{\partial p}{\partial \mu}\right)_{T,V} = z\left(\frac{\partial(\beta p)}{\partial z}\right)_{\beta} \tag{2.6.85}$$

is easily found, since

$$z\frac{dg_\ell(z)}{dz} = g_{\ell-1}(z) .$$

We get

$$n = a\left\{g_{\frac{d}{p}}(z) - \frac{d}{2p}\chi^2\left[g_{\frac{d}{p}-2}(z) + (1-2\alpha)g_{\frac{d}{p}-1}(z)\right]\right\} . \tag{2.6.86}$$

According to our general considerations, the chemical potential is restricted to negative values for the system we consider. Since

$$g_\ell(z) > g_{\ell'}(z) ,$$

when $\ell < \ell'$ and $z \to 1$, eq. (2.6.83) implies that the pressure becomes negative for μ sufficiently close to 0 (z sufficiently close to 1) if χ is real. This is not a signal of instability of the ideal q-gas, rather it signals the breakdown of the expansion in χ. From (2.6.77) it is clear that the effective expansion parameter is χ^2 times a positive power of $Z_0(z, x_i)$, and this grows large when $z \to 1$ and $x_i \to 0$. Although the problem of negative pressure is absent for imaginary χ, also in this case the expansion cannot be trusted as $z \to 1$. Thus it is not possible to address the interesting question how the phenomenon of Bose-Einstein condensation is affected by the deformation.

For low densities, however, the expansion in χ^2 is meaningful. We invert eq. (2.6.86) to obtain z as a series in n

$$z = a_1\left(\frac{n}{a}\right) + a_2\left(\frac{n}{a}\right)^2 + \cdots , \tag{2.6.87}$$

where

$$a_1 = 1 + \frac{d}{p}(1-\alpha)\chi^2 ,$$

$$a^2 = -\frac{1}{2^{d/p}}\left(1 - \frac{d}{p}\alpha\chi^2\right) .$$

Substituting (2.6.87) into (2.6.83) and expanding in powers of n, we obtain the virial expansion of the equation of state

$$\beta p = n(1 + Bn + \cdots) . \tag{2.6.88}$$

The second virial coefficient is given by

$$B = -\frac{1}{2^{d/p+1}a}(1 - \frac{d}{p}\chi^2)\,. \qquad (2.6.89)$$

Thus, for very low densities, the q-gas behaves as a classical ideal gas. The expression for the second virial coefficient is independent of the parameter α in the Hamiltonian, and shows that the deformation weakens the attraction between pairs of q-bosons producing an increase in the pressure, when χ is real, whereas the effect of the deformation is the opposite, when χ is imaginary. For details, we refer to [50].

For interesting physical applications of quantum groups in non-equilibrium statistical mechanics, the reader can consult [2] and refs. therein.

2.6.7 Nonlinear Regge trajectory and quantum dual string theory

The development of the string theory is, more or less, directly related to the Veneziano's discovery [232] (for a review and refs. see [112]) of a four-point crossing-symmetric scattering amplitude with linear Regge trajectories. Afterwards Fubini, Gordon and Veneziano [232] provided an elegant operator formalism for the dual amplitudes, which further led to the interpretation of the Veneziano model as a theory of interacting strings in physical space-time.

A q-deformed dual string amplitude was proposed in [47], which has not only the required properties of crossing symmetry and factorization, the correct pole structure and a suitable asymptotic behaviour, but also leads to a spin-(mass)^2 relation, which possesses features interesting from the physical point of view. One introduces an infinite number of q-oscillators which build up a Fock space with well-known properties [47]. However, for the specific values of the deformation parameter q being a root of unity, only a finite number of oscillators for each (harmonic) mode possesses non-vanishing norm. This is a fundamental property which is responsible for a drastic change in the high energy behaviour of the amplitudes [48] and, therefore, in the mass spectrum of the physical particles.

Let us consider first the usual Veneziano 4-point dual amplitude given by

$$A_4 = \int_0^1 z^{-\alpha(t)-1}(1-z)^{-\alpha(s)-1}dz, \qquad (2.6.90)$$

where $\alpha(s) = \alpha' s + \alpha_0$ is the linear Regge trajectory, with α' and α_0 the Regge slope and intercept, respectively. We will take our units so that $\alpha' = 1$.

The amplitude (2.6.90) can be factorized by introducing an infinite set of oscillators [232, 112] satisfying $[a_m^\mu, a_n^{\nu +}] = \delta_{mn} g_{\mu\nu}$; μ and ν are the Lorentz indices; $n, m = 1, 2, \ldots, \infty$ correspond to the different oscillator modes. We have then the identity

$$(1-z)^{-A\bar{A}} = \prod_{n=1}^{\infty} \langle 0| e^{\frac{A a_n}{\sqrt{n}}}\, z^{n a_n^+ a_n}\, e^{\frac{\bar{A} a_n^+}{\sqrt{n}}} |0\rangle \,, \tag{2.6.91}$$

where the contraction of the Lorentz indices is understood in all the scalar products. Here the four-vectors A_μ and $\bar{A}_\mu$ correspond to incoming and outgoing momenta, respectively. In what follows we shall denote $a = A\bar{A} \equiv \alpha(s)+1$ and $b \equiv \alpha(t)+1$, and omit the Lorentz indices.

To perform the q-deformation within the operator formalism, it seems most natural to use the q-deformed oscillators instead of the usual ones in the factorization procedure. Therefore, we replace the identity (2.6.91) by the q-deformed expression [48]

$$F(a,z) = \prod_{n=1}^{\infty} \langle 0| e_q^{\frac{A a_n}{\sqrt{n}}}\, z^{n N_n}\, e_q^{\frac{\bar{A} a_n^+}{\sqrt{n}}} |0\rangle = \prod_{n=1}^{\infty} \sum_{\ell=0}^{\infty} \left(\frac{a}{n}\right)^{\ell} \frac{z^{n\ell}}{[\ell]_q!} = \prod_{n=1}^{\infty} e_q^{\frac{a}{n} z^n} \,. \tag{2.6.92}$$

The q-oscillators entering in (2.6.92), satisfy (2.4.19),(2.4.20).

Notice that we have assumed the total energy operator of the genuinely free system to be $H = \sum_{n=1}^{\infty} n N_n$ instead of $H = \sum_{n=1}^{\infty} n a_n^+ a_n$, and therefore we have replaced $a_n^+ a_n$ by the number operator N_n in the exponent of z. As a consequence of this assumption, the amplitudes will exhibit poles in s when $\alpha(s)$ is a positive integer, or poles in t when $\alpha(t)$ is a positive integer as in the undeformed case.

In order to preserve duality, expressed by the symmetry $z \to 1-z$ in (2.6.90), we define then the q-deformed 4-point amplitude as

$$\begin{aligned} A_4^q &= \int_0^1 dz\, F(a,z) F(b,1-z) \\ &= \prod_{n=1}^{\infty} \langle 0| e_q^{\frac{A a_n}{\sqrt{n}}} \int_0^1 dz\, z^{n N_n} F(b, 1-z)\, e_q^{\frac{\bar{A} a_n^+}{\sqrt{n}}} |0\rangle \;. \end{aligned} \tag{2.6.93}$$

Inserting a complete set of orthonormal states $1 = \sum_\lambda |\lambda\rangle\, \langle\lambda|$, with $|\lambda\rangle = \prod_i \frac{(a_i^+)^{\lambda_i}}{\sqrt{[\lambda_i]_q!}} |0\rangle$ we find A_4^q in its factorized form

$$A_4^q = \sum_\lambda V(A,\lambda) D(\lambda, b) V^+(\bar{A}, \lambda) \,, \tag{2.6.94}$$

where

$$V(A,\lambda) = \langle 0| \prod_{n=1}^{\infty} e_q^{\frac{Aa_n}{\sqrt{n}}} |\lambda\rangle \qquad (2.6.95)$$

is the vertex operator (with only one leg off shell) and

$$D(\lambda,b) = \langle\lambda| \int_0^1 dz\ z^{\sum_{n=1}^{\infty} nN_n} F(b,1-z)|\lambda\rangle \qquad (2.6.96)$$

is the propagator.

The other vertex operator Γ with two legs off shell , which is needed for higher n-point functions ($n \geq 5$), is given now by

$$\Gamma(\lambda,\lambda',p,\bar{p}) = \langle\lambda| \prod_{n=1}^{\infty} e_q^{\frac{pa_n^+}{\sqrt{n}}} e_q^{\frac{\bar{p}a_n}{\sqrt{n}}} |\lambda'\rangle\ , \qquad (2.6.97)$$

where $p = \bar{p}$ is the momentum of the unexcited leg in the vertex. The 5-point q-deformed amplitude, e.g., now can be written as a Feynman-like diagram in terms of products of vertices and propagators in the tree approximation:

$$A_5^q = \sum_{\lambda,\lambda'} V(A,\lambda)D(\lambda,b_1)\Gamma(\lambda,\lambda',p,\bar{p})D(\lambda',b_2)V^+(A,\lambda'), \qquad (2.6.98)$$

where b_1, b_2 are the Mandelstam variables of the corresponding channels.

Notice that (2.6.95), (2.6.96) and (2.6.97) lead to the usual undeformed expressions for the vertices and the propagators in the limit $q \to 1$. Let us first examine the singularities of A_4^q as a function of s and t. Since the integral in (2.6.93) is symmetric in s and t, we need only to analyse the singularities in one of the two variables, e.g. in the variable t. We first observe that near the singular point $z = 0$ the integrand $F(b,1-z)$ in (2.6.93) can be approximated as

$$F(b,1-z) \sim z^{-b}, \qquad (2.6.99)$$

and therefore the integral (2.6.93) is not defined when $\alpha(t) \geq 0$ (or $\alpha(s) \geq 0$) . Following the standard procedure [112], one can define it by analytic continuation to show that the amplitude A_4^q exhibits poles in t at any integer value. The residue at the n-th pole in the t-channel is a polynomial in s of degree $J \leq n$, which when decomposed into spherical harmonics describes the exchange of a set of particles of spins $\leq J$.

Let us consider now the effects of the deformation in the high energy behaviour of A_4^q . In the undeformed case the behaviour of the 4-point amplitude (2.6.90) for large a can be easily obtained from the contour integral

$$A_4 \sim \oint_{z=0} \frac{dz}{2\pi i}(1 + az + \cdots + \frac{a(a+1)\cdots(a+n-1)}{n!}z^n + \cdots)z^{-b}, \quad (2.6.100)$$

leading to

$$A_4 \sim \frac{a(a+1)\cdots(a+b-2)}{\Gamma(b)} \sim \frac{a^{b-1}}{\Gamma(b)}, \quad (2.6.101)$$

where Γ is the gamma function.

In order to discuss the high energy behaviour of the q-deformed 4-point amplitude (2.6.93), we will follow a procedure similar to the one used for the undeformed case, namely, we write A_4^q as the contour integral

$$A_4^q \sim \oint_{z=0} \frac{dz}{2\pi i} F(a, z) z^{-b} \quad . \quad (2.6.102)$$

Consider now a generic term in the product (2.6.92), which can be written as

$$C_{n_1,\dots,n_m} a^{n_1+n_2+\cdots+n_m} z^{n_1+2n_2+\cdots+mn_m}, \quad (2.6.103)$$

where $m = 1, 2, \dots, \infty$ and $C_{n_1,\dots,n_m}$ are coefficients which depend on q. Since we are interested in the high energy behaviour of (2.6.102), i.e. in the case when a is large, we should pick in $F(a, z)$ the highest power of a which gives nonvanishing contribution to the contour integral (2.6.102). We are therefore led to solve the algebraic equation

$$n_1 + 2n_2 + \cdots + mn_m = b - 1, \quad (2.6.104)$$

with the condition that the spin

$$J = n_1 + n_2 + \cdots + n_m, \quad (2.6.105)$$

which is the power of a in (2.6.103), takes its maximum value.

In what follows we shall consider two cases:

(*i*) Assume that q is real. Since in this case the n_i in (2.6.104) range from 0 to ∞ , we obtain that the maximum value of J in (2.6.105) will be given by

$J = b - 1$ when $n_1 = b - 1, n_i = 0$ for $i \geq 2$. Thus we obtain from (2.6.103) that for high energies

$$A_4^q \sim a^{b-1}. \tag{2.6.106}$$

We notice that the high s-behaviour of the 4-point q-deformed amplitude is proportional to s^t, which leads to a mass spectrum with a linearly-rising Regge trajectory as in the *undeformed* case.

(ii) Let us consider now the case when the deformation parameter q is a p-th root of unity, i.e. $q = \exp(2i\pi/p)$. The q-analogs $[\ell]$ thus become $[\ell] = \sin(2\pi\ell/p)/\sin(2\pi/p)$. We should point out here the main difference with respect to the undeformed case: due to the truncation of the Fock space for each oscillator mode in the Fubini-Veneziano operator formulation, each term in the product (2.6.92) consists of a finite series ending at $\ell = \tilde{p}$, where

$$\tilde{p} = \begin{cases} p-1, & \text{for } p \text{ odd} \\ p/2-1, & \text{for } p \text{ even}. \end{cases} \tag{2.6.107}$$

In this case $0 \leq n_i \leq \tilde{p}$ and we have two possibilities:

If $b \leq \tilde{p} + 1$, then it is always possible to find a solution of (2.6.104) such that $n_1 = b - 1 \leq \tilde{p}$ and $n_i = 0, i \geq 2$, which gives the maximum value of J in (2.6.105). Then as before the high energy behaviour will be given by eq. (2.6.106) and, thus, *the trajectories will be linear.*

If $b > \tilde{p} + 1$, it is easy to see that for a generic m the highest power of a in (2.6.103) is $J = m\tilde{p}$ and is obtained when all the n_i are equal to their maximum value $\tilde{p}$. Then according to (2.6.104), J will satisfy the second-order algebraic equation

$$\frac{J(J+\tilde{p})}{2\tilde{p}} = b - 1, \tag{2.6.108}$$

which has the positive solution

$$J = \frac{\tilde{p}}{2}\left\{-1 + \sqrt{1 + \frac{8(b-1)}{\tilde{p}}}\right\}. \tag{2.6.109}$$

Thus finally we obtain that for large values of a and for $b > \tilde{p} + 1$

$$A_4^q \sim a^J, \tag{2.6.110}$$

where J is given by (2.6.109).

We notice now from (2.6.109) and (2.6.110) that the high s-behaviour of the q-deformed amplitudes is proportional to $s^{\sqrt{t}}$ which leads to the mass spectrum with a *square-root Regge trajectory* $\alpha(t) = \sqrt{t} + const.$ It appears, as was expected by physical arguments, that the crucial change in the high energy behaviour of A_4^q occurs due to the truncation in the series in (2.6.92) and as mentioned before, is a direct consequence of q being a root of unity. This change can be understood by the finite character of the Fock space as a consequence of the assumed values of q.

Thus quantum groups and the q-deformation provide with a new phenomenon of a linear Regge trajectory turning into a square-root trajectory for higher masses in the case when the deformation parameter is a root of unity. This case is of utmost physical interest and in particular, appears in rational conformal field theories. An ultimate aim to combine the quantum group ideas with the (super)string theory in order to obtain a q-deformed (super)string amplitude, can be pursued along similar lines.

2.6.8 q-Deformation of the Virasoro algebra

The conformal mappings of a complex plane $\mathbb{C}$ belong to a classical part of mathematics which has found various applications in physics. One of the recent promising applications is related to conformal field and string theories, leading to a better understanding and to new methods in 2D-quantum field theory (see [112]).

The conformal mappings of a complex plane $\mathbb{C}$

$$z \to z' = z \; - \; \sum_{n=0}^{\infty} a_n z^n \, , \qquad (2.6.111)$$

are generated in the neighborhood of identity mapping by differential operators

$$\mathcal{L}_n \;=\; -z^{n+1}\partial_z \; , \; n = 0, 1, 2, \ldots \; , \qquad (2.6.112)$$

acting on a suitable set of holomorphic functions $\Phi(z)$, $z \in \mathbb{C}$. They form a basis of the Witt algebra and satisfy the commutation relations

$$[\mathcal{L}_n, \mathcal{L}_m] \;=\; (n-m)\mathcal{L}_{m+n} \; , \; n, m = 0, 1, 2, \ldots \; . \qquad (2.6.113)$$

The Virasoro algebra [234] is obtained by extending the commutation relations to all $n, m \in \mathbb{Z}$, followed by a central extension of the algebra. Denoting the generators of the Virasoro algebra by L_n, $n \in \mathbb{Z}$, their defining relations are

$$[L_n, L_m] = (n-m)L_{m+m} + \frac{c}{12}(n^3 - n)\delta_{n+m,0}\ , \ n, m = 0, \pm 1, \pm 2, \ldots\ . \tag{2.6.114}$$

where c is the central element commuting with all L_n. We would like to stress that in any irreducible representation c can be put to be a constant, which is real in the unitary representation $L_n^+ = L_{-n}$.

The unitary irreducible representations of the Virasoro algebra are well known, [134, 106]. The simplest one is given by the Sugawara construction of L_n in terms of an infinite set of free bosonic operators $a_n, a_n^+ = a_{-n}$, $n = 1, 2, \ldots$, satisfying the commutation relations

$$[a_n, a_m] = n\delta_{n+m,0}\ , \tag{2.6.115}$$

(here we have excluded the zero modes inessential for us) . The Sugawara formula for L_n reads

$$L_n = -\frac{1}{2}\sum : a_k a_m : \delta_{k+m,n}\ , \tag{2.6.116}$$

where the summation goes over $k, m = \pm 1, \pm 2, \ldots$ (and this convention is used in what follows too). The Sugawara formula (2.6.116) leads to the unitary irreducible representation of the Virasoro algebra (2.6.114) with $c = 1$.

Introducing the free fields

$$X(z) = \sum \frac{i}{n}\, a_n\, z^{-n} \qquad \text{(string coordinate)}\ ,$$

$$\Phi(z) = \sum a_n\, z^{-n} \qquad \text{(string momentum)}\ , \tag{2.6.117}$$

one can straightforwardly show that

$$[L_n, X(z)] = (\mathcal{L}_n X)(z)\ ,$$

$$[L'_n, X(z)] = (\mathcal{L}'_n X)(z)\ , \tag{2.6.118}$$

where $\mathcal{L}_n$ is the differential operator given in (2.6.112) and $\mathcal{L}'_m = -z^{m+1}\partial_z + mz^m$. Both $\mathcal{L}_n$ and $\mathcal{L}'_n$ realize the centreless Virasoro algebra (2.6.114) (with $c = 0$) in terms of differential operators. However, the realization depends on the field in question.

The string momentum field $\Phi(z)$ allows to express the Virasoro current (related to the string energy - momentum) as

$$L(z) = \sum L_m\, z^{-m} = \frac{1}{2} : \Phi^2(z) :\ . \tag{2.6.119}$$

The proper q-deformed Virasoro algebra is a substantial tool for the construction of q-conformal field theory or q-deformed string theory. There are a few different q-deformations of the infinite algebras introduced above. Here we consider two of them.

The first one is related to the q-deformation of the Witt algebra (see [69, 43, 193, 44]) and is given by the formula

$$\hat{\mathcal{L}}_m = \frac{z^m}{\omega(q)}\left(D(q^2)-1\right), \quad m \in \mathbb{Z}, \tag{2.6.120}$$

where $\omega(q) = q - q^{-1}$ and $D(q)$ is the dilatation operator acting on a suitable set of functions as $D(q)\Phi(z) = \Phi(zq)$, q is real parameter. The relation $D(q^\alpha) = D^\alpha(q)$ holds. The set of operators $\hat{\mathcal{L}}_m$, $m \in \mathbb{Z}$, satisfies the q-commutation relations

$$q^{m-m'}\hat{\mathcal{L}}_m\hat{\mathcal{L}}_{m'} - q^{m'-m}\hat{\mathcal{L}}_{m'}\hat{\mathcal{L}}_m = [m-m']_q\hat{\mathcal{L}}_{m+m'}. \tag{2.6.121}$$

Alternatively, they can be realized with the help of q-oscillators satisfying (2.1.1), (2.1.2) as follows [42]

$$\hat{\mathcal{L}}_m = q^{-N}(a^+)^{m+1}a. \tag{2.6.122}$$

The central extension of the q-commutation relations (2.6.121) is known (see [44] and references therein), but the q-analogue of Sugawara construction is still missing.

However, the operators $\hat{\mathcal{L}}_m$, $m \in \mathbb{Z}$, do not close under the commutation and an additional index $\alpha \in \mathbb{Z}$ is needed. Putting

$$\hat{\mathcal{L}}^\alpha_m = \frac{z^m}{\omega(q)}\{q^{\frac{\alpha m}{2}}D(q^\alpha) - q^{-\frac{\alpha m}{2}}D(q^{-\alpha})\}, \quad m, \alpha \in \mathbb{Z}, \tag{2.6.123}$$

one obtains the commutation relations

$$[\hat{\mathcal{L}}^\alpha_m, \hat{\mathcal{L}}^{\alpha'}_{m'}] = \sum_{\sigma'}[\frac{1}{2}(\alpha m' - \sigma'\alpha' m)]_q\frac{[\alpha+\sigma'\alpha']_q}{[\alpha]_q[\sigma'\alpha']_q}\hat{\mathcal{L}}^{\alpha+\sigma'\alpha'}_{m+m'}, \tag{2.6.124}$$

where the summation runs over $\sigma' = \pm 1$. In [46] was found the central extension of the algebra (2.6.124) based on the q-analogue of Sugawara construction in terms of a system of standard fermionic oscillators $b_r, b^+_r = b_{-r}$, $r = 1/2, 3/2, \ldots$, satisfying anticommutation relations

$$[b_r, b_s]_+ = \delta_{r+s,0}. \tag{2.6.125}$$

The operators

$$L_m^\alpha = L_m^{-\alpha} = \frac{1}{2\omega(q^\alpha)} \sum q^{\alpha(s-r)/2} : b_r b_s : \delta_{r+s,m} , \qquad (2.6.126)$$

satisfy the commutation relations (2.6.124). Here the summation is over $r, s = \pm 1/2, \pm 3/2, \ldots$. Introducing the Virasoro currents

$$L^\alpha(z) = \sum L_m^\alpha z^{-m} , \qquad (2.6.127)$$

one obtains

$$L^\alpha(z) = -\frac{1}{2\omega(q^\alpha)} : \Psi(zq^{\alpha/2})\Psi(zq^{-\alpha/2}) : , \qquad (2.6.128)$$

where

$$\Psi(z) = \sum b_r z^{-r-\frac{1}{2}} , \qquad (2.6.129)$$

is a free fermionic field. We see that the additional index α directly measures the splitting of arguments of the fields entering the Virasoro currents. Recalling that the unit circle $|z| = 1$ is related to the light-cone local string world sheet parameters [112], the point-splitting (2.6.128) can be interpreted as the Schwinger point-splitting [217] of the currents in complex direction in the complexified 2D space-time (for details see [59] and refs. therein).

However, we need a bosonic realization of the q-analogue of Sugawara construction in order to perform a q-analogue of the program described above in the non-deformed case. Such a realization was proposed in [46] and it led to a central extension of a slightly more complicated q-Witt algebra as in (2.6.124).

The starting point was the infinite set of bosonic operators $a_n, a_n^+ = a_{-n}$, $n = 1, 2, \ldots$, satisfying the commutation relations

$$[a_n, a_m] = [n]_q \delta_{n+m,0} . \qquad (2.6.130)$$

From this system of oscillators the auxiliary fields, string coordinate and string momentum, are given as

$$X(z) = \sum \frac{i}{[n]_q} a_n z^{-n} ,$$

$$\Phi(z) = \sum a_n z^{-n-1} . \qquad (2.6.131)$$

Then a bosonic realization of the q-Sugawara generators defined by

$$L_m^\alpha = L_m^{-\alpha} = \frac{1}{2} \sum q^{\alpha(n-k)/2} : a_k a_n : \delta_{k+n,m} . \qquad (2.6.132)$$

satisfy the commutation relations

$$[L^{\alpha}_{m}, L^{\alpha'}_{m'}] = \frac{1}{4}\sum_{\sigma,\sigma'}[\frac{1}{2}(m - m' - m'\sigma\alpha + m\sigma'\alpha']_q \, L^{\sigma\alpha+\sigma'\alpha'+1}_{m+m'} + C.T., \quad (2.6.133)$$

where the symbol $C.T.$ denotes the central term which is inessential for our purposes (it can be found in [46]). The summation in (2.6.133) runs over $\sigma, \sigma' = \pm 1$. It is important to mention that the introduction of the index α is inevitable. For $\alpha \in \mathbb{Z}$ the algebra (2.6.133) closes (the minimal set for which (2.6.133) closes is formed by α - odd integer). For the Virasoro currents (2.6.127) one can straightforwardly derive the q-analogue of (2.6.119):

$$L^{\alpha}(z) = \frac{1}{2} : \Phi(zq^{\alpha/2})\Phi(zq^{-\alpha/2}) : . \quad (2.6.134)$$

The additional index α again directly measures the splitting of arguments of the fields entering the Virasoro currents, see [59].

The problem of co-algebra structure for the infinite q-commutator algebra (2.6.121) with generators $\hat{\mathcal{L}}_m$, $m \in \mathbb{Z}$, remains open. The generators L^{α}_{m}, $m, \alpha \in \mathbb{Z}$, with an additional index α form a Lie (commutator) algebra which has a trivial (co-commutative) co-algebra structure. It is not clear whether they can posses nontrivial co-algebra structures too.

Finally, there are different approaches motivated by the theory of q- deformation of infinite dimensional affine algebras [89, 107, 22, 108, 61], in which the canonical realizations contain an infinite set of oscillators satisfying commutation relations similar to (21) but with more complicated functions on r.h.s. This could lead to new q-deformations of Virasoro algebra possesing a nontrivial co-algebra structures. The program of finding proper q-deformed Virasoro algebra is still not completed.

Chapter 3

q-Deformation of Space-Time Symmetries

The construction of relativistic quantum theory in the frame of "quantum geometry", i.e. geometry with non-commuting and/or discrete coordinates is an old dream of physicists (about old attempts see, e.g. [221, 26] and refs. therein), the reason being the hope to obtain in this way quantum theory with improved ultraviolet properties and new insight on the underlying space-time geometry. The attempts to construct such models were not very successful because of lacking of corresponding deformed relativistic and/or gauge transformations. In recent years this problem has received a revival and considerable interest, due to appearance of quantum groups.

At present there is a well-developed theory of quantum semisimple matrix Lie groups and algebras described in previous chapters. But as is well known in the classical case, Minkowski geometry is intrinsically connected with inhomogeneous Poincaré group. In this Chapter we will describe some approaches for the construction of q-deformation of inhomogeneous groups. But before that, in Sections 1 and 2, we shall consider general relations between q-deformation and space discretization.

In one-dimensional space q-derivative can be represented by Jackson difference operator [127],[99]. In Section 3.1 we will show how this provides , in turn, with the description of a quantum mechanical particle on a one-dimensional lattice [77]. Thus the q-deformation of a differential calculus apparently leads to space discretization.

To construct lattice-like regularization one definitely needs the multidimensional finite difference calculus. In Section 3.2 we consider $GL_q(N)$-invariant q-spaces with commuting coordinates and q-deformed differential calculus and construct multidimensional analog of the Jackson calculus. Using explicit for-

mulas for a two-dimensional q-space we study a quantum mechanical particle and simplest field theoretical model in the deformed space-time.

In Section 3.3 we consider projective approach to the construction of inhomogeneous q-groups, discuss the reason for the appearance of additional dilatations in the q-deformed case and the way to avoid them.

To illustrate more on the physical and mathematical meaning of noncommutativity of space-time coordinates, in Section 3.4 we discuss another limiting case, opposite to those considered in Section 3.2. This time we describe realistic model of 4-dimensional quantum Minkowski space with q-Poincaré group acting on it, while the latter is taken to be a pure twisted one, i.e. it can be obtained from the usual Poincaré group by twist, and so the q-deformation of its differential calculus prove to be trivial. Though it is clear that such a space-time cannot lead to ultimate aim of the space-time q-deformation, to improve ultraviolet properties of field theory, we use this relatively simple case of q-symmetry to study various properties of q-deformation.

Jordan-Schwinger construction discussed in Section 2.4 is applied in Section 3.5 for the construction of space-time algebras. Another topic of this section is a construction of inhomogeneous space-time q-groups via contraction of semisimple ones both on the level of q-algebras and q-groups.

General classification of quantum Poincaré groups according to [192] and elements of general theory of quantum inhomogeneous groups from [191] are presented in Section 3.6.

3.1 One-dimensional lattice and q- deformation of differential calculus

In this section we clarify the relation between q-derivatives and quantum mechanics on a lattice following [77].

Let us start from usual discrete derivatives (finite difference operators) on regular (equidistant) lattice with spacing $a \in \mathbb{R}$

$$\delta f(x) := \frac{1}{a}[f(x+a) - f(x)] , \tag{3.1.1}$$

$$\bar{\delta} f(x) := \frac{1}{a}[f(x) - f(x-a)] . \tag{3.1.2}$$

They satisfy the relation

$$\bar{\delta} f(x) = \delta f(x-a) \tag{3.1.3}$$

and the following deformed Leibnitz rules

$$\begin{aligned} \delta f(x)g(x) &= (\delta f(x))g(x) + f(x+a)\delta g(x) \\ &= (\delta f(x))g(x+a) + f(x)(\delta g(x)) , \end{aligned} \tag{3.1.4}$$

$$\begin{aligned} \bar{\delta} f(x)g(x) &= (\bar{\delta} f(x))g(x) + f(x-a)\bar{\delta} g(x) \\ &= (\bar{\delta} f(x))g(x-a) + f(x)(\bar{\delta} g(x)) . \end{aligned} \tag{3.1.5}$$

An analog of Stokes formula (cf. Sections 1.6 and 2.2) leads to the following definite "integral"

$$\int_{x_0-ma}^{x_0+na} dx\, f(x) := a \sum_{k=-m}^{n} f(x_0+ka) . \tag{3.1.6}$$

An inner product of two complex functions f and g can be defined by

$$< f, g >:= \int_{x_0-\infty}^{x_0+\infty} dx\, f(x)^* g(x) . \tag{3.1.7}$$

This inner product makes the set of functions on a one-dimensional lattice $x_0 \pm m$, $m = 0, 1, 2, \ldots$ which satisfy normalizability condition, into a Hilbert space $\mathcal{H} = \ell_2(\mathbb{Z})$.

It is convenient to introduce the shift operators

$$A := 1 + a\delta , \qquad \bar{A} := 1 + a\bar{\delta} , \tag{3.1.8}$$

which satisfy

$$(Af)(x) = f(x+a), \qquad (\bar{A}f)(x) = f(x-a) , \tag{3.1.9}$$

and thus also

$$A\bar{A} = \bar{A}A = 1 , \tag{3.1.10}$$

$$AX = (X+a)A , \qquad \bar{A}X = (X-a)\bar{A} , \tag{3.1.11}$$

where X is the position operator: $(Xf)(x) := xf(x)$. The operators $\bar{A}, A$ are well defined on the Hilbert space $\mathcal{H}$ and are hermitian conjugate with respect to the scalar product (3.1.7)

$$\bar{A} = A^\dagger . \tag{3.1.12}$$

Together with (3.1.10) this shows that A is a unitary operator. The natural generalization of a (self-adjoint) momentum operator is

$$P := \frac{1}{2ia}(A - \bar{A}) , \tag{3.1.13}$$

with the action on a function of the form

$$(Pf)(x) = \frac{1}{2ia}[f(x+a) - f(x-a)] . \tag{3.1.14}$$

Another self-adjoint operator

$$H := \frac{1}{2a^2}(A + \bar{A} - 2) = \frac{1}{2a}(\delta - \bar{\delta}) = -\frac{1}{2}\delta\bar{\delta} , \tag{3.1.15}$$

(the latter equality follows from (3.1.10)) tends in the limit $a \to 0$ to the Hamiltonian of a free particle (with unit mass) and so can serve as a Hamiltonian on a lattice.

The functions $f_k = \exp(ikx)$ are simultaneous eigenfunctions of P and H,

$$Pf_k = \frac{1}{a}\sin(ka)f_k , \quad Hf_k = \frac{1}{a^2}[1-\cos(ka)]f_k , \qquad k \in \left(-\frac{\pi}{a}, \frac{\pi}{a}\right) . \tag{3.1.16}$$

In the limit $a \to 0$, one recovers the spectra of the corresponding operators of the continuum theory.

One can easily see that the CR of $A, \bar{A}$ and X coincide with the relations for q-oscillator operators b^+, b, N in the degenerate representation (see Section 2.1) up to the rescaling $N \sim X/a$. Thus we already know the representation for $A, \bar{A}, X$: representation space is spanned by the vectors $|x_0, n\rangle := \bar{A}^n|x_0\rangle$, $|x_0, -n\rangle := A^n|x_0\rangle$, $n \in \mathbb{Z}_+$, where $|x_0\rangle := |x_0, 0\rangle$ is an eigenstate of X with the eigenvalue x_0:

$$X|x_0\rangle = x_0|x_0\rangle .$$

The spectrum of X forms a lattice with the spacing a:

$$X|x_0, n >= (x_0 + na)|x_0, n > , \quad n \in \mathbb{Z} ,$$

and the choice of x_0 with $0 \leq x_0 < a$, determines different (inequivalent) representations and can be interpreted as the position of the lattice on a real line.

To transform this discrete calculus into q-calculus, one has to introduce new coordinate

$$y := q^{x/(q-1)} , \quad q := 1 + a . \tag{3.1.17}$$

In the continuum limit this becomes exponential map: $y = e^x$ as $a \to 0$. Using the definition of discrete derivative (3.1.1), one finds

$$\delta y = y ,$$

and thus for deformed differentials one has

$$dy = (dx)y \ .$$

Hence $df = dx\delta f = dyDf = dxyDf$ **(**D **is the derivative with respect to the coordinate** y**) which implies**

$$D = y^{-1}\delta \ .$$

Together with (3.1.1) this gives the usual expression for the Jackson derivative

$$(Df)(y) = \frac{f(qy) - f(y)}{(q-1)y} \ . \tag{3.1.18}$$

Similarly,

$$(\bar{D}f)(y) = \frac{f(y) - f(q^{-1}y)}{(1-q^{-1})y} \ . \tag{3.1.19}$$

The derivatives $D, \bar{D}$ **together with the position operator** Y **satisfy the CR**

$$DY - qYD = 1 \ , \tag{3.1.20}$$

$$\bar{D}Y - q^{-1}Y\bar{D} = 1 \ , \tag{3.1.21}$$

$$\bar{D}D - qD\bar{D} = 0 \ . \tag{3.1.22}$$

Expressed in terms of the coordinates y**, the shift operators** A **and** $\bar{A}$ **take the form**

$$A = 1 + (q-1)YD \ , \qquad \bar{A} = 1 - (1-q^{-1})Y\bar{D} \ , \tag{3.1.23}$$

and act on functions of y **as follows:**

$$(Af)(y) = f(qy) \ , \qquad (\bar{A}f)(y) = f(q^{-1}y) \ . \tag{3.1.24}$$

The operators $A, \bar{A}, Y$ **have CR**

$$AY = qYA \ , \qquad \bar{A}Y = q^{-1}Y\bar{A} \ . \tag{3.1.25}$$

The inner product introduced on an equidistant lattice is transformed into Jackson integral considered in Section 1.6

$$< f, g >= \int_0^{q\infty} dy y^{-1} f(y)^* g(y) = (q-1) \sum_{k=-\infty}^{\infty} f(q^k y_0)^* g(q^k y_0) \ .$$

As the operator Y is self-adjoint and $A^{\dagger} = \bar{A}$, using (3.1.23) one has

$$D^{\dagger} = -\bar{D} + Y^{-1} . \tag{3.1.26}$$

The irreducible representations of the algebra $A, \bar{A}, Y$ are obtained as follows. Let $|y_0>$ be an eigenstate of Y with eigenvalue y_0. Then the states

$$|y_0, n> := \bar{A}^n |y_0> , \qquad |y_0, -n> := A^n |y_0> , \quad n \in \mathbb{Z}_+ , \tag{3.1.27}$$

span the irrep of the algebra and are eigenstates of Y:

$$Y|y_0, n> = q^n y_0 |y_0, n> . \tag{3.1.28}$$

Thus quantum mechanics on an equidistant lattice can be described equivalently as a kind of q-deformed quantum mechanics with the CR (3.1.20)-(3.1.22), with Y being the position operator and $D, \bar{D}$ being the building blocks for the construction of a deformed counterpart of a momentum operator in analogy with (3.1.13) (see also next Section).

Another approach to q-deformation of one-dimensional Heisenberg algebra with self-adjoint q-derivative (momentum operator) and two mutually conjugate coordinate operators (the physical position operators being their self-adjoint combination) is considered in [215, 119].

3.2 Multidimensional Jackson calculus and particle on two-dimensional quantum space

We have found out in the preceding Section that Jackson derivatives are suitable for description of a quantum mechanical particle on a one-dimensional lattice. To consider quantum mechanics on higher dimensional lattices, one needs multidimensional generalization of the Jackson calculus. Such calculus invariant with respect to the q-group $GL_r(N) := GL_{r,1}(N)$, can be constructed in the space $C_r^N[x^i]$ with commuting coordinates x^i.

As it was shown in Section 1.5, commuting coordinates differ from non-commuting ones by the non-commuting factors e^i. The situation reminds a transition from usual three-dimensional Euclidian coordinates to well known quaternions with basis $\{\sigma_i\}_{i=1}^3$: sometimes it is convenient to put into correspondence the coordinates $\{x^i\}_{i=1}^3$ with non-commutative quaternions $\hat{x}^i := x^i \sigma_i$ (no summation). Though this transformation brings about new algebraic structure and permits to express three-dimensional rotations in pure algebraic way, the underlying geometrical and physical structure remain the same.

This analogy leads to the conclusion that one can freely choose the most convenient quantum space among the set of twisted q-spaces. In particular, in the case of spaces $C^N_{r,q_{ij}}[x^i]$, the simplest choice is the spaces $C^N_r[x^i]$ with commuting coordinates. The CR for coordinates and derivatives on this space are the following [209]:

$$x^i x^j = x^j x^i , \quad \forall\, i,j \,, \quad \partial_i x^i = 1 + r x^i \partial_i + (r-1) \sum_{l=i+1}^{N} x^l \partial_l \,,$$
$$\partial_i \partial_k = \frac{1}{r} \partial_k \partial_i \,, \quad \partial_i x^k = r x^k \partial_i \,, \quad \partial_k x^i = x^i \partial_k \,, \qquad (3.2.1)$$
$$i < k, \; i,k = 1,...,N \,.$$

To develop Jackson calculus, one defines the operators of finite dilatations

$$A_i = 1 + (r-1) \sum_{j=i}^{N} x^j \partial_j \,, \qquad (3.2.2)$$

with the commutation relations

$$\begin{aligned} A_i A_k &= A_k A_i \,, \quad \forall\, k,i \,, \\ A_k x^i &= x^i A_k \,, \quad k > i \,, \\ A_i x^k &= r x^k A_i \,, \quad k \geq i \,. \end{aligned} \qquad (3.2.3)$$

Note that the operators A_i are analogous to operators $Y^i{}_j$ of vector fields on a simple quantum group, introduced in Section 1.6. The relations (3.2.2) permit to express the q-derivatives in terms of A_i

$$\partial_i = (1-r)^{-1} (x^i)^{-1} (A_{i+1} - A_i) \,, \quad i = 1,...,N \,, \qquad (3.2.4)$$
$$A_{N+1} := 1 \,.$$

The relations (3.2.3) and (3.2.4) lead to the following realization of the q-derivatives in the space of functions of N commuting variables

$$\partial_i f(x^1,...,x^N) = \frac{f(x^1,...,x^i, r x^{i+1},...,r x^N) - f(x^1,...,x^{i-1}, r x^i,...,r x^N)}{(1-r) x^i} \,. \qquad (3.2.5)$$

One can check easily that the finite differences (3.2.5) indeed satisfy the $GL_r(N)$ invariant CR (3.2.1). These differences look like natural multidimensional generalization of the Jackson derivative.

As is shown in [239] there are two types of CR for q-derivatives and coordinates, which are invariant with respect to q-deformed groups. The first possibility is presented in (3.2.1), the second one is the following

$$
\begin{aligned}
\tilde{\partial}_i x^i &= 1 + \frac{1}{r} x^i \tilde{\partial}_i + (\frac{1}{r} - 1) \sum_{l=1}^{i-1} x^l \tilde{\partial}_l \, , \\
\tilde{\partial}_i \tilde{\partial}_k = \frac{1}{r} \tilde{\partial}_k \tilde{\partial}_i \, , \quad \tilde{\partial}_i x^k &= x^k \tilde{\partial}_i \, , \quad \tilde{\partial}_k x^i = \frac{1}{r} x^i \tilde{\partial}_k \, , \\
& i < k, \; i, k = 1, \ldots, N \, .
\end{aligned}
\tag{3.2.6}
$$

For these CR, it is natural to introduce the operators

$$
\tilde{A}_i = 1 + (\frac{1}{r} - 1) \sum_{j=1}^{i} x^j \tilde{\partial}_j \, , \tag{3.2.7}
$$

which commute with each other and have the following CR with the coordinates

$$
\begin{aligned}
\tilde{A}_k x^i &= r^{-1} x^i \tilde{A}_k \, , \quad k > i \, , \\
\tilde{A}_i x^k &= x^k \tilde{A}_i \, , \quad k \geq i \, .
\end{aligned}
\tag{3.2.8}
$$

These relations permit to construct the realization of the q-derivatives $\tilde{\partial}_i$ in terms of the finite differences

$$
\begin{aligned}
&\tilde{\partial}_i f(x^1, \ldots, x^N) \\
&= \frac{f(r^{-1}x^1, \ldots, r^{-1}x^i, x^{i+1}, \ldots, x^N) - f(r^{-1}x^1, \ldots, r^{-1}x^{i-1}, rx^i, \ldots, rx^N)}{(r^{-1} - 1)x^i} \, ,
\end{aligned}
$$

which is the generalization of the Jackson derivative of the second type (3.1.19).

Now we apply the above considered formulas for construction of quantum mechanics on a two-dimensional quantum plane. For convenience, we rewrite the CR (3.2.1), (3.2.3), (3.2.6) and (3.2.8) in this particular case denoting $x^1 = z$, $x^2 = \bar{z}$, $\partial_1 = \partial$, $\partial_2 = \bar{\partial}$, $\tilde{\partial}_1 = \tilde{\partial}$, $\tilde{\partial}_2 = \tilde{\bar{\partial}}$, $A_1 = A$, $A_2 = \bar{A}$, $\tilde{A}_1 = \tilde{A}$, $A_2 = \tilde{\bar{A}}$

$$
z\bar{z} = \bar{z}z \, , \quad \partial\bar{\partial} = \frac{1}{r}\bar{\partial}\partial \, , \quad \tilde{\partial}\tilde{\bar{\partial}} = \frac{1}{r}\tilde{\bar{\partial}}\tilde{\partial} \, , \tag{3.2.9}
$$

$$
\begin{aligned}
\partial z = 1 + rz\partial + (r-1)\bar{z}\bar{\partial} \, , \qquad & \bar{\partial}\bar{z} = 1 + r\bar{z}\bar{\partial} \, , \\
\partial\bar{z} = r\bar{z}\partial \, , \qquad & \bar{\partial}z = z\bar{\partial} \, ,
\end{aligned}
\tag{3.2.10}
$$

$$
Az = rzA \, , \quad A\bar{z} = r\bar{z}A \, , \quad \bar{A}z = z\bar{A} \, , \quad \bar{A}\bar{z} = r\bar{z}\bar{A} \, , \tag{3.2.11}
$$

$$\tilde{\partial} z = 1 + \frac{1}{r} z \tilde{\partial}\,, \qquad \tilde{\bar{\partial}} \bar{z} = 1 + (\frac{1}{r} - 1) z \tilde{\partial} + \frac{1}{r} \bar{z} \tilde{\bar{\partial}}\,,$$

$$\tilde{\partial} \bar{z} = \bar{z} \tilde{\partial}\,, \qquad \tilde{\bar{\partial}} z = \frac{1}{r} z \tilde{\bar{\partial}}\,, \tag{3.2.12}$$

$$\tilde{A} z = \frac{1}{r} z \tilde{A}\,, \quad \tilde{A} \bar{z} = \bar{z} \tilde{A}\,, \quad \tilde{\bar{A}} z = \frac{1}{r} z \tilde{\bar{A}}\,, \quad \tilde{\bar{A}} \bar{z} = \frac{1}{r} \bar{z} \tilde{\bar{A}}\,. \tag{3.2.13}$$

All A-operators commute with each other and satisfy the relations

$$\tilde{\bar{A}} A = 1\,, \qquad \tilde{\bar{A}} \bar{A} = \tilde{A}\,, \qquad \tilde{A} A = \bar{A}\,. \tag{3.2.14}$$

The simplest way to find these relations, is to derive them from the action of these operators on an arbitrary function $f(z, \bar{z})$. The CR between different types of the q-derivatives have the form

$$\begin{aligned} \tilde{\partial} \partial &= r \partial \tilde{\partial}\,, & \tilde{\bar{\partial}} \bar{\partial} &= r \bar{\partial} \tilde{\bar{\partial}}\,, \\ \tilde{\partial} \bar{\partial} &= \bar{\partial} \tilde{\partial}\,, & \tilde{\bar{\partial}} \partial &= r^2 \partial \tilde{\bar{\partial}}\,. \end{aligned} \tag{3.2.15}$$

Now we must define $*$-involution in the algebra of the operators which enter the relations (3.2.9)-(3.2.15).

We want to consider the parameter r as a lattice spacing and hence, it must be a real number. The appropriate involution for $GL_r(2)$ in this case is the following [210]

$$T^* = CTC\,, \quad C = \begin{pmatrix} 0 & 1 \\ 1 & 0 \end{pmatrix}.$$

This means that

$$T^* = \begin{pmatrix} a^* & b^* \\ c^* & d^* \end{pmatrix} = \begin{pmatrix} d & c \\ b & a \end{pmatrix},$$

and $z^* = \bar{z}$. It is not difficult to see that the CR for the A-operators and the coordinates are consistent with the involution

$$A^* = \tilde{\bar{A}}\,, \qquad \bar{A}^* = \tilde{A}\,. \tag{3.2.16}$$

This gives the involution rules for the derivatives

$$\partial^* = -\tilde{\bar{\partial}} + (\frac{1}{r} - 1) \frac{z}{\bar{z}} \tilde{\partial} + \frac{1}{\bar{z}}\,, \qquad \bar{\partial}^* = -\tilde{\partial} + \frac{1}{z}\,.$$

The next step is to construct the representation of the operators in a Hilbert space so that the involution would coincide with hermitian conjugation (therefore in the following we will denote the involution by a sign of hermitian conjugation). To construct a convenient basis in the Hilbert space, one needs hermitian operators. Combinations of the coordinates of the form

$$x = (z + \bar{z})/2 , \qquad y = (z - \bar{z})/2i ,$$

are not convenient as they have identical CR with a part of A-operators and rather cumbersome CR with another part. A-operators play the role of conjugate momenta on a lattice (cf. [77]). A natural choice for position operators follows from the observation that the A-operators generate finite dilatations rather than translations. This implies the introduction of the operators

$$\rho = \sqrt{\bar{z}z} , \qquad \rho^* = \rho , \tag{3.2.17}$$

and

$$\Phi = \sqrt{\bar{z}z^{-1}} , \qquad \Phi^*\Phi = 1 . \tag{3.2.18}$$

As follows from (3.2.16) and (3.2.14), we have also the operators A and $B := \bar{A}\bar{A}$ such that

$$A^*A = 1 , \qquad B^* = B . \tag{3.2.19}$$

These operators have the following CR

$$\begin{aligned} A\rho &= r\rho A , & B\Phi &= r\Phi B , \\ A\Phi &= \Phi A , & B\rho &= \rho B . \end{aligned} \tag{3.2.20}$$

All other operators can be expressed in terms of these four ones.

The CR (3.2.20) play the role of canonical commutation relations (q-analog of two-dimensional Heisenberg algebra). As is seen from (3.2.20), we have two mutually commuting subalgebras generated by the pairs of the operators A, ρ and B, Φ (analog of canonically conjugate momenta and coordinates). Their matrix representations are constructed on a common domain D_{ρ_0,b_0} consisting of all linear combinations of vectors $| N, m\rangle_{\rho_0,b_0}$

$$\begin{aligned} \rho \mid N, m\rangle_{\rho_0,b_0} &= \rho_0 r^{-N} \mid N, m\rangle_{\rho_0,b_0} , \\ B \mid N, m\rangle_{\rho_0,b_0} &= b_0 r^{-m} \mid N, m\rangle_{\rho_0,b_0} , \end{aligned} \tag{3.2.21}$$

$$\begin{aligned} A \mid N, m\rangle_{\rho_0,b_0} &= \mid N+1, m\rangle_{\rho_0,b_0} , \\ \Phi \mid N, m\rangle_{\rho_0,b_0} &= \mid N, m-1\rangle_{\rho_0,b_0} . \end{aligned} \tag{3.2.22}$$

The constants ρ_0 and b_0 mark the different representations and from the eigenvalues of ρ and B it follows that in the ranges $[\rho_0, r\rho_0)$ and $[b_0, rb_0)$ the representations are inequivalent. The matrices ρ, B are hermitian, while A, Φ are unitary with respect to the scalar product defined by

$${}_{\rho_0,b_0}\langle N, m \mid N', m'\rangle_{\rho_0,b_0} = \delta_{NN'}\delta_{mm'} .$$

As usual, D_{ρ_0,b_0} can be completed to a Hilbert space $\mathcal{H}_{\rho_0,b_0}$

$$\mathcal{H}_{\rho_0,b_0} = \left\{ \sum_{N,m\in\mathbb{Z}} C_{Nm} \mid N,m\rangle_{\rho_0,b_0} : \sum_{N,m\in\mathbb{Z}} \mid C_{Nm} \mid^2 < \infty \right\} .$$

The operators ρ and B are essentially self-adjoint in the Hilbert space, their self-adjoint extension is defined on the domain

$$D^+_{\rho_0,b_0} = \left\{ \sum_{N,m\in\mathbb{Z}} C_{Nm} \mid N,m\rangle_{\rho_0,b_0} : \quad \sum_{N,m\in\mathbb{Z}} \mid C_{Nm} \mid^2 r^{-2N} < \infty \, , \right.$$
$$\left. \sum_{N,m\in\mathbb{Z}} \mid C_{Nm} \mid^2 r^{-2m} < \infty \right\} .$$

The operators A, ϕ can be easily extended to unitary operators with the domain $\mathcal{H}_{\rho_0,b_0}$. Therefore in what follows we will denote the involution by a sign of hermitian conjugation.

We will consider one of the representations labeled by ρ_0, b_0 and for brevity put $\rho_0 = b_0 = 1$ (one can always achieve these values by appropriate rescaling of the operators $\rho \to \rho_0\rho$ and $B \to b_0 B$, the defining CR (3.2.20) being invariant with respect to this transformation).

So, starting from $GL_r(2)$ invariant differential calculus on a quantum plane, we have come naturally to the polar coordinates, the operator ρ being the operator of radial coordinates and the operator Φ being of the form $\Phi = e^{-i\phi}$, where ϕ is (hermitian) operator of an angle coordinate. The structure of the algebra involution leads to the mixed representation with one coordinate (ρ) and one momentum diagonal operators. In classical case a multiplication of a function by Φ shifts its Fourier component numbers by minus unity. This corresponds to the action (3.2.22) of the operator Φ on vectors of $\mathcal{H}$ and implies that B is connected with an angular momentum operator. In fact, it is an analog of the operator of the form

$$\exp\{i\phi_0 \frac{\partial}{\partial\phi}\} = \exp\{\phi_0 M\} \, , \tag{3.2.23}$$

in the case of quantum mechanics on usual continuous plane, where ϕ_0 is some fixed angle and $i\partial/\partial\phi$ is (two-dimensional) angular momentum operator. Eigenfunctions of this operator are periodic functions $\exp\{-in\phi\}$ with eigenvalues $\exp\{n\phi_0\} = (e^{\phi_0})^n$. The latter expression coincides with eigenvalue of operator B if one equates $r = \exp\{\phi_0\}$.

Thus the operator ρ defines the values of radial coordinate, while the operator A shifts them (play the role of conjugate momentum). Analogously, the operator B defines angular momentum values, while the operator Φ of the conjugate coordinate shifts its eigenvalues.

Now we can consider a q-subgroup Λ of the $GL_r(2,\mathbb{R})$ of matrices of the form

$$T = \begin{pmatrix} a & 0 \\ 0 & d \end{pmatrix} ,$$

with $ad = da$. Because of the latter relation, it looks like an ordinary group isomorphic to a multiplicative group of complex numbers. But one must remember about CR with the generators $Y^i_{\ j}$ of the corresponding quantum universal enveloping algebra. For left-invariant and right-covariant generators, they have the following general form (cf. Section 1.6.1)

$$Y^i_{\ j} T^k_{\ s} = T^k_{\ l} Y^m_{\ n} (\tilde{R}_{21})^{il}_{\ \ mt} (\tilde{R}_{12})^{nt}_{\ \ js} , \tag{3.2.24}$$

where $\tilde{R}_{12}$ and $\tilde{R}_{21}$ are properly normalized R-matrices of the $GL_r(2,\mathbb{R})$. The CR (3.2.24) gives for the subgroup Λ (i.e., if $b = c = 0$)

$$\begin{array}{ll} Y^1_{\ 1} a = r a Y^1_{\ 1} , & Y^1_{\ 1} d = d Y^1_{\ 1} , \\ Y^2_{\ 2} a = a Y^2_{\ 2} , & Y^2_{\ 2} d = r d Y^2_{\ 2} . \end{array} \tag{3.2.25}$$

It is easy to see that CR for $\mathcal{D} := Y^1_{\ 1} Y^2_{\ 2}$, $\bar{\mathcal{D}} := Y^2_{\ 2}$, a and d are the same as for the $A, \bar{A}, z, \bar{z}$ (recall also that according to the chosen involution, $d = a^*$). So the algebra (3.2.11) on the quantum plane is isomorphic to that of the q-subgroup Λ and we can identify the q-plane with this subgroup. The operators $\tilde{A}, \bar{\tilde{A}}$ correspond to right-invariant and left-covariant generators (Section 1.6.1). From the other hand, q-subgroup Λ can play the role of symmetry group, the whole $GL_r(2,\mathbb{R})$ group being the group of linear canonical transformations (of CR (3.2.10),(3.2.12)).

The coaction of Λ on the coordinates is

$$\begin{array}{ll} z \to z' = a \otimes z , & \bar{z} \to \bar{z}' = d \otimes \bar{z} , \\ \rho \to \rho' = a_\rho \otimes \rho , & \Phi \to \Phi' = a_\Phi \otimes \Phi , \\ a_\rho = \sqrt{da} = \sqrt{a^* a} , & a_\Phi = \sqrt{a/d} = \sqrt{a/a^*} . \end{array} \tag{3.2.26}$$

The coordinates ρ', Φ' have a representation in the Hilbert space $\mathcal{H} \otimes \mathcal{H}$ (i.e., the cross product of the same Hilbert spaces since in the considered case the comodule coincides with the symmetry q-group) which has a subspace (diagonal) isomorphic to $\mathcal{H}$. The latter corresponds to the representation of CR based on the coordinates ρ', Φ' as primary ones.

All A-operators and thus the operator B are invariant with respect to the coaction (3.2.26). If the subgroup Λ is considered as a symmetry group, it is reasonable to construct a Hamiltonian out of the invariant operators A and B. In particular, for a two-dimensional quantum free particle one can try the Hamiltonian

$$\tilde{H} = \frac{\Omega}{(1-r)^2}\left[(1-A)(1-A^\dagger) + (1-B)^2\right] , \qquad (3.2.27)$$

where Ω is a constant with the dimension of energy. In order to find eigenfunctions and eigenvalues of the Hamiltonian, introduce the operators

$$\rho^{iP} = \exp\{iP\ln\rho\} ,$$

and the states (cf. Section 1.6.3)

$$\mid \mathbf{I}, m\rangle = \sum_{N=-\infty}^{\infty} \mid N, m\rangle ,$$

$$\mid P, m\rangle = \rho^{iP} \mid \mathbf{I}, m\rangle ,$$

with the properties

$$A \mid \mathbf{I}, m\rangle = \mid \mathbf{I}, m\rangle ,$$
$$A \mid P, m\rangle = r^{iP} \mid P, m\rangle ,$$

where P is a real number: $0 \leq P \leq \pi/\chi$, $\chi := \ln r$. For eigenvalues of the Hamiltonian (3.2.27) we obtain

$$\tilde{H} \mid P, m\rangle = \xi_{P,m} \mid P, m\rangle ,$$

$$\xi_{P,m} = \Omega\left(\frac{(1-\cos\chi P)}{(1-r)^2} + [m;r]^2\right) \underset{r\to 1}{\longrightarrow} \Omega(P^2+m^2) .$$

Thus the operator $\tilde{H}$ has the correct continuous limit but its eigenvalues are not invariant with respect to the reflection $m \longrightarrow -m$. This means that left and right modes have different properties and positive modes have decreasing spectrum which can lead to additional divergences in the corresponding field theories. These properties are caused by exponent-like form of the operator B analogous to (3.2.23), as we discussed above. So it is natural to consider the Hamiltonian

$$H = \frac{\Omega}{(1-r)^2}\left[(1-A)(1-A^\dagger) + \ln^2 B\right] ,$$

with the eigenvalues $\Omega\lambda_{P,m}$, where

$$\lambda_{P,m} = \left(\frac{(1-\cos\chi P)}{(1-r)^2} + \frac{\chi^2}{(1-r)^2}m^2\right) \underset{r\to 1}{\longrightarrow} (P^2+m^2) \,. \tag{3.2.28}$$

To construct a two dimensional quantum field theory, we need a kind of integral over the variable ρ. It can be defined analogously to that in Section 1.6.3.

For m-th component $f_m(\rho)$ of "Fourier expansion" we define

$$\begin{aligned}\int_0^K d_r\rho\, f_m(\rho) &:= \langle K,m \mid \partial_\rho^{-1} f_m(\rho) \mid \mathbf{1},m\rangle \\ &= (1-r)\langle K,m \mid (1-A)^{-1}\rho f_m(\rho) \mid \mathbf{1},m\rangle \\ &= (1-r)\sum_{l=-\infty}^{K} r^{-l} f_m(r^{-l}) \,,\end{aligned} \tag{3.2.29}$$

where the derivative ∂_ρ is defined as in (3.2.4)

$$\partial_\rho := \frac{1}{(1-r)\rho}(1-A) \,.$$

The last expression for the integral in (3.2.29) has the form of a usual Jackson integral.

To construct the action, we note that the two-dimensional space with the coordinates $z, \bar{z}$ and the symmetry transformations (3.2.26) can be considered as a result of the conformal map

$$u \to z = e^u \,, \qquad \bar{u} \to \bar{z} = e^{\bar{u}} \,, \tag{3.2.30}$$

of a cylinder with the coordinates $\bar{u}, u$. This map is well known and is used widely in field theoretical context, in particular, in conformal and string theories (in frame of the so-called radial quantization, see e.g.[112]). A coordinate along the cylinder is associated with a time coordinate and space coordinate takes values on a circle. After the conformal mapping the coordinate ρ plays the role of time and Φ of space coordinate. It is easy to see that in coordinates $u, \bar{u}$ the transformations (3.2.26) become translations, while the time coordinate $\tau := Re\, u$ takes values on an equidistant lattice with a spacing $\ln r$. In continuous case a field theory free action for a scalar real field Ψ in polar coordinates on the z-plane has the form

$$\begin{aligned} S_0 &= \int_0^\infty\int_0^{2\pi} \frac{d\rho}{\rho} d\phi\, \Psi(\rho,\phi)(\rho\partial_\rho\rho\partial_\rho + \partial_\Phi^2)\Psi(\rho,\phi) \\ &= \sum_{m=-\infty}^{\infty}\int_0^\infty \frac{d\rho}{\rho}\, \Psi_{-m}(\rho)(\rho\partial_\rho\rho\partial_\rho - m^2)\Psi_m(\rho) \,,\end{aligned} \tag{3.2.31}$$

This expression is explicitly invariant with respect to dilatations of ρ and translations of ϕ (continuous analog of (3.2.26)).

A quantum plane analog has the following form

$$S_0^r = \lim_{K\to\infty} \sum_{m=-\infty}^{\infty} \langle K, m \mid (1-A)^{-1}\Psi_{-m}(\rho)G^{-1}\Psi_m(\rho) \mid \mathbf{1}, m\rangle \,.$$

Here the operator G^{-1} has a form similar to the Hamiltonian H

$$G^{-1} = \frac{H}{\Omega}\,,$$

though its meaning is quite different, of course. The finite difference operator G^{-1} in this action has the eigenvalues $\lambda_{P,m}$ presented in (3.2.28). In the case of interacting Ψ^4 theory a one-loop correction to, for example, mass term is proportional to the trace of the operator G

$$\sum_{m=-\infty}^{\infty} \int_0^{-\pi/\ln r} \frac{dP}{\lambda_{P,m}}\,,$$

which has no divergences. Note, however, that summation over orbital number m is the same as in the continuous case. Such partial discretization of the space-time is enough for the regularization of two-dimensional models but in higher dimensions such a field theory can be ultraviolet divergent even in q-space-time. Another lesson from the considered models is that q-derivatives or finite differences constructed with the use of operators of the form (3.2.2) are connected with dilatations and not with translations. This implies that corresponding coordinates are related with usual ones (in which derivatives generate translations) by nonlinear exponent-like map of the type (3.2.30). It is clear, that symmetry transformations, e.g. of Lorentz group, have quite different form in non-linearly transformed coordinate frames.

Unfortunately, many important quantum groups, such as orthogonal ones, have no twisted counterparts with commuting coordinates of the corresponding quantum space. But even in this case, the freedom can be used for choice of the most convenient q-coordinate frame. Note also, that linear groups play important role as groups of space-time symmetries. It is enough to mention $SL(2,C)$ as universal covering of Lorentz group and $SU(2,2)$ as covering of four-dimensional conformal group. We shall consider their quantization in the subsequent sections.

3.3 Projective construction of quantum inhomogeneous groups

The first step in the construction of q-deformations of inhomogeneous groups was made in [245] and [230], where a q-analog of the group of 2D Euclidean transformations had been found. As we have shown in Section 2.2, this q-group defines the differential calculus on the complex q-plane that was used for the construction of the Bargmann-Fock representation for the q-oscillator and the corresponding q-deformed path integral. However, this case is rather degenerate because the coaction $\delta(z), \delta(\bar{z})$ on the complex coordinates $z, \bar{z}$ of q-plane has the form

$$\delta(z) = a \otimes z + w \otimes 1,$$
$$\delta(\bar{z}) = \bar{a} \otimes \bar{z} + \bar{w} \otimes 1,$$

where a corresponds to quantum U(1)-rotations and $w, \bar{w}$ are translation parameters. Thus this transformation is factorized and in fact becomes one-dimensional. The problem arises if one tries to define inhomogeneous transformations of higher dimensional spaces or of the 2D q-plane with real coordinates. In these cases, one is forced to introduce additional scaling transformations [212] which has no analog in classical limit $q \rightarrow 1$. But as is well known the symmetry of physical systems with respect to groups with dilatations is very restrictive: the invariant length on the corresponding space-time is absent, all particles must have zero masses, the inhomogeneous group $IGL(N)$ has no invariants at all etc. So it is important to understand the nature of the problem and to find possible ways to overcome it.

There is the same problem in attempts to construct q-Poincaré algebra in which q-translations are generated by q-derivations [185] or to find QUE Poincaré algebra as a Hopf subalgebra of $\mathcal{U}_q(O(4,2))$ or $\mathcal{U}_q(SU(2,2))$ [80, 164]: one has to add generator of dilatations to define a self-consistent coproduct. In [80] the general consideration of the problem based on Cartan automorphisms and Bruhat decomposition of non-compact algebras was done in the context of canonical procedure for q-deformation of non-compact Lie algebras. In particular, it was shown that Poincaré containing Hopf subalgebra of the quantum conformal algebra has to include dilatations, i.e. to be the 11-generator Weyl algebra (parabolic-type subalgebra).

In this Section we give explanation of the appearance of additional dilatations in q-affine groups and suggest a method for the construction of inhomogeneous q-groups without them.

Let $\mathbb{C}P^N$ be a usual (non-deformed) projective space with homogeneous coordinates $X = (X^1, ..., X^{N+1}) \in \mathbb{C}^{N+1}$ (see e.g. [123]) and with the equivalence relation $X \sim \lambda X, 0 \neq \lambda \in \mathbb{C}$. The projective space is topologically nontrivial and can be covered by $N+1$ patches $U_1, ..., U_{N+1}$ defined by the conditions

$$X \in U_k \Longleftrightarrow X^k \neq 0.$$

On the patch U_{N+1}, one can introduce inhomogeneous coordinates $x^m = X^m/X^{N+1}, m = 1, ..., N$. The action of the group $GL(N+1)$ is deduced from its action in $\mathbb{C}^{N+1}$

$$X \rightarrow X' = TX \; , \qquad T \in GL(N+1). \tag{3.3.1}$$

On the patch U_{N+1}, it is useful to write T in a block-diagonal form

$$T = \begin{pmatrix} A & \vec{b} \\ \vec{c}^{\mathsf{T}} & d \end{pmatrix} , \tag{3.3.2}$$

where $A \in GL(N)$, $\vec{b}$ and $\vec{c}$ are N-dimensional vectors and d is a number.

Let us denote by G the subgroup of $GL(N+1)$-matrices with $\vec{c} = 0$. The requirement $T \in GL(N+1)$, then gives

$$T = \begin{pmatrix} A & \vec{b} \\ 0 & d \end{pmatrix} \in G \Longleftrightarrow A \in GL(N), \; d \neq 0. \tag{3.3.3}$$

Then the action (3.3.1) on the inhomogeneous coordinates $\vec{x} = (x^1, ..., x^m)^{\mathsf{T}}$ reduces to $IGL(N)$ transformations of U_{N+1} given by

$$\vec{x} \rightarrow \vec{x}\,' = d^{-1}(A\vec{x} + \vec{b}) = K\vec{x} + \vec{t}, \tag{3.3.4}$$

where $K = d^{-1}A \in GL(N)$ and $\vec{t} = d^{-1}\vec{b}$.

We now identify U_{N+1} with the affine space, and the mapping (3.3.4) with the inhomogeneous (affine) transformations. We note that even if we restrict originally the homogeneous part A of the affine transformations to a smaller group, e.g. to $SL(N)$, then K is extended by the dilatations. So we are forced to include dilatations into the affine transformations.

Thus the problem of the appearance of additional dilatations in q-inhomogeneous groups can be reduced to a simpler question of the possibility of quantum deformation of the subgroup G of the matrices (3.3.3) satisfying the additional condition $d = 1$. Now we shall show that if there is a coordinate among $X^1, ..., X^N$ which does not commute with X^{N+1}, then $d \neq 1$ in the quantum deformation of (3.3.3),(3.3.4). Let the coordinates X^i of the multiparametric quantum space $\mathbb{C}^{N+1}_{q,r}$ obey the relations

$$X^i X^j = q_{ij} X^j X^i \; , \qquad i < j,$$

where q_{ij} are the parameters of the quantum deformation. The q-matrix T analogous to that defined in (3.3.1),(3.3.2) transforms these coordinates via the coaction

$$\delta : A \to T \otimes A : \delta(x^i) = \sum_j T^i_j \otimes x^j$$

The commutation relations for the matrix elements ${T^i}_j$ are given by RTT-relation (1.4.10). For elements of any 2×2 submatrix of T

$$\begin{pmatrix} {T^i}_j & {T^i}_l \\ T^k_j & T^k_l \end{pmatrix} \equiv \begin{pmatrix} a & b \\ c & d \end{pmatrix}, \quad \forall i,j,k,l\,, \tag{3.3.5}$$

(the second matrix is a simple notation for the first one) one has from (1.4.10), (1.5.32), (3.3.5)

$$\begin{aligned} ab &= \frac{r^2}{q_{cn}} ba\,, & cd &= \frac{r^2}{q_{cn}} dc, \\ bc &= \frac{q_{rn} q_{cn}}{r^2} cb\,, & ad &= \frac{q_{rn}}{q_{cn}} da + \frac{r^2 - 1}{q_{cn}} bc\,, \\ ac &= q_{rn} ca\,, & bd &= q_{rn} db\,. \end{aligned} \tag{3.3.6}$$

Here the indices of the parameters are the numbers of the two rows (rn) or the two columns (cn) of the 2×2 submatrix. Now let $j = N+1$ and i correspond to the coordinates X^i such that $q_{i(N+1)} \neq 1$. The corresponding submatrix has the form

$$\begin{pmatrix} a & b \\ 0 & d \end{pmatrix}, \tag{3.3.7}$$

and it is immediately seen, e.g. from (3.3.6) that ${T^{N+1}}_{N+1} = d \neq 1$ in this case. Thus one has to introduce nontrivial d-transformations. If X^{N+1} is an invertible element of the algebra $\mathbb{C}^{N+1}_{q,r}$, then it is possible to define inhomogeneous coordinates also in the quantum space:

$$x^i = X^i (X^{N+1})^{-1}.$$

In this case the coactions

$$\begin{aligned} \delta(X^i) &= {X'}^i = \sum_{j=1}^{N} {T^i}_j \otimes X^j + {T^i}_{N+1} \otimes X^{N+1}\,, \\ \delta(X^{N+1}) &= {X'}^{N+1} = d \otimes X^{N+1}\,, \end{aligned}$$

become

$$\delta(x^i) = {x'}^i \equiv {X'}^i \left({X'}^{N+1}\right)^{-1} = \left(\sum_{j=1}^{N} {T^i}_j d^{-1} \otimes X^j + {T^i}_{N+1} d^{-1} \otimes 1 \right).$$

This coincides with the transformation which is called q-inhomogeneous. But as we have mentioned above, they are not reduced to ordinary affine group in the limit $q \to 1$ because of additional scalings.

From the above discussion it is clear how to avoid the appearance of additional dilatations in the inhomogeneous quantum transformations. We have to require that the last coordinate X^{N+1} of the initial projective space commutes with all the others. If we put $q_{i(N+1)} = 1$ $(i = 1, ..., N)$, then for the elements of q-matrix

$$\begin{pmatrix} A & \vec{t} \\ 0 & d \end{pmatrix}, \tag{3.3.8}$$

we have from (3.3.6) the following commutation relations (hereafter $i, j, k, l = 1, ..., N$):

- For $T^i{}_j$, we have usual CR of the homogeneous linear q-group $GL_{q,r}(N)$;
- For t^i :

$$t^i t^j = q_{ij} t^j t^i, \qquad i < j, \tag{3.3.9}$$

$$T^i{}_j t^i = r^2 t^i T^i{}_j,$$

$$T^i{}_j t^k = q_{ik} t^k T^i{}_j + (r^2 - 1) t^i T^k{}_j, \qquad i < k, \tag{3.3.10}$$

$$t^k T^i{}_j = \frac{q_{ki}}{r^2} T^i{}_j t^k, \qquad i > k; \tag{3.3.11}$$

- For d :

$$T^i{}_j d = d T^i{}_j,$$

$$t^i d = d t^i.$$

It is easy to verify straightforwardly that the conditions $T^{N+1}{}_i = 0$, do not contradict any of the commutation relation (3.3.6). Thus the element d and coordinate X^{N+1} commute with all quantities and can be put equal to unity without any contradictions. From the other hand, one cannot define (at least in this way) the pure translational group as is seen from (3.3.10) - (3.3.11) (it is impossible to put all $T^i{}_i$ equal to unity and $T^i{}_j = 0$ $(i \neq j)$).

Now to be sure that it is possible to construct by this method the inhomogeneous group without any dilatations (and for the construction of the antipode), we have to show that $\det_q T$ can be made a central element of the quantum algebra for certain choice of the deforming parameters q_{ij} and r^2. In the case of the multiparametric deformation of the $GL(N+1)$, one has (1.5.35)

$$T^i{}_k \det_q T = \frac{\prod_{\alpha=1}^{k-1} q_{\alpha k} \prod_{\delta=i+1}^{N+1} q_{i\delta}}{\prod_{\beta=1}^{i-1} q_{\beta i} \prod_{\gamma=k+1}^{N+1} q_{k\gamma}} (r^2)^{i-k} (\det_q T) T^i{}_k \equiv C_{ik} (\det_q T) T^i{}_k . \quad (3.3.12)$$

Thus $[T^i{}_i, \det_q T] = 0$, $[T^i{}_k T^k{}_i, \det_q T] = 0$, $C_{ik} = C_{in} C^{-1}_{nk}$ and it is sufficient to check the centrality for T^i_{N+1} only, i.e. in our case for t^i. Taking into account the conditions $q_{i(N+1)} = 1$, one obtains from (3.3.12) N equations for $[N(N-1)/2] + 1$ parameters

$$\left(\prod_{\alpha=1}^{i-1} q^{-1}_{\alpha i} \prod_{\beta=i+1}^{N} q_{i\beta} \right) (r^2)^{i-N-1} = 1 \, , \qquad i = 1, ...N. \quad (3.3.13)$$

As a result one has $ISL(N)_q$ with

$$\frac{N(N-3)}{2} + 1$$

free parameters. It is interesting that for $N = 2$ there are no free continuous parameters, while nontrivial solutions do exist. The equations in this case become very simple

$$q = r^4 \, , \qquad q = r^{-2} \, ,$$

and there are two nontrivial solutions (for $n = 1, 2$)

$$r^2 = exp\{\frac{2\pi i n}{3}\} \, , \qquad q = exp\{\frac{4\pi i n}{3}\} \, .$$

The form of the antipode S for the elements of the homogeneous submatrix was given in Section 1.5. For the translations, it has the form

$$S(t^i) = -S(M^i{}_j) t^j \, .$$

The form of counity ε is obviously

$$\varepsilon \begin{pmatrix} A & \vec{t} \\ 0 & 1 \end{pmatrix} = \begin{pmatrix} 1 & 0 \\ 0 & 1 \end{pmatrix} .$$

One more advantage of treating inhomogeneous quantum groups as q-subgroups of $GL(N+1)_q$ is that differential calculus which is invariant with respect to the former follows directly from that for the latter. Indeed, to extract $IGL(N)_q$ invariant differential calculus from the corresponding formulas for the $GL(N+1)_q$, one has to put simply

$$dX^{N+1} = 0 \quad \text{and} \quad \partial_{N+1} = 0 \quad \text{(in operator sense)}\,,$$

according to our condition $X^{N+1} = 1$. Note that the commutation relations for the differentials dX^k with coordinates X^i

$$X^i dX^i = r^2 (dX^i) X^i\,,$$
$$X^i dX^k = q_{ik}(dX^k)X^i + (r^2 - 1)(dX^i)X^k\,, \qquad i < k,$$
$$(dX^k)X^i = \frac{q_{ki}}{r^2} X^i dX^k\,, \qquad i > k\,,$$

reproduce the commutation relations (3.3.9)-(3.3.11) for the parameters of translations and coordinates. This corresponds to the usual interpretation of differentials as shifts of coordinates.

3.4 Twisted Poincaré group and geometry of q-deformed Minkowski space

Here we apply the general idea of inhomogeneous q-group construction developed in previous section for deformation of physically important Poincaré group [57]. To avoid q-group contractions, we use as a starting point the twistor approach (see, e.g. [189]) for the definition of Minkowski space and Poincaré group. This leads us to a very simple q-deformation, the QUE algebra deformation parameter r^2 being trivial, $r^2 = 1$. This means that we obtain twisted q-group and q-Lie algebra with diagonal R-matrix, i.e. triangular QUEA.

One more important problem that can be solved due to geometrical twisting procedure, concerns the construction of q-Poincaré group representations. As is well known, representations of classical Poincaré group are constructed through induction from its $SU(2) \otimes T_4$ subgroup (in massive case), where T_4 is the group of translations. But the point is that this q-subgroup (or Lie q-subalgebras) does not exist for the known q-deformed Poincaré groups (q-algebras) . In other words comultiplication for the generators of this subgroup contains other q-Poincaré generators (see next Section). This problem becomes even more serious in our twisted case: our q-group has not even $SU(2)_q$ subgroup. So this example is very appropriate to the studying of the problem.

We shall show that one has to use twisted q-subgroup for induced representation construction. Simultaneously, our approach solves another important problem - how to construct the representations in the Wigner basis. This in turn permits to understand the relation between q-deformation of the Poincaré group and q-deformation of algebra of creation and annihilation operators. We shall argue that q-wave functions play the role of q-beins in representation spaces, so that the creation and annihilation operators become the usual ones, satisfying CR of ordinary oscillator algebra.

3.4.1 Quantum deformation of the Poincaré group

Let us remind some basic facts about the Grassmann manifold $G(2,4)$ and its transformations by conformal and Poincaré group in classical case [189]. $G(2,4)$ is a set of 2-dimensional complex subspaces $\mathbb{C}^2$ of 4-dimensional space $\mathbb{C}^4$. The homogeneous coordinates of $G(2,4)$ form $2\otimes 4$ matrix Z with matrix elements $Z^j{}_\alpha(\alpha = 1,2; j = 1,...,4)$. Grassmann manifold is topologically nontrivial and can be covered by six patches (big cells) $U_{ij}(i < j;\ i,j = 1,...,4)$ defined by the conditions

$$Z \in U_{ij} \Longleftrightarrow \det Z_{(ij)} \neq 0.$$

The $2\otimes 2$ submatrix $Z_{(ij)}$ is formed by i and j rows of the matrix Z. On any big cell U_{ij} one can introduce inhomogeneous coordinates $z = Z_{(kl)}Z^{-1}_{(ij)}$ (here $k,l \neq i,j$). In fact, each U_{ij} is a complexified Minkowski space with coordinates $x^\mu \in \mathbb{C}$ defined by the relation $z = x^\mu\sigma_\mu$ (σ_μ are the Pauli matrices). If $z = z^+$, then $x^\mu \in \mathbb{R}$ are the coordinates of the usual Minkowski space $\mathcal{M}$. The action of the Poincaré group is deduced from the action of $SU(2,2)$ on $G(2,4)$:

$$Z \to Z' = TZ, \qquad T \in SU(2,2). \tag{3.4.1}$$

Let us choose the big cell U_{34} for definiteness, and write the matrix $T \in SU(2,2)$ in a block form,

$$T = \begin{pmatrix} A & B \\ C & D \end{pmatrix}, \tag{3.4.2}$$

where A,B,C,D are $2\otimes 2$ matrices and choose the hermitian metric Φ of the group $SU(2,2)$ in the form

$$\Phi = \begin{pmatrix} 0 & i\mathbf{I}_2 \\ -i\mathbf{I}_2 & 0 \end{pmatrix}. \tag{3.4.3}$$

Matrices T obey the unitarity condition

$$T^{+}\Phi T = \Phi \ . \tag{3.4.4}$$

The Poincaré subgroup of the $SU(2,2)$ is defined by

$$C = 0 \ ,$$
$$\det A = 1 \ . \tag{3.4.5}$$

Note that such simple and convenient form of conditions distinguishing Poincaré group from the conformal one, is possible due to the antidiagonal form of the metric Φ. In this case we have from (3.4.4) and (3.4.5)

$$B^{+}D = D^{+}B \ , \tag{3.4.6}$$

$$A^{+} = D^{-1} \ , \tag{3.4.7}$$

$$\det A = \det D = 1 \ . \tag{3.4.8}$$

The group action (3.4.1) on the inhomogeneous coordinates z takes the form

$$z \to z' = AzA^{+} + BA^{+}. \tag{3.4.9}$$

The hermitian matrices $z \in \mathcal{M}$ are transformed by (3.4.9) to hermitian ones because of conditions (3.4.6),(3.4.7) and the Minkowski length $ds^2 = \det(z_1 - z_2)$ is invariant. Thus (3.4.9) are completely equivalent to the usual Poincaré group of transformations.

To "quantize" this construction [57], we have to find, first of all, a real form $SU_{q_{ij},r}(2,2)$ of $SL_{q_{ij},r}(4,C)$, corresponding to the hermitian metric (3.4.3). In other words, we must find a multiparametric R-matrix consistent with the involution,

$$T^{*} = \Phi(T^{-1})^{t}\Phi, \tag{3.4.10}$$

followed from the unitarity condition (3.4.4). Here $(T^{-1})_{ij}$ is the antipode of T_{ij} defined in a usual way for $SL_{q_{ij},r}(4,C)$ group. Multiparametric R-matrix for $GL_{q_{ij},r}(4,C)$ has the form (1.5.32). As usual, for the $SU(m,n)$ case we put $q_{ij}, r^2 \in \mathbb{R}$ and thus from the relations which define T_{ij},

$$R_{ik,rs}T_{rv}T_{sw} = T_{kb}T_{ia}R_{ab,vw} \ , \tag{3.4.11}$$

we obtain the equations

$$R_{ik,rs}T^{*}_{sw}T^{*}_{rv} = T^{*}_{ia}T^{*}_{kb}R_{ab,vw},$$

which can be written, due to (3.4.10), in the form

$$\tilde{R}_{sr,ki}T_{rv}T_{sw} = T_{kb}T_{ia}\tilde{R}_{wv,ab}, \tag{3.4.12}$$

where

$$\tilde{R}_{sr,ki} = \Phi_{sa}\Phi_{rb}R_{ab,cd}\Phi_{ck}\Phi_{di}. \tag{3.4.13}$$

Comparison of (3.4.11) and (3.4.12) yields the identity

$$\tilde{R}_{sr,ki} = R_{ik,rs} \ ,$$

which in turn gives for the deformation parameters the conditions

$$r^2 = 1, \ q_{ij} = q_{\hat{j}\hat{i}} \qquad (\hat{i} = 5 - i, \hat{j} = 5 - j),$$

followed from the explicit form of the R-matrix. In our case of $GL_{q_{ij},r}(4, C)$ group, this means

$$r^2 = 1, \ q_{12} = q_{34}, \ q_{13} = q_{24} \ (q_{ij} \in \mathbb{R}). \tag{3.4.14}$$

We would like to stress that the condition $r^2 = 1$ is a consequence of the antidiagonal form of the metric Φ, which is very appropriate for picking up the Poincaré subgroup. In the case of other classically equivalent hermitian metrics, we would have other reality conditions for parameters of deformation. For example, for diagonal metric $\Phi'_{ij} = \epsilon_i\delta_{ij}$ ($\epsilon_i = 1$ for $i = 1, 2$; $\epsilon_i = -1$ for $i = 3, 4$) one has for real parameters

$$q_{ij}^{-1} = q_{ij}r^{-2},$$

so that $q_{ij} = r$, $\forall i, j$. This confirms once more the fact that "quantization" of the group removes its "degeneracy", i.e. classically equivalent groups become inequivalent after deformation.

To construct the q-Poincaré subgroup $\mathcal{P}_q$ of SU(2,2)$_q$, we have to require centrality of $\det_q A$ and $\det_q D = \det_q(A^+)^{-1}$ according to condition (3.4.8), i.e.

$$[T_{ij}, \det_q A] = [T_{ij}, \det_q D] = 0 \ , \tag{3.4.15}$$

where

$$\det_q A = A_a A_d - r^2 q_{12}^{-1} A_b A_c \ ,$$
$$\det_q D = D_a D_d - r^2 q_{34}^{-1} D_b D_c \ ,$$

and we have written T-matrix (3.4.2) with $C = 0$ in the form

$$T = \begin{pmatrix} A_a & A_b & B_a & B_b \\ A_c & A_d & B_c & B_d \\ 0 & 0 & D_a & D_b \\ 0 & 0 & D_c & D_d \end{pmatrix} .$$

Using commutation relations (3.4.11) for T_{ij} defined by the R-matrix, one has from (3.4.15) the conditions

$$q_{12} = q_{34} = 1, \quad r^2 = 1,$$

and equations

$$q_{14}q_{24} = 1, \quad q_{13}q_{14} = 1,$$
$$q_{13}q_{23} = 1, \quad q_{23}q_{24} = 1.$$

This leads to the equalities

$$q_{12} = q_{34} = 1, \quad q_{14} = q_{23} \equiv q, \quad q_{13} = q_{24} \equiv q^{-1}, \tag{3.4.16}$$

which do not contradict the real form conditions (3.4.14). Taking the deformation parameters according to (3.4.16), we can put $\det A = \det D = 1$ to obtain a consistent q-Poincaré subgroup $\mathcal{P}_q$ of $SU_q(2,2)$ with a structure of real Hopf algebra inherited from $SU_q(2,2)$.

We can further restrict our q-group to obtain Lorentz subgroup formed by the matrices

$$T = \begin{pmatrix} A & 0 \\ 0 & (A^+)^{-1} \end{pmatrix}. \tag{3.4.17}$$

Matrix elements of A, A^+ belong to *-Hopf subalgebra of $\mathcal{P}_q$, so matrices (3.4.17) define q-Lorentz subgroup of $\mathcal{P}_q$. A very unusual property of our deformation is the commutativity of elements of A with each other and of A^+ with each other. The only nontrivial commutation relations are those between elements of A and A^+. This is the simplest possible deformation of Lorentz group.

Constructed q-deformation contradicts the condition

$$A = (A^+)^{-1},$$

that picks up the $SO(3)_q$ rotation subgroup, since elements of A between themselves and with those of A^+ have different CR. So there is no q-rotation subgroup of q-Lorentz group in our case. The absence of this subgroup is not very important from the physical point of view because, as we hope, q-Poincaré symmetry has meaning at extremely high energy, so that $SO(3)_q$ subgroup cannot play essential physical role. Of course, this fact could have strong influence on q-Poincaré representation theory since the subgroup of rotations is a small group of fixed momentum for massive states in the $q = 1$ case and thus, is an important ingredient of induced representations of the Poincaré group. The $U(1)_q$ subgroup, however, exists. But representations of

Poincaré-like groups are induced from the subgroups of the form $K \otimes T_4$, where K is some subgroup of the Lorentz group. Unfortunately, $(U(1) \otimes T_4)_q$ is not subgroup of the $\mathcal{P}_q$. Indeed, conditions which pick up the subgroup include, e.g. $T_{11} = T_{33}$. But the matrix elements T_{11} and T_{33} have different CR with, for example, T_{23}. So we have no appropriate subgroup of $\mathcal{P}_q$ and the possibility of the induced representations is questionable. There is a similar problem in other deformations of the Poincaré group too (see next section). Below we shall suggest the solution of the problem at least in the simple twisted case considered here.

We would like to note that it is very natural to look for q-Poincaré group as a q-subgroup of the q-conformal one at least for two reasons: *i*) to start from $SU(2,2)$ is desirable from the physical point of view because of the well-known important role of the conformal group at high energy and small distances, where we hope that non-commutative geometry becomes essential (direct construction of the q-Poincaré group does not automatically imply that it is a q-subgroup of some quantum deformation of conformal group); *ii*) conformally invariant field theory has problems because of ultraviolet divergences, but the most attractive aim of q-deformation is to remove these divergences, and so to start from q-conformal symmetry is logically a self-consistent approach. Important mathematical background for the construction of the q-deformed conformal field theory has been developed in [81, 82].

3.4.2 Quantum Minkowski geometry

The commutation relations for homogeneous coordinates Z are the same as for the two last columns of the matrix T in (3.4.2). In this case they are invariant with respect to transformations. Moreover, due to the centrality of $\det Z_{(34)}$ (in analogy with $\det D$) we can put $\det Z_{(34)} = 1$ and define inhomogeneous coordinates

$$z = Z_{(12)} Z^{-1}_{(34)} \equiv \begin{pmatrix} z^1 & z^4 \\ z^2 & z^3 \end{pmatrix} .$$

Using once more the CR (3.4.11), we obtain

$$z^i z^j = Q_{ij} z^j z^i , \tag{3.4.18}$$

where

$$\begin{cases} Q_{ij} &= q^2 \quad \text{if} \quad j = i + 1 (mod\ 4) , \\ Q_{ij} &= 1 \quad \text{if} \quad j = i + 2 (mod\ 4) \quad \text{or} \quad i = j , \\ Q_{ji} &= Q_{ij}^{-1} . \end{cases}$$

To obtain the deformed Minkowski length l_q which is invariant under q-Poincaré transformations, we have to take into account that the matrix elements of submatrix A do not commute with those of A^+ and thus we cannot put $l_q = \det z$, though this is a central element of the algebra, but assume a more general expression $l_q = \det_{q^n} z \equiv z_1 z_4 - q^n z_2 z_3$, with some integer n (l_q is central for any n). Straightforward calculations show that if $n = 2$ then the determinant factorizes

$$\det_{q^2}(AzA^+) = (\det A)(\det_{q^2} z)(\det A^+) = \det_{q^2} z \ ,$$

and the deformed Minkowski length

$$l_q = \det_{q^2} z = g^q_{ij} z^i z^j \ ,$$

is invariant under homogeneous transformations due to the conditions $\det A = \det A^+ = 1$. Here g^q_{ij} is q-analog of classical $SO(2,2)$ metric g_{ij}:

$$g^q_{ij} = \begin{pmatrix} 0 & 0 & 1 & 0 \\ 0 & 0 & 0 & -q^2 \\ 1 & 0 & 0 & 0 \\ 0 & -q^2 & 0 & 0 \end{pmatrix} . \tag{3.4.19}$$

Note that we consider the quantum analog $\mathcal{M}_q$ of usual Minkowski space $\mathcal{M}$ but for the sake of convenience we use, instead of real coordinates, the coordinates with the reality conditions of the form

$$(z^1)^* = z^1, \quad (z^3)^* = z^3, \quad (z^2)^* = z^4, \quad (z^4)^* = z^2, \tag{3.4.20}$$

in contrast to the usual pure real ones $(x^\mu)^* = x^\mu$. One can straightforwardly check that the involution (3.4.20) is consistent with the defining relations (3.4.18). Then the inhomogeneous transformations have the form

$$z \to (z')^\alpha{}_\beta = (AzA^+)^\alpha{}_\beta + (BA^+)^\alpha{}_\beta \ , \tag{3.4.21}$$

and a direct inspection gives

$$[(AzA^+)^\alpha{}_\beta, (BA^+)^\alpha{}_\beta] = 0, \quad \forall \alpha, \beta = 1, 2. \tag{3.4.22}$$

From this relation we can deduce the affine structure of the quantum Minkowski space: the identity (3.4.22) implies that we can consider a set of coordinates $(z_t)^\alpha{}_\beta$, where t is a discrete or continuous index distinguishing different copies of q-space-time and put

$$z^i_t z^j_{t'} = Q_{ij} z^j_{t'} z^i_t \ , \qquad \forall\, t,\, t' \ . \tag{3.4.23}$$

Thus again as in Section 2.3.1 the CR for coordinates of different copies of quantum spaces do not depend on the indices which distinguish the copies. This time the guiding principle to choose such CR is the property (3.4.22) of the q-Poincaré group. Now we can define the invariant of the complete Poincaré group:

$$ds_q^2 = \det_{q^2}(z_t - z_{t'}). \tag{3.4.24}$$

The coordinates of q-Minkowski space $\mathcal{M}_q$ form the vector representation of $\mathcal{P}_q$. The spinorial representations of q-Lorentz subgroup can be derived rather easily in analogy with the usual ones (see e.g. [183]). A spinor ξ_α and its conjugate $\bar{\xi}^{\dot\alpha}$ ($\alpha = 1, 2$) are transformed by q-Lorentz submatrix according to

$$\begin{pmatrix} \xi' \\ \bar{\xi}' \end{pmatrix} = \begin{pmatrix} A & 0 \\ 0 & (A^+)^{-1} \end{pmatrix} \begin{pmatrix} \xi \\ \bar{\xi} \end{pmatrix} .$$

As we have already mentioned the matrix elements of A commute with each other and hence $\xi_1\xi_2 = \xi_2\xi_1$. The same is true for $(A^+)^{-1}$ and for the components of $\bar{\xi}$. So q-Lorentz invariants $\xi_\alpha\xi_\beta\epsilon^{\alpha\beta}$ and $\bar{\xi}^{\dot\alpha}\bar{\xi}^{\dot\beta}\epsilon_{\dot\alpha\dot\beta}$ can be constructed with the help of usual antisymmetric tensors $\epsilon_{\alpha\beta}$ and $\epsilon_{\dot\alpha\dot\beta}$. Thus lowest spinors do not differ from the undeformed ones. The q-relations between ξ and $\bar{\xi}$ are the same as those for the matrix elements of any column of matrix Z or T

$$\xi_\alpha\bar{\xi}^{\dot\beta} = q_{\alpha(\beta+2)}\bar{\xi}^{\dot\beta}\xi_\alpha \, , \tag{3.4.25}$$

or, using (3.4.16),

$$\begin{aligned} \xi_1\bar{\xi}^{\dot 1} &= q^{-1}\bar{\xi}^{\dot 1}\xi_1 \, , \quad & \xi_1\bar{\xi}^{\dot 2} &= q\bar{\xi}^{\dot 2}\xi_1 \, , \\ \xi_2\bar{\xi}^{\dot 1} &= q\bar{\xi}^{\dot 1}\xi_2 \, , \quad & \xi_2\bar{\xi}^{\dot 2} &= q^{-1}\bar{\xi}^{\dot 2}\xi_2 \, . \end{aligned}$$

We can generalize the relations (3.4.25) as in the case of coordinates, and introduce the relations for different spinors

$$\xi_\alpha\bar{\eta}^{\dot\beta} = q_{\alpha(\beta+2)}\bar{\eta}^{\dot\beta}\xi_\alpha \, ,$$

$$\xi_\alpha\eta_\beta = \eta_\beta\xi_\alpha \, , \qquad \bar{\xi}^{\dot\alpha}\bar{\eta}^{\dot\beta} = \bar{\eta}^{\dot\beta}\bar{\xi}^{\dot\alpha} \, .$$

Now the higher spin representations can be constructed by a method similar to the one in the classical case [183, 13]. Let us consider non-normalized symmetric spinors

$$F_{\sigma\dot\rho}(j_1, j_2) = \xi_1^{j_1+\sigma}\xi_2^{j_1-\sigma}\bar{\eta}_{\dot 1}^{j_2+\dot\rho}\bar{\eta}_{\dot 2}^{j_2-\dot\rho} \, , \tag{3.4.26}$$

where $-\infty < j_1, j_2 < \infty$, $\sigma \leq | j_1 |$, $\rho \leq | j_2 |$. They form a representation space V_{j_1,j_2} of q-Lorentz group of dimension $(2j_1+1)(2j_2+1)$. The transformations of these spinors $F(j_1,j_2)$ are defined by the transformations of spinors ξ and $\bar{\eta}$

$$\begin{aligned} F'_{\sigma\dot{\rho}}(j_1,j_2) &= (A\xi)_1^{j_1+\sigma}(A\xi)_2^{j_1-\sigma}(A^*\bar{\eta})_{\dot{1}}^{j_2+\dot{\rho}}(A^*\bar{\eta})_{\dot{2}}^{j_2-\dot{\rho}} \\ &\equiv \mathcal{D}(A)_{\sigma}{}^{\lambda}F_{\lambda\dot{\tau}}(j_1,j_2)\mathcal{D}((A^+)^{-1})^{\dot{\tau}}{}_{\dot{\rho}} , \end{aligned}$$

the representation matrices $\mathcal{D}(A)$, $\mathcal{D}((A^+))$ being the same as in the classical case due to commutativity of elements of matrix A with each other. This is similar to the homogeneous part of coordinates transformations (3.4.21). Note, however, that to obtain non-isotropic Minkowski vector, one has to use two different spinors (two-columns matrix Z) and their conjugates [189]. So coordinates z^i correspond to nonsymmetric spinors. Quantization leads to the deformation of invariant metric on the space $V_{j_1 j_2}$ in the same way as on the q-Minkowski space. Consider the spinors with upper indices

$$\Phi^{\sigma\dot{\rho}}(j_1,j_2) = (\zeta^1)^{j_1+\sigma}(\zeta^2)^{j_1-\sigma}(\bar{\theta}^{\dot{1}})^{j_2+\dot{\rho}}(\bar{\theta}^{\dot{2}})^{j_2-\dot{\rho}} .$$

Taking into account the CR between spinors and conjugate spinors, one can show that the invariant metric has the form

$$< F, \Phi > = \sum_{\sigma,\, \dot{\rho}} a^q_{\sigma\dot{\rho}}(j_1,j_2)F_{\sigma\dot{\rho}}(j_1,j_2)\Phi^{\sigma\dot{\rho}}(j_1,j_2) ,$$

where

$$a_q(j_1,j_2;\sigma,\dot{\rho}) = \frac{(2j_1)!(2j_2)!}{(j_1+\sigma)!(j_1-\sigma)!(j_2+\rho)!(j_2-\rho)!}q^{-4\sigma\rho} .$$

In the classical limit $q=1$, one obtains the usual normalization factor $\sqrt{a(j_1,j_2)}$ of symmetric spinors.

3.4.3 q-tetrades and transformation to commuting coordinates

The obtained q-Poincaré group [57] has a diagonal 16×16 R-matrix with the following explicit form

$$R = diag\{1,1,q,q^{-1},1,1,q^{-1},q,q^{-1},q,1,1,q,q^{-1},1,1\} . \qquad (3.4.27)$$

Such groups are called the pure twisted ones. According to Section 1.5.3, let us introduce a q-tetrade in the q-space-time $\mathcal{M}_q$, i.e. four q-vectors $e^n{}_\alpha$, so that

$$e^n{}_\alpha e^m{}_\beta = Q_{nm}e^m{}_\beta e^n{}_\alpha , \qquad (m,n,\alpha,\beta = 1,...,4). \qquad (3.4.28)$$

One can show that the q-determinant defined by

$$e \equiv \epsilon^q_{mnpk} e^m{}_1 e^n{}_2 e^p{}_3 e^k{}_4 = \epsilon^{\alpha\beta\gamma\delta} e^1{}_\alpha e^2{}_\beta e^3{}_\gamma e^4{}_\delta \tag{3.4.29}$$

is the q-Poincaré invariant and central element of the algebra. Here ϵ^q_{ijkl} is the q-deformed antisymmetric Levi-Cevita tensor: $\epsilon^q_{1234} = 1$ and $\epsilon^q_{ijkl} = -Q_{ji}\epsilon^q_{jikl} = -Q_{kj}\epsilon^q_{ikjl} = -Q_{lk}\epsilon^q_{ijlk}$. Now it is possible to introduce commuting coordinates $\tilde{z}^\alpha$ on $\mathcal{M}_q$ through the relations

$$z^m = e^m_\alpha \otimes \tilde{z}^\alpha , \tag{3.4.30}$$

or

$$\tilde{z}^\alpha = e^\alpha{}_m \otimes z^m , \tag{3.4.31}$$

where $e^\alpha{}_m$ are the elements of the inverse matrix of the q-tetrade defined with the help of q-determinant e (3.4.29),

$$e^\alpha{}_m = (e^{-1})\epsilon^q_{l\ldots rmp\ldots k} e^l{}_1 \ldots e^r_{a-1} e^p_{a+1} \ldots e^k{}_4 Q(l\ldots r, m) , \tag{3.4.32}$$

(generalization of the inverse matrix formula in usual case). It is easy to see that

$$e^\alpha{}_m e^\beta{}_n = Q_{mn} e^\beta{}_n e^\alpha{}_m , \tag{3.4.33}$$

and that $\tilde{z}^\alpha$ $(a = 1, \ldots, 4)$ are commuting coordinates.

In vector notations the q-Lorentz transformations have the form

$$z'^m = (\Lambda_q)^m_n \otimes z^n . \tag{3.4.34}$$

Substituting (3.4.30) in (3.4.34), one obtains

$$\tilde{z}'^a = \tilde{\Lambda}^a{}_b \otimes \tilde{z}^b ,$$

with

$$\tilde{\Lambda}^a{}_b = (e')^a{}_m \otimes (\Lambda_q)^m_n \otimes e^n_b .$$

One can check that the matrix $\tilde{\Lambda}$ has commuting elements. From the other hand, we can construct the Λ-matrix from commuting classical Lorentz matrix $\tilde{\Lambda}$

$$\Lambda^m_n = (e)^m_\alpha \otimes \tilde{\Lambda}^\alpha_\beta \otimes e^\beta{}_n .$$

3.4.4 Twisted Poincaré algebra and induced representations of the q-group

For the representations of the generators of q-translations it is natural to use q-derivatives. Thus first of all, we have to derive formulae for differential calculus on our Minkowski space [57]. Simple calculations give the result

$$\partial_i \partial_j = Q_{ij} \partial_j \partial_i \,, \tag{3.4.35}$$

$$dz^i dz^j = -Q_{ij} dz^j dz^i \,,$$

$$\partial_j z^i = \delta^i_j + Q_{ij} z^i \partial_j \tag{3.4.36}$$

To find q-Poincaré algebra, we can start from a general anzatz for the action of q-Lorentz generators L on Minkowski coordinates,

$$L z^i = \alpha_i z^i L + A^i_j z^j \,. \tag{3.4.37}$$

The first term in the rhs is diagonal because of the diagonal R-matrix. The unknown matrix elements A^i_j and α_i can be defined from the condition of metric invariance

$$L l_q = l_q L \,, \tag{3.4.38}$$

and from the definition of invariance

$$L(z^i z^j - Q_{ij} z^j z^i) = (z^i z^j - Q_{ij} z^j z^i) L \,. \tag{3.4.39}$$

One can show that these conditions lead to a unique solution for q-generators (3.4.37) (i.e. unique solution for α_i and A^i_j). Obviously, it would be desirable to find a form of q-Lie algebra CR analogous to the covariant tensorial form of nondeformed Lorentz Lie algebra ($SO(2,2)$ in our case):

$$[M^{mn}, M^{pk}] = g^{mk} M^{np} + g^{np} M^{mk} - g^{mp} M^{nk} - g^{nk} M^{mp} \,, \tag{3.4.40}$$

where g^{mk} is inverse to (3.4.19) at $q = 1$. This can be achieved most easily by constructing the spinless representation of the q-Lie algebra by q-differential operators, i.e. by the deformation of usual Killing vectors

$$M^{mn} = z^m \partial^n - z^n \partial^m, \quad \partial^m = g^{mn} \partial_n \,,$$

on commuting Minkowski space. For this aim, we introduce the two-index q-Lorentz generators $L^{mn} = -Q_{mn} L^{nm}$ and put

$$L^{mn} = z^m \partial^n - Q_{mn} z^n \partial^m, \quad \partial^m = g_q^{mn} \partial_n \,. \tag{3.4.41}$$

One can check that (3.4.41) satisfies (3.4.38),(3.4.39) and is in one-to-one correspondence with the unique solution of the anzatz (3.4.37). For the CR of L^{mn} one can deduce the desirable form:

$$[L^{mn}, L^{pk}]_{Q(mn,pk)} = g_q^{mk} Q_{mn} Q_{pk} L^{np} + g_q^{np} L^{mk} - g_q^{mp} Q_{mn} L^{nk} - g_q^{nk} Q_{pk} L^{mp} \,, \tag{3.4.42}$$

where we use the symbol

$$Q(ijk..., lmn...) \equiv Q_{il} Q_{im} Q_{in} ... Q_{jl} Q_{jm} Q_{jn} ... Q_{kl} Q_{km} Q_{kn} ... \,.$$

Here the factors in rhs contain all ordered pairs of indices from two sets in lhs divided by a comma. In particular, $Q_{mn,pk} \equiv Q_{mp} Q_{mk} Q_{np} Q_{nk}$.

Now we can add the generators of q-translations, i.e. q-derivatives, to complete the q-Poincaré Lie algebra:

$$[\partial^k, L^{mn}]_{Q(k,mn)} = g_q^{km} \partial^n - g_q^{km} Q_{mn} \partial^m \,. \tag{3.4.43}$$

Coproduct Δ for generators of q-Poincaré algebra can be read off from the action on monoms $m(A,B,C,D) \equiv (z^1)^A (z^3)^B (z^2)^C (z^4)^D$ (A, B, C, D are arbitrary integer numbers). For example, $\Delta(L^{23}) = L^{23} \otimes 1 + q^{2(q^2 L^{24} - L^{13})} \otimes L^{23}$; $\Delta(\partial_1) = \partial_1 \otimes 1 + q^{2q^2 L^{24}} \otimes \partial_1$. Counity ϵ is defined, as usual, $\epsilon(L^{mn}) = \epsilon(\partial_k) = 0$, and antipode S is defined from the property (1.2.9).

The introduction of the q-tetrade permits to establish the relation of our twisted algebra with classical one very easily. Indeed, using differential operator representation (3.4.41), one has

$$L^{mn} = e_a{}^m z^a e_b{}^n \partial^b - Q_{mn} e_b{}^n z^b e_a{}^m \partial^a = e_a{}^m e_b{}^n (z^a \partial^b - z^b \partial^a) = e_a{}^m e_b{}^n M^{ab} \tag{3.4.44}$$

where we use

$$\partial^m = e_a{}^m \partial^a \,.$$

Obviously one can define the inverse transformation

$$M^{ab} = e_n{}^b e_m{}^a L^{mn} \,.$$

Now all the properties of the q-algebra including commutators, Casimir operators, etc. can be derived from the well-known properties of the usual Poincaré algebra of (M^{ab}, ∂^c). In particular, q-deformed Casimir operators have the form

$$C_M^q = g_q^{mn} \partial_m \partial_n = \partial_1 \partial_3 - q^{-2} \partial_2 \partial_4 \,, \tag{3.4.45}$$

$$C_{PWL}^q = g_q^{mn} \omega_m^q \omega_n^q \equiv \omega_q^2 \,, \tag{3.4.46}$$

where

$$\omega_i^q = \frac{1}{4}\epsilon_{ijkl}^q L^{jk}\partial^l \,, \tag{3.4.47}$$

are components of the deformed Pauli-Wigner-Lubanski vector.

Now we turn to the representation theory. Two obvious and most serious problems, as we mentioned already, are the following.

First of all, induced representations are constructed in the space of the functions on the group (more precisely, on the coset space) [13]. For q-groups this means that in general one must use the functions of non-commuting variables. Another problem is that in the classical case the Poincaré group representations are induced from those of the appropriate subgroup ($SU(2)\otimes \mathcal{T}_4$ in massive case and $E(2)\otimes \mathcal{T}_4$ in massless case). But our $\mathcal{P}_q$ has neither $(SU(2)\otimes \mathcal{T}_4)_q$, nor $SU(2)_q$, nor even $(U(1)\otimes \mathcal{T}_4)_q$ subgroups. We stress once more that similar situation occurs in other deformations of the Poincaré-like groups which we shall consider in the next Section. Nonunitary representations of the q-Poincaré transformations in the space of functions $\psi(z)$ of noncommuting coordinates (i.e. fields from the physical point of view) can be obtained rather easily. Let $\psi_{\sigma\dot\rho} \in V_{j_1 j_2}$, and thus it is transformed by q-Lorentz group as multispinor (3.4.26). Then q-Poincaré transformations can be realized on the maps $\psi : \mathcal{M}_q \longrightarrow V_{j_1 j_2}$, in a way quite analogous to the classical case (cf. [13])

$$\psi'_{\alpha\dot\rho}(z) = \mathcal{D}(A)_\sigma{}^\lambda \psi_{\lambda\dot\tau}(z'')\mathcal{D}((A^+)^{-1})^{\dot\tau}{}_{\dot\rho} \,, \tag{3.4.48}$$

where z'' is the coordinate matrix transformed by the antipode A^{-1}

$$z'' = A^{-1}z(A^+)^{-1} - A^{-1}B \,.$$

But due to the existence of the q-tetrade, we can use the commuting coordinate $\tilde z$ as the fields arguments. In our pure twisted case the full differential calculus becomes commutative after the transition to commutative coordinates, so that we can easily make the next step and introduce Fourier transformation of the fields $\psi(\tilde z)$. Let us consider the momentum space with coordinates

$$p^i = e^i{}_\alpha \tilde p^\alpha \,,$$

where in general the components of momentum p^i have the CR of the q-derivatives ∂^i. Then the Fourier transformation can be defined as usual

$$\psi(\tilde p) = (2\pi)^{-3/2}\int d\tilde z\; e^{i\tilde p\tilde z}\psi(\tilde z) \,.$$

At this point our pure twisted case is essentially different from the general one. In the general case, one cannot make the coordinates and the momentum components (derivatives) simultaneously commutative. So that the construction of the integral calculus and Fourier analysis are complicated problems. But an appropriate choice of the coordinates within equivalence class of twisted frames can help to solve the problems at least in two respects: *i)* it is possible to choose the system with commuting derivatives and hence to construct the states with definite value of momentum components; *ii)* some coordinate systems can be more convenient for construction of the integral calculus and Fourier transformations than the standard ones.

As usual we consider eigenstates of the Casimir operator C_M, so that $\psi(\tilde{p})$ actually depends on only three independent momentum components, $\psi = \psi(\tilde{\mathbf{p}})$. These Fourier transformed fields are in one-to-one correspondence with the so-called spinor basis of the Poincaré group representations [13]. Considering, for example, the negative frequency part of $\psi(\tilde{p})$, one can divide it into the wave function and creation operators

$$\psi^M(\tilde{\mathbf{p}}) = v^M_\alpha(\tilde{\mathbf{p}}) \otimes (a^+)^\alpha(\tilde{\mathbf{p}})\,. \tag{3.4.49}$$

This relation is quite analogous to (3.4.30), the wave function $v^M_\alpha(\tilde{\mathbf{p}})$ playing the role of the q-bein in the space of the representation. So the CR for creation and annihilation operators can be completely different from those for the components of the field $\psi^M(\tilde{\mathbf{p}})$. This in turn means that q-deformation of the small subgroup, transforming creation and annihilation operators and particle state basis, is different from the whole $\mathcal{P}_q$. In particular, in our case taking the appropriate CR for wave functions we can choose $a^{\dagger\,\alpha}$, a^α so that they are transformed by the usual nondeformed $SU(2)$-group. The $\mathcal{P}_q$ transformations for $\psi(\tilde{\mathbf{p}})$ has the form

$$\psi'^{\,M}(\tilde{\mathbf{p}}) = e^{i\tilde{p}'\tilde{t}} D^M_N \otimes \psi^N(\tilde{\mathbf{p}}')\,, \quad \tilde{p}' = \tilde{\Lambda}\tilde{p}\,.$$

Here $\tilde{t}^\alpha$ are the commuting parameters of translations. Working analogously to the case of coordinates, we have for the transformations

$$v'^{\,M}_{\ \alpha}(\tilde{\mathbf{p}}) \otimes (a^+)'^{\,\alpha}(\tilde{\mathbf{p}}) = e^{i\tilde{p}'\tilde{t}} D^M_N(\Lambda) \otimes v^N_\beta(\tilde{\mathbf{p}}') \otimes (a^\dagger)^\beta(\tilde{\mathbf{p}}')\,.$$

Assuming the orthogonality relation

$$v^M_{\ \alpha} G_{MN} v^N_{\ \beta} = \delta_{\alpha\beta}\,,$$

one obtains

$$\begin{aligned}(a^+)'^{\,\alpha}(\tilde{\mathbf{p}}) &= e^{i\tilde{p}'\tilde{t}} \left[v'^{\,M\alpha}(\tilde{\mathbf{p}}) \otimes G_{MN} D^M_N(\Lambda) \otimes v^N_\beta(\tilde{\mathbf{p}}') \right] \otimes a^{\dagger\,\beta}(\tilde{\mathbf{p}}') \\ &\equiv e^{i\tilde{p}'\tilde{t}} \mathcal{D}^\alpha_\beta(\Lambda) \otimes (a^+)^\beta(\tilde{\mathbf{p}}')\,,\end{aligned}$$

and analogous transformations for the annihilation operators a^α.

We illustrate the above described general scheme with the example of vector massive representation. In this case

$$\psi^m(\tilde{\mathbf{p}}) = v^m_\alpha(\tilde{\mathbf{p}})a^\alpha(\tilde{\mathbf{p}}) \qquad (\alpha = -1, 0, 1) .$$

The wave function v^m_α is constructed from the corresponding classical wave function $u^a{}_\alpha(\tilde{\mathbf{p}})$ [13]

$$u^a{}_\alpha(\tilde{\mathbf{p}}) = u^a{}_\alpha(0) - \frac{\tilde{p}^a + g^{0a}m}{\tilde{p}^0 + m} \cdot \frac{\tilde{p}_b u^b{}_\alpha(0)}{m} ,$$

where

$$u^a{}_1(0) = \frac{1}{\sqrt{2}}(0, 1, i, 0) , \quad u^a{}_0(0) = (0, 0, 0, -1) , \quad u^a{}_{-1}(0) = \frac{1}{\sqrt{2}}(0, -1, i, 0) ,$$

so that

$$v^m_\alpha = e^m{}_a u^a{}_\alpha . \tag{3.4.50}$$

Hence, v^m_α satisfies the usual Proca-type wave equation

$$p_m v^m = \tilde{p}_a u^a = 0 .$$

From the normalization condition for the classical wave function

$$g_{ab} u^{\dagger\, a}{}_\alpha(\tilde{\mathbf{p}}) u^b{}_\beta(\tilde{\mathbf{p}}) = -\delta_{\alpha\beta} ,$$

one has an analogous condition for v^m_α

$$g^q_{ab} v^{\dagger\, a}{}_\alpha(\tilde{\mathbf{p}}) v^b{}_\beta(\tilde{\mathbf{p}}) = -\delta_{\alpha\beta} .$$

Thus from the q-Poincaré transformations of the field ψ^m

$$\psi'^{\, m}(p) = e^{ip'^t} \Lambda^m{}_n \psi^n(p') ,$$

we have the following transformations for the creation operator

$$(a^+)'^{\,\alpha}(\tilde{\mathbf{p}}) = e^{i\tilde{p}'\tilde{t}} \left[v'^{\, m\alpha}(\tilde{\mathbf{p}}) \otimes g^q_{ml} \Lambda^l{}_n \otimes v^n{}_\beta(\tilde{\mathbf{p}}') \right] \otimes (a^+)^\beta(\tilde{\mathbf{p}}') , \tag{3.4.51}$$

and hence, the corresponding unitary transformations for one particle states. The CR for the wave functions $v^n{}_\alpha$ are defined by (3.4.50). The CR for field components ψ^m are defined in general by its spinor content (3.4.26) and in the considered case coincide with those for the wave functions. This leads to usual (non-deformed) CR for operators a^α, $a^{\dagger\,\beta}$ and $SU(2)$ transformations (3.4.51) with commuting matrix elements. So the small subgroup which defines the particle states transformations, is twisted with respect to the initial q-Lorentz group. This opens the way for construction of the particle states representations in the Poincaré-like groups in the case of physically nontrivial deformations.

3.4.5 Twisted Minkowski space in the case of related q and $\hbar$ constants

As we discussed in Section 2.3, the physical meaning of a deformation crucially depends on the relation between q-parameter and Planck constant. Above, in the consideration of the q-Minkowski space we assumed that q is independent of $\hbar$. According to Section 2.3 there is another possibility: $q = q(\hbar)$, e.g. $q(\hbar) = e^{i\gamma\hbar}$. In this case the notion of q-bein becomes more involved: as they have now CR containing $\hbar$, we have to consider them as dynamical variables. This implies development of gravitational theory in q-space. This is an interesting possibility which has still to be explored. To avoid the consideration of q-gravitational theory and as a preliminary step to understand the physical meaning of the q-space with $q = q(\hbar)$, one can consider the quasi-classical limit $\hbar \to 0$ directly in q-commuting coordinate frame (i.e. satisfying the CR (3.4.18)) [71, 72].

The limit $q = e^{\gamma\hbar} \to 1$, $\hbar \to 0$, is governed by the correspondence (1.1.30),(2.3.36), provided that the R-matrix is normalized as

$$R = 1 + \gamma\hbar r + \mathcal{O}(\hbar^2) \quad , \tag{3.4.52}$$

r being their quasi-classical counterparts (do not confuse with r-parameter of multiparametric deformation!). The Poisson bracket (PB) of x^i and p_i (given by the limit of the commutation rules for coordinates and derivatives) which now are c-numbers, become noncanonical, the departure from the canonical ones being governed by the new constant γ (cf. (2.3.15)). For the case of $\mathcal{M}_q$, which we consider in this section, $r^{(1)} = r^{(4)} = 0$ and $r^{(2)} = r^{(3)} = r = diag(1, 0, 0, 1)$.

The PB for coordinates z and momenta p (which transforms as the matrix of q-derivatives) are obtained from the quasi-classical limits of (3.4.18) and (3.4.36). In this way, for $q = 1$ the latter reduces to $[x^i, p_j] = i\hbar\delta^i_j$. The quasi-classical limits now give

$$\begin{aligned} \{z_1, z_2\}_p &= i\gamma[z_1 r z_1, \sigma] \ , \\ \{p_1, p_2\}_p &= -i\gamma[p_1 r p_1, \sigma] \ , \\ \{p_1, z_2\}_p &= i\gamma[p_1 z_2, r] - \sigma \ . \end{aligned} \tag{3.4.53}$$

Here indices have the same meaning as in RTT-relation (cf., e.g. Section 1.1) i.e. z_1 acts in $\mathbb{C}^2 \otimes \mathbb{C}^2$ as z in the first copy of $\mathbb{C}^2$ and unity in the second one etc., σ is the permutation matrix in $\mathbb{C}^2 \otimes \mathbb{C}^2$. These PB are invariant under the usual Lorentz transformations provided that the entries of A and $A^\dagger$ have zero Poisson brackets with those of z and p (which has to be the case since they already commuted in the q-case) and the elements of the usual Lorentz group have nontrivial PB which is the quasi-classical limit of (3.4.11)

$$\{A_1, A_2^\dagger\}_p = i\gamma(A_1 r^{(2)} A_2^\dagger - A_2^\dagger r^{(2)} A_1) \quad . \tag{3.4.54}$$

(so one obtains a Lie-Poisson group rather than a Lie group together with its homogeneous Poisson space).

To discuss the Hamiltonian dynamics of a particle, one can use the mass-shell constraint $\varphi = p^2 - m^2 = (p_1 p_3 - q^{-2} p_2 p_4) - m^2$ as usual, and introduce the gauge $\varphi' = (z^1 - \tau)$ to eliminate the unwanted degree of freedom and separate the evolution parameter (for gauge or "parametric" invariant consideration of relativistic particle see, e.g. [112]). This gauge means that light-cone-like variables are used as suggested by the real elements of Z. In this picture, the 2×2 PB matrix C of the constraint and the gauge is specially simple, $C = \{\varphi, \varphi'\}_p(i\sigma_2) = -ip_3\sigma_2$. The new (Dirac) brackets are obtained from

$$\{A, B\}_D = \{A, B\}_p - (\{A, \varphi\}_p, \{A, \varphi'\}_p) C^{-1} (\{\varphi, B\}_p, \{\varphi', B\}_p)^\top , \tag{3.4.55}$$

for any observables A, B. The momentum component $p_1 = (m^2 + |p_2|^2)/p_3$ is considered as the energy in the quasi-classical limit. Then, from the standard Hamiltonian equation

$$\dot{A} = \{H, A\}_p ,$$

one obtains, with p_1 as Hamiltonian, the expression for components v^i of particle velocity

$$\begin{pmatrix} 0 & v^4 \\ v^2 & v^3 \end{pmatrix} = \{p_1, Z\}_D = \begin{pmatrix} 0 & -p_2/p_3 - i\gamma p_1 z^4 \\ -p_4/p_3 + i\gamma p_1 z^2 & p_1/p_3 \end{pmatrix} \tag{3.4.56}$$

Thus, even in this simple case, one finds that for $\gamma \neq 0$ the velocities $v^{2,4}$ depend on the coordinates. Evolution of the momenta is defined by the equations

$$\dot{P} = \begin{pmatrix} \dot{p}_1 & \dot{p}_2 \\ \dot{p}_4 & \dot{p}_3 \end{pmatrix} = \{p_1, P\}_p = i\gamma \begin{pmatrix} 0 & -p_1 p_2 \\ p_1 p_4 & 0 \end{pmatrix} . \tag{3.4.57}$$

Hence,

$$\begin{gathered} p_1(\tau) = \text{const} \quad , \quad p_3(\tau) = \text{const} \; , \\ p_2(\tau) = \exp(-i\gamma p_1 \tau) p_2(0) = (p_4(\tau))^* , \end{gathered} \tag{3.4.58}$$

($p_2 = (p_4)^*$ is not conserved) and

$$\begin{aligned} z^3(\tau) &= \frac{p_1}{p_3}\tau + z^3(0) , \\ z^2(\tau) &= (z^2(0) - \frac{p_4(0)}{p_3}\tau)\exp(i\gamma p_1\tau) = (z^4(\tau))^* . \end{aligned} \tag{3.4.59}$$

Although the limit $\gamma = 0$ reproduces the standard constant momenta and linear evolution of coordinates with respect to the parameter τ, the γ-deformed behaviour is, even in this simple case, strongly oscillating in the (x, y) plane and may relate points separated by space-like intervals.

In general, one could expect a nontrivial behaviour of the dynamical variables as the general consideration in Section 2.3 shows that the $q = q(\hbar)$ case corresponds to systems with curved phase spaces and hence to nontrivial dynamics.

3.5 Jordan-Schwinger construction for q-algebras of space-time symmetries and contraction of quantum groups

As we saw in Section 2.4, the use of the q-analog of the harmonic oscillators in the Jordan-Schwinger (JS) construction is very useful in building up the q-deformation of Lie algebras. In this section we will introduce this method for the Lorentz and anti-de Sitter space-time symmetries [52, 45]. The former construction leads to QUEA of quantum Lorentz group which in a sense is opposite to those considered in the preceding section. Recall that there we considered the group constructed from two copies of $SL(2,\mathbb{C})$ group with non-trivial (q-deformed) CR only between elements which belong to *different* copies. The q-Lorentz algebra in the first part of this section consists of two standard $SU_q(2)$ QUEA commuting with each other. The second construction (q-anti-de Sitter QUEA) is interesting because its special contraction (rather its q-deformed version [163, 166]) leads to another example of q-Poincaré algebra. In the last part of this section we will describe dual pattern of the q-contraction for an algebra of function on the anti-de Sitter quantum group. As we noted already, projective space approach discussed in Section 3.3 cannot be applied directly to (pseudo)orthogonal groups and (pseudo)Euclidian spaces. To circumvent the problem, we used in the preceding section the twistor formalism for Minkowski space and the embedding of the q-Poincaré in the q-conformal group. Quantum version of a group contraction [253] gives another method of derivation of inhomogeneous q-deformed (pseudo)Eucledian groups from higher dimensional orthogonal ones.

3.5.1 Fock space representation of the q-Lorentz algebra

We start with short reminding of the JS construction for classical (non-deformed) Lorentz group. Choosing two commuting sets of $SU(2)$ generators

$$\left[J_i^{(m)}, J_j^{(p)}\right] = i\delta^{mp}\varepsilon_{ijk}J_k^{(p)}, \tag{3.5.1}$$

where $p, m = 1, 2$ enumerate two copies of $SU(2)$ and $i, j, k = 1, 2, 3$, we construct the generators

$$\begin{aligned} M_i &= J_i^{(1)} + J_i^{(2)} && \text{(rotations)}, \\ L_i &= -i(J_i^{(1)} - J_i^{(2)}) && \text{(boosts)}, \end{aligned} \tag{3.5.2}$$

which satisfy CR

$$\begin{aligned} [L_i, L_j] &= -i\varepsilon_{ijk}M_k , \\ [M_i, M_j] &= i\varepsilon_{ijk}M_k , \\ [M_i, L_j] &= i\varepsilon_{ijk}L_k . \end{aligned} \tag{3.5.3}$$

These clearly generate $so(1,3)$ Lorentz group. So combining this with the standard JS map for $SU(2)$, we can write the Lorentz group generators in terms of the bosonic oscillator operators as follows,

$$\begin{aligned} &M_+ = a_0^+ b_0 + c_0^+ d_0 , && L_+ = i(c_0^+ d_0 - a_0^+ b_0) , \\ &M_- = a_0 b_0^+ + d_0^+ c_0, && L_- = i(d_0^+ c_0 - b_0^+ a_0), \\ &M_3 = \tfrac{1}{2}(N_a - N_b + N_c - N_d) , && L_3 = \tfrac{i}{2}(N_c - N_d - N_a - N_b) , \end{aligned} \tag{3.5.4}$$

where $(a_0^+, a_0), (b_0^+, b_0), (c_0^+, c_0), (d_0^+, d_0)$ are four pairs of independent (usual, nondeformed) oscillator operators, $N_a = a_0^+ a_0$, $N_b = b_0^+ b_0$, etc. Clearly

$$M_{\pm,3}^\dagger = M_{\pm,3} , \qquad L_{\pm,3}^\dagger = -L_{\pm,3} ,$$

with

$$M_\pm := \frac{1}{\sqrt{2}}(M_1 \pm M_2) , \qquad L_\pm := \frac{i}{\sqrt{2}}(L_1 \pm L_2) .$$

The fact that not all the generators are hermitian is expected since the realization of the Lorentz algebra described above will lead to its *finite* representations and $so(1,3)$ is not compact and so the representation cannot be a unitary one. Using the commutation relations of the oscillator operators $\left[a_0, a_0^+\right] = 1$, $[a_0, a_0] = \left[a_0^+, a_0^+\right] = 0$ etc., it is easy to check that the generators (3.5.4) satisfy the commutation relations of $so(1,3)$

$$\begin{aligned}
[M_+, M_-] &= 2M_3 , & [M_3, M_\pm] &= \pm M_\pm , \\
[L_+, L_-] &= -2M_3 , & [L_3, L_\pm] &= \mp M_\pm , \\
[M_+, L_-] &= 2L_3 , & [M_-, L_+] &= -2L_3 , \\
[M_3, L_\pm] &= \pm L_\pm , & [L_3, M_\pm] &= \pm L_\pm .
\end{aligned} \tag{3.5.5}$$

Using the CR (1.1.36) for $su(2)_q$ and the relations of the type (3.5.2) with obvious substitution $\{J_1^{(m)}, J_2^{(m)}, J_3^{(m)}\} \longrightarrow \{X_{(m)}^\pm, H_{(m)}\}$, one finds CR for $so_q(1,3)$ [52]

$$\begin{aligned}
[M_+, M_-] &= [M_3 + iL_3]_q + [M_3 - iL_3]_q , \\
[M_3, M_\pm] &= \pm M_\pm , \\
[L_+, L_-] &= -([M_3 + iL_3]_q + [M_3 - iL_3]_q) , \\
[L_3, L_\pm] &= \mp M_\pm , \\
[M_+, L_-] &= -i([M_3 + iL_3]_q - [M_3 - iL_3]_q) , \\
[M_-, L_+] &= i([M_3 + iL_3]_q - [M_3 - iL_3]_q) , \\
[M_3, L_\pm] &= \pm L_\pm , \\
[L_3, M_\pm] &= \pm L_\pm .
\end{aligned} \tag{3.5.6}$$

Notice that the consistency of the hermiticity properties and (3.5.6) (for instance, $M_+^\dagger = M_-$ and the commutator $[M_+, M_-]$) requires that both copies of $su_q(2)$ algebras be deformed using the same q.

The Hopf algebra structure of $so_q(1,3)$ is induced by the Hopf algebra structure of the two chiral $su_q(2)$'s and using the relation (3.5.2) between chiral and q-Lorentz generators one can easily derive comultiplication rules for the latter. We leave this as an exercise for the reader. The commutativity

$$\Delta(J_i^{(1)})\Delta(J_j^{(2)}) = \Delta(J_j^{(2)})\Delta(J_i^{(1)}) ,$$

guarantees that Δ is the $so_q(1,3)$ algebra homomorphism. The antipodes are also constructed from those for $su_q^{(1),(2)}(2)$.

The finite representations of $so_q(1,3)$ may be obtained following a procedure similar to the one used for derivation of the finite representations of Lorentz algebra. This takes advantage of the fact that the representations of $su_q(2)$ are already known.

Thus, proceeding as in the $q = 1$ case, the Fock space states which define the basis of the finite representations of $so_q(1,3)$ are given by

$$\begin{aligned} |j,j',m',m\rangle &= \left([j'+m']_q!\,[j'-m']_q!\right)^{-1/2} \\ &\quad \times (a^+)^{j+m}(b^+)^{j-m}(c^+)^{j'+m'}(d^+)^{j'-m'}\,|0\rangle\,, \qquad (3.5.7) \\ j,j' = 0,1/2,1,\ldots\,; &\qquad m = -j,-j+1,\ldots,j\,,\quad m' = -j',-j'+1,\ldots,j'\,. \end{aligned}$$

Then the action of the q-deformed generators on (3.5.7) is given by

$$\begin{aligned} M_\pm|j,j',m,m'\rangle &= \sqrt{[j\mp m]_q\,[j\pm m+1]_q}\;|j,j',m\pm 1,m'\rangle \\ &+ \sqrt{[j'\mp m']_q\,[j'\pm m'+1]_q}\;|j,j',m,m'\pm 1\rangle\,, \\ L_\pm|j,j',m,m'\rangle &= -i\sqrt{[j\mp m]_q\,[j\pm m+1]_q}\;|j,j',m\pm 1,m'\rangle \\ &+ i\sqrt{[j'\mp m']_q\,[j'\pm m'+1]_q}\;|j,j',m,m'\pm 1\rangle\,, \qquad (3.5.8) \end{aligned}$$

$$\begin{aligned} M_3|j,j',m,m'\rangle &= (m+m')|j,j',m,m'\rangle\,, \\ L_3|j,j',m,m'\rangle &= -i(m-m')|j,j',m,m'\rangle\,. \end{aligned}$$

Eqs. (3.5.8) define the irreducible finite representations of $so_q(1,3)$. As in the $q = 1$ case, they are characterized by two numbers (j, j'), and their dimension is $(2j+1)(2j'+1)$.

It is easy to find the two Casimir operators of this q-Lorentz algebra in such a way that they lead to the familiar expressions $\vec{M}^2 - \vec{L}^2$ and $\vec{M}\vec{L}$ in the classical limit. For this purpose, it is sufficient to take $C_q^1 = 2(C_{q(1)} + C_{q(2)})$, $C_q^2 = -i(C_{q(1)} - C_{q(2)})$, where $C_q(1)$, $C_{q(2)}$ are the Casimir operators (1.1.39) of the two quantum $su_q(2)$ algebras.

3.5.2 q-Deformed anti-de Sitter algebra and its contraction

The classical ($q = 1$) anti-de Sitter algebra is defined by

$$[M_{ab}, M_{cd}] = i(g_{ab}M_{cd} + g_{bc}M_{ad} - g_{ac}M_{bd} - g_{bd}M_{ac})\,,$$

where $g_{ab} = diag\{+1,-1,-1,-1,+1\}$. Let us introduce for convenience the generators

$$\begin{aligned} M_i &:= \frac{1}{2}\varepsilon_{ijk}M_{jk}\ , \quad i,j,k=1,2,3\ , \\ L_i &:= M_{i4}\ , \\ P_\mu &:= M_{0\mu}\ , \quad \mu=1,2,3,4\ . \end{aligned} \tag{3.5.9}$$

These generators can be expressed in terms of *two* pairs $(a_0^+, a_0), (b_0^+, b_0)$ of usual oscillator operators

$$\begin{aligned} \vec{L} &= -\tfrac{i}{4}(\xi^\dagger\vec{\sigma}\epsilon\xi^\dagger + \xi^\dagger\epsilon\vec{\sigma}\xi)\ , & \vec{M} &= \tfrac{1}{2}\xi^\dagger\vec{\sigma}\xi\ , \\ \vec{P} &= \tfrac{1}{4}(\xi^\dagger\vec{\sigma}\epsilon\xi^\dagger - \xi^\dagger\epsilon\vec{\sigma}\xi)\ , & \vec{P}_4 &= \tfrac{1}{2}(\xi^\dagger\xi + 1)\ , \\ \xi &:= \begin{pmatrix} a_0 \\ b_0 \end{pmatrix}\ , & \xi^\dagger &:= (a_0^+, b_0^+)\ , \end{aligned}$$

where σ^i are the Pauli matricies and $\epsilon := i\sigma^2$. To quantize the algebra, it is convenient to use the Chevalley basis of $so(3,2)$ and apply the Drinfel'd-Jimbo prescription (cf. Section 1.4). In this basis $so_q(5,\mathbb{C})$ is defined by the relations (1.3.13),(1.3.17),(1.3.18) and (1.3.21) with $i,j=1,2$ and the Cartan matrix

$$a_{ij} = \begin{pmatrix} 2 & -2 \\ -1 & 2 \end{pmatrix}\ .$$

Cartan-Weyl basis has in addition

$$\begin{aligned} X_3^+ &:= X_1^+X_2^+ - qX_2^+X_1^+\ , \\ X_4^+ &:= [X_1^+, X_2^+]\ , \\ X_3^- &:= X_2^-X_1^- - q^{-1}X_1^-X_2^-\ , \\ X_4^- &:= [X_3^-, X_1^-]\ . \end{aligned}$$

Consider the case $q \in \mathbb{C}$, $|q| = 1$. Then the real form $so_q(3,2)$ is defined by the involution (denoted by hermitian conjugation sign)

$$\begin{aligned} H_i^\dagger &= H_i\ , \\ (X_1^+)^\dagger &= X_1^-\ , \\ (X_i^+)^\dagger &= -X_i^-\ , \quad i=2,3,4\ . \end{aligned} \tag{3.5.10}$$

These generators can be expressed in terms of q-oscillators [45]

$$\begin{aligned} H_1 &= \tfrac{1}{2}(N_a - N_b)\,, & H_2 &= N_b + \tfrac{1}{2}\,, \\ X_1^+ &= \frac{1}{\sqrt{[2]_q}} a_{(1)}^+ a_{(2)}\,, & X_2^+ &= \frac{i}{[2]_q}(a_{(2)}^+)^2\,, \\ X_3^+ &= \frac{i}{\sqrt{[2]_q}} a_{(1)}^+ a_{(2)}^+ q^{-N_b}\,, & X_4^+ &= \frac{i}{[2]_q}(a_{(1)}^+)^2 q^{-2N_b}\,. \end{aligned} \tag{3.5.11}$$

(other generators are defined by (3.5.10)). Using the expressions for anti-de Sitter generators (3.5.9) in terms of the generators in Cartan-Chevalley basis

$$\begin{aligned} M_1 &= \frac{1}{\sqrt{2}}(X_1^+ + X_1^-), & L_1 &= \tfrac{1}{2}(X_4^+ - X_4^- - X_2^+ + X_2^-)\,, \\ M_2 &= \frac{-i}{\sqrt{2}}(X_1^+ - X_1^-), & L_2 &= \frac{-i}{2}(X_4^+ + X_4^- + X_2^+ + X_2^-)\,, \\ M_3 &= H_1, & L_3 &= \frac{-1}{2}(X_3^+ - X_3^-)\,, \\ P_1 &= \tfrac{i}{2}(X_4^+ + X_4^- - X_2^+ - X_2^-), & P_2 &= \tfrac{1}{2}(X_4^+ - X_4^- + X_2^+ - X_2^-)\,, \\ P_3 &= \frac{-i}{\sqrt{2}}(X_3^+ + X_3^-), & P_4 &= H_1 + H_2\,, \end{aligned} \tag{3.5.12}$$

one finally obtains the realization of anti-de Sitter algebra in terms of q-oscillators. Using CR for q-oscillators (or directly CR for $so_q(3,2)$ in Cartan-Weyl basis), one finds CR for the anti-de Sitter generators [52]:

a) Rotations:

$$[M_+, M_-] = 2\,[M_3]_{q^2}\;, \quad [M_3, M_\pm] = \pm M_\pm\;; \tag{3.5.13}$$

b) Boosts:

$$\begin{aligned} [L_3, L_+] &= -\frac{1}{2}q^{-2P_4}(1 + q^{2M_3})M_+ \\ &- \frac{1}{4}(1 - q^2)(K_-(N_- - L_+) - (N_- + L_+)K_+)\,, \\ [L_-, L_+] &= [M_3 + P_4]_{q^4} + [M_3 - P_4]_{q^4} \\ &+ \frac{1}{2}(q^2 - 1)(K_-^2 - q^{-2}K_+^2)\,, \\ [L_3, L_+] &= -\frac{1}{2}M_- q^{2P_4}(1 + q^{-2M_3}) \\ &+ \frac{1}{4}(1 - q^2)(K_+(N_+ - L_-) - (N_+ + L_-)K_-)\,, \end{aligned} \tag{3.5.14}$$

where $K_\pm := \mp L_3 + iP_3$, $M_\pm := P_2 \pm iP_1$;

c) Rotations $\leftrightarrow$ boosts:

$$
\begin{aligned}
[M_3, L_\pm] &= \pm L_\pm \ , \qquad [M_3, l_3] = 0 \ , \\
[M_-, L_3] &= \frac{1}{2}(N_+ + L_-) - \frac{1}{2}(N_+ - L_-)q^{2M_3} \ , \\
[M_-, L_+] &= q^{2M_3}K_+ - K_- + \frac{1}{2}(1 - q^{-2})M_-(L_+ - N_-) \ , \\
[M_-, L_-] &= \frac{1}{2}(1 - q^2)M_-(N_+ + L_-) \ , \\
[M_+, L_3] &= -\frac{1}{2}(N_- + L_+) + \frac{1}{2}q^{-2M_3}(N_- - L_+) \ , \\
[M_+, L_+] &= \frac{1}{2}(1 - q^2)M_+(N_- + L_+) \ , \\
[M_+, L_-] &= q^{-2M_3}K_- - K_+ + \frac{1}{2}(1 - q^2)(N_+ - L_-)M_+ \ ;
\end{aligned}
\tag{3.5.15}
$$

d) Curved translations:

$$
\begin{aligned}
[P_4, P_3] &= iL_3 \ , \qquad [P_4, P_2] = iL_2 \ , \qquad [P_4, P_1] = iL_1 \ , \\
[P_3, P_2] &= \frac{i}{4}\left(M_- q^{2P_4}(1 + q^{-2M_3}) + q^{-2P_4}(1 + q^{2M_3})M_+\right) \\
&- \frac{i}{8}(q^2 - 1)\Big(K_+(L_- - N_+) + (N_- + L_+)K_+ \\
&+ K_-(L_+ - N_-) + (N_+ + L_-)K_-\Big) \ , \\
[P_3, P_1] &= \frac{1}{4}\left(M_- q^{2P_4}(1 + q^{-2M_3}) - q^{-2P_4}(1 + q^{2M_3})M_+\right) \\
&- \frac{1}{8}(q^2 - 1)\Big(K_+(L_+ - N_-) + (N_- + L_+)K_+ \\
&- (L_- + N_+)K_- + K_+(N_+ - L_-)\Big) \ , \\
[P_2, P_1] &= \frac{i}{2}\left([P_4 + M_3]_{q^2} - [P_4 - M_3]_{q^2}\right) + \frac{i}{4}(1 - q^2)(K_-^2 - q^{-2}K_+^2) \ ;
\end{aligned}
\tag{3.5.16}
$$

e) Curved translations $\leftrightarrow$ Lorentz generators:

$$
\begin{aligned}
[P_4, M_\pm] &= 0 \ , \qquad [P_4, M_3] = 0 \ , \\
[P_4, L_\pm] &= \pm N_\mp \ , \qquad [P_4, L_3] = -iP_3 \ ,
\end{aligned}
$$

$$
\begin{aligned}
[P_1, M_+] &= \frac{i}{2}\left(K_+ + q^{-2M_3}K_-\right) \\
&+ \frac{i}{4}(q^2-1)\Big(M_+(N_- + L_+) + (N_+ - L_-)M_+\Big) \\
[P_1, M_-] &= -\frac{i}{2}\left(K_- + K_+ q^{2M_3}\right) \\
&- \frac{i}{4}(1-q^{-2})\Big(M_-(N_- - L_+) + (N_+ + L_-)M_-\Big), \\
[P_1, M_3] &= -iP_2, \\
[P_1, L_+] &= -\frac{i}{2}\left([P_4 + M_3]_{q^2} + [P_4 - M_3]_{q^2}\right) \\
&- \frac{i}{4}(1-q^{-2})\left(2q^{-2(P_4-M_3)}M_+^2 + K_+^2 + q^2K_-^2\right), \\
[P_1, L_-] &= -\frac{i}{2}\left([P_4 + M_3]_{q^2} + [P_4 - M_3]_{q^2}\right) \\
&+ \frac{i}{4}(q^2-1)\left(2M_-^2 q^{2(P_4-M_3)} + K_-^2 + q^{-2}K_+^2\right), \\
[P_1, L_3] &= \frac{i}{4}\Big((1-q^{2M_3})q^{-2P_4}M_+ + M_- q^{2P_4}(1-q^{-2M_3})\Big) \\
&+ \frac{i}{8}(q^2-1)\Big(K_+(N_+ - L_+) + (N_- + L_+)K_+ \\
&+ K_-(N_- - L_+) + (N_+ + L_-)K_-\Big), \\
[P_2, M_+] &= -\frac{1}{2}\left(K_+ + q^{-2M_3}K_-\right) \\
&+ \frac{1}{4}(q^2-1)\Big(M_+(N_- + L_+) - (N_+ - L_-)M_+\Big) \\
[P_2, M_-] &= -\frac{1}{2}\left(K_- + K_+ q^{2M_3}\right) \\
&+ \frac{1}{4}(1-q^{-2})\Big((N_+ + L_-)M_- - M_-(N_- - L_+)\Big), \\
[P_2, M_3] &= iP_1, \qquad\qquad (3.5.17) \\
[P_2, L_+] &= \frac{1}{2}\left([P_4 + M_3]_{q^2} + [P_4 - M_3]_{q^2}\right) \\
&- \frac{1}{4}(1-q^{-2})\left(2q^{-2(P_4-M_3)}M_+^2 - K_+^2 - q^2K_-^2\right), \\
[P_2, L_-] &= -\frac{1}{2}\left([P_4 + M_3]_{q^2} + [P_4 - M_3]_{q^2}\right) \\
&- \frac{1}{4}(q^2-1)\left(2M_-^2 q^{2(P_4-M_3)} - K_-^2 - q^{-2}K_+^2\right),
\end{aligned}
$$

$$
\begin{aligned}
[P_2, L_3] &= \frac{1}{4}\Big((1-q^{2M_3})q^{-2P_4}M_+ + M_-q^{2P_4}(1-q^{-2M_3})\Big) \\
&+ \frac{1}{8}(q^2-1)\Big((N_+ + L_+)K_+ - (N_+ + L_-)K_- \\
&- K_+(N_+ - L_-) + K_-(N_- - L_+)\Big) , \\
[P_3, M_+] &= \frac{i}{2}\Big((N_- + L_+) + q^{-2M_3}(N_- - L_+)\Big) , \\
[P_3, M_-] &= \frac{i}{2}\Big((N_+ + L_-) + (N_+ - L_-)q^{2M_3}\Big) , \\
[P_3, M_3] &= 0 , \\
[P_3, L_+] &= \frac{i}{2}(1-q^{2M_3})q^{-2P_4}M_+ \\
&- \frac{i}{4}(q^2-1)\Big((N_- + L_+)K_+ + K_-(N_- - L_+)\Big) , \\
[P_3, L_-] &= \frac{i}{2}M_-q^{2P_4}(1-q^{-2M_3}) \\
&- \frac{i}{4}(1-q^{-2})\Big(K_-(N_+ + L_-) + (N_+ - L_-)K_+\Big) , \\
[P_3, L_3] &= -i\,[P_4]_{q^2} .
\end{aligned}
$$

It is well known that the Poincaré algebra can be obtained as a contraction of the anti-de Sitter one (as well as from the de Sitter algebra) in the limit in which the radius of the de Sitter universe, R, becomes infinite. The contraction is performed by rescaling of the generators: $M_{\mu\nu} \longrightarrow M_{\mu\nu}$, $P_\mu \longrightarrow RP_\mu$, and then taking the limit $R \longrightarrow \infty$. As is seen by inspecting (3.5.14), the commutators of the Lorentz sector *in the deformed case* do not close and the limit $R \longrightarrow \infty$ would lead to a contradiction. This shows that any attempt aimed at extending the above contraction to the q-deformed $so_q(3,2)$, has to involve *also* the deformation parameter q, since (3.5.13),(3.5.14) close only in the $q = 1$ limit, in which it reproduces the Lorentz algebra. A contraction limit with $R \longrightarrow \infty$ was obtained in [163, 166] by means of above redefinition, $P_\mu \longrightarrow RP_\mu$, *and* by setting

$$
q = \exp\{\frac{i}{2\kappa R}\} ,
$$

where $0 < \kappa < \infty$ is a parameter with dimension of mass in the $so_q(3,2)$ space, leading to

$$\begin{aligned} [M_+, M_-] &= 2M_3 , & [P_\mu, P_\nu] &= 0 , \\ [M_3, M_\pm] &= \pm M_\pm , & [M_i, P_4] &= 0 , \\ [M_i, P_i] &= i\varepsilon_{ijk} P_k , \end{aligned} \tag{3.5.18}$$

with other commutation relations $[L, L]$, $[M, L]$ and $[P, L]$ being deformed

$$\begin{aligned} [L_-, L_+] &= 2M_3 \cos\frac{P_4}{\kappa} - \frac{1}{\kappa}(P_3 L_3 + L_3 P_3) + \frac{1}{2\kappa^2} P_3^2 , \\ [M_\pm, L_\pm] &= \frac{1}{2i\kappa} M_\pm P_\mp , \\ [L_\pm, P_1] &= i\kappa \sin\frac{P_4}{\kappa} \pm \frac{1}{2\kappa} P_3^2 . \end{aligned} \tag{3.5.19}$$

Since in the $\kappa \longrightarrow \infty$ limit the CR (3.5.18),(3.5.19) reproduce the Poincaré algebra, they may be considered as a deformation of the Poincaré algebra. It should be noticed that the deformation parameter in this case, κ, is a dimensional one, and that the CR (3.5.18) in the rotation and translation sectors are undeformed. However, this algebra is non-cocommutative, so must correspond to non-commutative geometry of corresponding quantum group and q-Minkowski space. In the next subsection we will see that this is indeed the case and that non-trivial comultilication leads to non-commutative algebra of functions on corresponding $SO_q(3,2)$.

To obtain the Casimir operator, one has to add to the straightforward generalization of classical expression

$$\tilde{C}_1 = M_1^2 + M_2^2 + [M_3]_q^2 - \vec{L}^2 - \vec{P}^2 + [P_4]_q^2 = \sum_{\alpha=1}^{4} X_\alpha^+ X_\alpha^- + [H_1]^2 + [H_3]^2 ,$$

suitable terms which provide its invariance in the Fock space of the states built out of the q-oscillators. They prove to be

$$C_1 = \tilde{C}_1 + \frac{(q-q^{-1})^2}{(q+q^{-1})}\left([M_3]_q^2 + [P_4]_q^2\right) + (q-q^{-1})^2 \frac{q^2+q^{-2}}{q+q^{-1}} [M_3]_q^2 [P_4]_q^2 ,$$

and C_1 has the value

$$C_1 \mid_{\text{Fock space}} = -\frac{4}{(q+q^{-1})^2}\left(2 - \frac{3}{(q^{1/2}+q^{-1/2})^2}\right) .$$

The same $R \longrightarrow \infty$ contraction gives

$$C_1 \longrightarrow P^2 := \lim_{R\to\infty} \frac{1}{R^2} C_1 = -\vec{P}^2 + 4\kappa^2 \sin^2\left(\frac{P_4}{2\kappa}\right) ,$$

which is the mass operator for the deformed Poincaré algebra since it commutes with all its generators. In the q-oscillator Fock space the Casimir operator P^2 has zero eigenvalue, which means that the oscillator realization describes massless states.

3.5.3 Quantum inhomogeneous groups from contraction of q-deformed simple groups

In the classical contraction scheme[125], the $N-1$ dimensional space–time is identified with a neighbourhood of a particular point on the $N-1$ sphere (or hyperbola if the signature is Minkowskian), in the limit of infinite radius. In the quantum space $O_q^N(\mathbb{R})$ one considers [253, 254] a subspace of dimension $N-1$ characterized by the condition $x^T C x = const$ (see Section 1.5 Eq.(1.5.26)). This corresponds to the de Sitter sphere in the classical Euclidean contraction. For the contraction, it is more convenient to choose a real set of generators for the quantum space, $z_i = M_{ij} x_j = z_i{}^*$, with the matrix and its inverse

$$\begin{aligned} M &= \frac{1}{\sqrt{2}} \sum_{i=1}^{N} (\alpha_i e_{ii} + \beta_i e_{i'i}) , \\ M^{-1} &= \frac{1}{\sqrt{2}} \sum_{i=1}^{N} (\gamma_i e_{ii} + \delta_i e_{i'i}) , \end{aligned}$$

where e_{ii} are defined in (1.5.7), and

$$(\alpha_1, \ldots, \alpha_n) = (1, \ldots, 1) ,$$

$$(\alpha_{n'}, \ldots, \alpha_N) = (-i\varepsilon_n q^{\rho_n}, \ldots, -i\varepsilon_1 q^{\rho_1}) ,$$

$$\beta_j = i\alpha_j, \qquad \gamma_j = \frac{1}{\alpha_j}, \qquad \delta_j = \frac{1}{\beta_{j'}} \qquad \text{for } j \neq j' ,$$

$$\alpha_{\frac{N+1}{2}} = \beta_{\frac{N+1}{2}} = \gamma_{\frac{N+1}{2}} = \delta_{\frac{N+1}{2}} = \frac{1}{\sqrt{2}} ,$$

ε_i, ρ_i are defined in (1.5.20). Recall that $i' = N+1$, $N = 2n+1$ for $SO_q(2n+1)$; $N = 2n$ for $SO_q(2n)$. Accordingly, we take new real generators $V = (v_{ij}) = MTM^{-1}$ for the algebra $Fun(SO_q(N, \mathbb{R}))$, which satisfy slightly different orthogonality conditions

$$
\begin{array}{lll}
V\hat{C}V^T = \hat{C} , & \text{with} & \hat{C} = MCM^T , \\
V^T\tilde{C}V = \tilde{C} , & \text{with} & \tilde{C} = M^{-1T}CM^{-1} .
\end{array}
\tag{3.5.20}
$$

The comultiplication and counity are of the similar form, as in Section 1.5.2 and the antipode now is

$$
S(V) = \hat{C}V^T\tilde{C} . \tag{3.5.21}
$$

In this real basis, the quantum space relations become

$$
\begin{aligned}
z_i z_j - q z_j z_i - z_{i'} z_{j'} + q z_{j'} z_{i'} &= i\left(z_{i'} z_j - q z_j z_{i'} + z_i z_{j'} - q z_{j'} z_i\right) , \\
&\quad \text{if } i < j, i < i', j < j' , \\
z_{j'} z_{i'} - q z_{i'} z_{j'} - z_j z_i + q z_i z_j &= -i\left(z_{j'} z_i - q z_i z_{j'} + z_j z_{i'} - q z_{i'} z_j\right), \\
&\quad \text{if } i < j, i > i', j < j' , \\
z_{i'} z_j - q z_j z_{i'} + z_i z_{j'} - q z_{j'} z_i &= -i\left(z_i z_j - q z_j z_i - z_{i'} z_{j'} + q z_{j'} z_{i'}\right), \\
&\quad \text{if } i < j, i < i', j > j' ,
\end{aligned}
\tag{3.5.22}
$$

$$
\varepsilon_i[z_i, z_{i'}] = i\frac{q^2-1}{q^2+1}\sum_{k=i+1}^{M}\left(\frac{1+q^2}{2}\right)^{k-i}\varepsilon_k(z_k^2+z_{k'}^2)\overset{odd}{+}i\frac{q^2-1}{q+1}\left(\frac{1+q^2}{2}\right)^{M-i}z^2_{\frac{N+1}{2}}
$$

and the quadratic form is diagonal

$$
z^T\tilde{C}z = \frac{1+q^{2-N}}{1+q^2}\sum_{k=1}^{M}\left(\frac{1+q^2}{2}\right)^{k}\varepsilon_k(z_k^2+z_{k'}^2)\overset{odd}{+}\frac{1+q^{2-N}}{1+q}\left(\frac{1+q^2}{2}\right)^{M}z^2_{\frac{N+1}{2}}
\tag{3.5.23}
$$

(the last terms in the two above formulas exist only for odd-dimensional spaces, i.e. for the groups $SO_q(2n+1)$). From this expression, the meaning of $D = \mathrm{diag}(\varepsilon_1, \ldots, \varepsilon_N)$ as the signature of metric is clear, particularly in the limit $q \to 1$.

On the invariant $N-1$ dimensional sphere $z^T\tilde{C}z = const$, one selects a particular point $(z_i) = (R, 0, \ldots, 0)$ around which an expansion in R is performed. In the quantum case, this means that Z_1 has an expansion as

$$
z_1 = R\left(1 - \frac{\varepsilon_1}{2R^2}\sum_{a=2}^{N}\varepsilon_a z_a^2 + O(R^{-3})\right) . \tag{3.5.24}
$$

As we explained in the preceding subsection the contraction amounts to take simultaneously $R \to \infty$ and $q \to 1$ by letting $q = \exp(\gamma/R)$, with γ a finite constant. Inserting the expansion (3.5.24) in the relations (3.5.22), the limit

$R \to \infty$ is well defined because all the divergent terms cancel, and we are left with the unique constraint

$$[z_a, z_N] = -i\gamma z_a \ , \qquad a = 2, \dots, N-1 \ . \tag{3.5.25}$$

We therefore define the quantum space–time as the algebra generated by the z_a subject to the above constraint (3.5.25). The generators of $Fun(SO_q(N, \mathbb{R}))$ also must be expanded in the contraction parameter R

$$v_{ij} = \sum_{n=0}^{\infty} \frac{v_{ij}^n}{R^n}, \tag{3.5.26}$$

and from consistency requirements it is possible to derive all the necessary relations characterizing the algebra $Fun(\mathcal{P}_\gamma(N-1))$. The requirements are

1. The elements z_a remain of order 1 in the limit $R \to \infty$ and z_1 is of order R. This leads to $v_{a1}^0 = 0$.

2. Consistency of (3.5.24) under coaction δ gives

$$v_{11}^0 = 1 \ , \quad v_{11}^1 \otimes \mathbf{I} + v_{1a}^0 \otimes z_a = 0 \quad \Rightarrow \quad v_{1a}^0 = 0 \ , \ v_{11}^1 = 0 \ .$$

Thus dividing z_1 by R, one has in the $R \to \infty$ limit of $\delta(z) = V \otimes z$

$$\begin{aligned} \delta(\mathbf{I}) &= \mathbf{I} \otimes \mathbf{I} \ , \\ \delta(z_a) &= v_{a1}^1 \otimes \mathbf{I} + v_{ab}^0 \otimes z_b \ . \end{aligned} \tag{3.5.27}$$

From this form of the coaction, we see that v_{a1}^1 play the role of translations and v_{ab}^0 the role of Lorentz transformations. It is then natural to take the elements $\mathbf{I}, u_{ab} = v_{ab}^0$ and $u_a = v_{a1}^1$ as the generators of $Fun(\mathcal{P}_\gamma(N-1))$, the algebra of functions on the quantum Poincaré group $\mathcal{P}_\gamma(N-1)$. The expansion of the orthogonality relations (3.5.20) at zeroth order in $1/R$, gives

$$\begin{aligned} v_{ab}^0 \varepsilon_b v_{cb}^0 &= \varepsilon_a \delta_{ac} \ , \\ v_{ba}^0 \varepsilon_b v_{bc}^0 &= \varepsilon_a \delta_{ac} \ , \end{aligned} \tag{3.5.28}$$

and at first order, the relations are

$$\begin{aligned} &(v_{ji}^1 \varepsilon_i v_{ki}^0 + v_{ji}^0 \varepsilon_i v_{ki}^1 + \gamma v_{ji}^0 \varepsilon_i \theta_i \rho_i v_{ki}^0 + i\gamma v_{ji}^0 \varepsilon_i \rho_i v_{ki'}^0) e_{jk} \\ &\qquad = \gamma \varepsilon_i \theta_i \rho_i e_{ii} + i\gamma \varepsilon_i \rho_i e_{ii'} \ , \\ &(v_{ij}^1 \varepsilon_i v_{ik}^0 + v_{ij}^0 \varepsilon_i v_{ik}^1 - \gamma v_{ij}^0 \varepsilon_i \theta_i \rho_i v_{ik}^0 - i\gamma v_{ij}^0 \varepsilon_i \rho_i v_{i'k}^0) e_{jk} \\ &\qquad = -\gamma \varepsilon_i \theta_i \rho_i e_{ii} - i\gamma \varepsilon_i \rho_i e_{ii'} \ , \end{aligned} \tag{3.5.29}$$

where $\theta_i = 1$, if $i \leq n$ and $\theta_i = -1$, if $i > n$. These constraints are necessary for computing the antipode and the commutation relations.

To determine the commutation relations among the generators that follow from the contraction of the constraint $R_v V_1 V_2 = V_2 V_1 R_v$, one needs to expand that expression up to order R^{-2}. Higher order terms ($R^{-n}, n \geq 3$) will always contain elements v_{ij}^n of that order which by definition do not belong to the quantum Poincaré algebra, and thus do not yield new constraints on the set of generators. Performing the expansion, one gets

$$[u_{ab}, u_{cd}] \equiv [v_{ab}^0, v_{cd}^0] = 0 \ , \tag{3.5.30}$$

$$[u_a, u_{cd}] \equiv [v_{a1}^1, v_{cd}^0] = i\gamma\left((u_{Nd} - \delta_{Nd})\varepsilon_1\varepsilon_a\delta_{ac} + (u_{cN} - \delta_{cN})u_{ad}\right) \ , \tag{3.5.31}$$

$$[u_a, u_b] \equiv [v_{a1}^1, v_{b1}^1] = i\gamma(\delta_{Na}u_b - \delta_{Nb}u_a) \ . \tag{3.5.32}$$

The rest of the Hopf algebra structure is obtained by contraction of the comultiplication $\Delta(V) = V \dot{\otimes} V$, which yields

$$\Delta(u_{ab}) = u_{ac} \otimes u_{cb} \ , \qquad \Delta(u_a) = u_a \otimes 1 + u_{ab} \otimes u_b \ , \tag{3.5.33}$$

the counit $\varepsilon(V) = \mathbf{I}$:

$$\varepsilon(u_{ab}) = \delta_{ab} \ , \qquad \varepsilon(u_a) = 0 \ , \tag{3.5.34}$$

and the antipode

$$\begin{aligned} S(u_{ab}) &= \varepsilon_a \varepsilon_b u_{ba} \ , && \text{no sum on } a, b \ , \\ S(u_a) &= -\textstyle\sum_{b=2}^{N} \varepsilon_a u_{ba} \varepsilon_b u_b \ , && \text{no sum on } a \ . \end{aligned} \tag{3.5.35}$$

One readily checks that the commutation relations (3.5.30)–(3.5.32) satisfy the Jacobi identity and as they originate from the contraction of $Fun(SO_q(N, \mathbb{R}))$, it is natural to take them as the definition of the quantum Poincaré group $Fun(\mathcal{P}_\gamma(N-1))$. Furthermore, this definition is consistent with the ones obtained from quantization of the classical Poisson structure on the Poincaré group[252, 165].

Looking closer at (3.5.28) and (3.5.30), one sees that $U = (u_{ab})$ actually describes an ordinary orthogonal matrix (with commuting entries) which preserves the metric $\eta_{ac} = \varepsilon_1\varepsilon_a\delta_{ac}$, *i.e.* (3.5.28) become $U^T\eta U = \eta$.

There are good reasons to believe that the quantum Poincaré group is dual to the κ-Poincaré Hopf algebra. The approach proposed here is very reminiscent of the contraction used in deriving κ–Poincaré QUAE discussed in the preceding subsection and the deformation parameter q is treated in the same way. But the rigorous proof of this fact exists only in the two dimensional case[254].

Note that CR (3.5.32) for translation operators u_a and CR (3.5.25) for the coordinates of the quantum spaces have Lie algebra form. The interesting problem is to construct (if it does exist) a differential calculus for this space invariant with respect to the coaction of $\mathcal{P}_\gamma$.

Classical Lie algebras can be used directly for the construction of a kind of non-commutative geometry (see [173, 92, 94, 114, 115] and refs therein). This is a quite expected fact in view of the well known relation between Lie algebras and quantum mechanics on manifolds (coadjoint orbits) [19]. In particular, in [115] (see also refs. therein) a simple two dimensional field theoretical model on the so-called *fuzzy sphere* [173] was considered. A fuzzy sphere is an underlying object (in the sense of non-commutative geometry) of the family of non-commutative algebras A_j given by finite sums

$$A_j := [0] \oplus [1] \oplus \cdots \oplus [j] \, ,$$

of irreducible representations of Lie algebra $su(2)$. The algebra A_j are considered as the truncation (quantization) of the algebra $Fun(S^2)$ of functions on usual continuous sphere S^2. As all unitary representations of $su(2)$ are finite dimensional, a field theory defined on a fuzzy sphere, considered as a configuration space, has no ultraviolet divergences.

It is instructive to consider the phase space of the model. Recall that the $su(2)$ algebra derivatives (which can be considered as canonically conjugate momenta), also form $su(2)$ Lie algebra (see, e.g. [13]) and together with the initial "configuration" $su(2)$, they form the algebra $so(4) \sim su(2) \otimes su(2)$. Taking into account the existence of two Casimir operators in the $so(4)$ algebra, we get that the phase space has a topology $S^2 \hat{\otimes} S^2$. So the whole phase space of the model is compact and thus infinite momenta (and, hence, ultraviolet problems) do not appear. Notice, however, that in this phase space there exists, obviously, two commutative operators (from the different $su(2)$ subalgebras) which seems worth to be called "coordinates of the configuration subspace" of the phase space. In this way, one obtains usual quantum mechanics on the phase space $S^2 \hat{\otimes} S^2$. Thus the question about the exact physical meaning of models of this type is still open.

3.6 Elements of general theory of q-inhomogeneous groups and classification of q-Poincaré groups and q-Minkowski spaces

In this book we have met already many examples which confirm and illustrate the general statement: "quantization removes the degeneracy". For noncommutative geometry, this means that classically equivalent objects (groups, algebras, spaces) become different after the quantization.

In this situation the classification of possible quantum spaces and the corresponding q-groups of their symmetries becomes very important. Possible quantizations of Minkowski space, Lorentz and Poincaré groups of special relativity have been classified in [249, 192, 70]. Below we present the results of this classification together with elements of general theory of inhomogeneous quantum groups [191].

3.6.1 Classification of q-Lorentz groups and q-Minkowski spaces

To classify possible quantizations of Lorentz group Woronowicz and Zakrewski [249] have used the method which is already well familiar to us : most important properties of classical object are considered as defining ones in quantum deformed case. Following this principle, they have classified all quantum groups of 2×2 matrices satisfying three specific properties which characterize $SL(2,\mathbb{C})$ among the classical groups. These three properties are the following:

- a tensor square of fundamental representations has a trivial representation as a direct sum component;
- a tensor product, independent of the order of factors, of fundamental representation and its conjugate is irreducible;
- the group is a maximal group with the above two properties.

All such q-groups are determined by two matrices (intertwiners)

$$E : \mathbb{C} \longrightarrow M \otimes M \; ,$$

and

$$X : M \otimes \bar{M} \longrightarrow \bar{M} \otimes M \; ,$$

where M is (two-dimensional) fundamental representation with conjugate $\bar{M}$.

More precisely they have defined the quantum Lorentz group as follows:

Definition 3.1 *$\mathcal{L}_q$ is a quantum Lorentz group if $Fun_q(\mathcal{L})$ satisfies the conditions*

i) $Fun_q(\mathcal{L})$ is a Hopf *-algebra generated (as *-algebra) by matrix elements of a two-dimensional representation $M_{\alpha\beta}$, $(\alpha, \beta = 1,2)$;

ii) $M \textcircled{T} M \simeq I \oplus M^1$, where M^1 is any representation and I is a trivial representation;

iii) the representation $M \textcircled{T} \bar{M} \simeq \bar{M} \textcircled{T} M$ is irreducible;

iv) if $Fun'_q(\mathcal{L})$ satisfy i)–iii) and there exists Hopf *-algebra epimorphism $\rho : Fun'_q(\mathcal{L}) \longrightarrow Fun_q(\mathcal{L})$ (i.e., $\operatorname{im}\rho = Fun_q(\mathcal{L})$ such that $\rho(M') = M$, then ρ is an isomorphism (the universality condition).

Following Woronowicz's notation, we use here the sign $\textcircled{T}$ for the product of a q-group corepresentations:

$$(v \textcircled{T} w)_{ij,kl} := v_{ik} w_{jl} \; .$$

Comultiplication Δ is defined as usual for matrix q-groups

$$\Delta M_{\alpha\beta} := M_{\alpha\gamma} M_{\gamma\beta} \; .$$

According to [249], $Fun_q(\mathcal{L})$ is generated by the $M_{\alpha\beta}$ satisfying

$$(M \textcircled{T} M)E = E \; , \tag{3.6.1}$$

$$E'(M \textcircled{T} M) = E', \tag{3.6.2}$$

$$X(M \textcircled{T} \bar{M}) = (\bar{M} \textcircled{T} M)X \; , \tag{3.6.3}$$

where E, E' have *two* possible forms:

1) $E = e_1 \otimes e_2 - q e_2 \otimes e_1$, $E' = -q^{-1} e^1 \otimes e^2 + e^2 \otimes e^1$, $q \in \mathbb{C} \setminus \{0, i, -i\}$, ($E'E \neq 0$ which means $q \neq \pm i$)

or

2) $E = e_1 \otimes e_2 - e_2 \otimes e_1 + e_1 \otimes e_1$, $E' = -e^1 \otimes e^2 + e^2 \otimes e^1 + e^2 \otimes e^2$,

$$e_1 = \begin{pmatrix} 1 \\ 0 \end{pmatrix} , \quad e_2 = \begin{pmatrix} 0 \\ 1 \end{pmatrix} , \quad e^1 = (1,0) \; , \quad e^2 = (0,1) \; .$$

Matrix X defines CR for M and its conjugate. In general they also depend on continuous parameters of deformation (cf. preceding Section), and their possible explicit forms for the above cases 1) and 2) have been found in [249]. We do not reproduce them here, since physically interesting q-Lorentz groups must be included into the corresponding q-Poincaré group. As is shown in [192] this is not possible for all q-Lorentz groups. We will present their results on classification of q-Poincaré groups in subsection 3.5.3 together with corresponding explicit forms of the matrices X. We use notation M for q-Lorentz matrices instead of A used in Section 3.4, to stress that here we present the analysis of general situation and A is an example of one concrete quantization.

One can check that (3.6.1),(3.6.2) for the above case 1) is equivalent to the CR for the standard quantum group $SL(2,\mathbb{C})$ (cf. Sections 1.1. and 1.5). This shows, in particular, that one can use this q-group as a starting point and try to find different quantizations of Lorentz group within R-matrix approach with CR of "RTT"-type (see [70, 71] and refs. therein) for two copies M and $\tilde{M}$ of $SL_q(2,\mathbb{C})$ matrices

$$\begin{aligned} R^{(1)}M_1M_2 &= M_2M_1R^{(1)}\,, & R^{(2)}M_2\tilde{M}_1 &= \tilde{M}_1M_2R^{(2)}\,, \\ R^{(3)}M_1\tilde{M}_2 &= \tilde{M}_2M_1R^{(3)}\,, & R^{(4)}\tilde{M}_2\tilde{M}_1 &= \tilde{M}_1\tilde{M}_2R^{(4)}\,. \end{aligned} \tag{3.6.4}$$

Using the permutation operator σ, the first equation may be rewritten as

$$\sigma(R^{(1)})^{-1}\sigma M_1M_2 = M_2M_1\sigma(R^{(1)})^{-1}\sigma\,, \tag{3.6.5}$$

(and similarly for $R^{(4)}$ and tilded $\tilde{M}'s$). It follows from (3.6.4) that we may take $R^{(4)} = R^{(1)\dagger}$ or $R^{(4)} = \sigma(R^{(1)-1})^\dagger\sigma$ and that $R^{(2)} = \sigma R^{(3)}\sigma$ and $R^{(2,3)\dagger} = \sigma R^{(2,3)}\sigma$. Imposing the reality condition $\tilde{M}^{-1} = M^\dagger$, so that M and $\tilde{M}$ are copies of the same quantum group, one obtains $R^{(1)\dagger} = \sigma R^{(1)}\sigma$.

Relations (3.6.4) can be treated simultaneously by using a four dimensional q-Dirac spinorial realization in terms of the 4×4 matrices

$$\Upsilon \equiv \begin{pmatrix} M & 0 \\ 0 & \tilde{M} \end{pmatrix}\,, \tag{3.6.6}$$

and by introducing the R-matrix

$$R := \begin{pmatrix} R^{(1)} & 0 & 0 & 0 \\ 0 & R^{(3)} & 0 & 0 \\ 0 & 0 & R^{(2)\,-1} & 0 \\ 0 & 0 & 0 & \sigma R^{(4)}\sigma \end{pmatrix}\,. \tag{3.6.7}$$

In this way, the set of relations (3.6.4) defining a deformed Lorentz group can be written in "RTT"-form

$$R\Upsilon_1\Upsilon_2 = \Upsilon_2\Upsilon_1 R\,, \tag{3.6.8}$$

where the 16×16 matrices Υ_1 and Υ_2 are defined in block form by

$$\Upsilon_1 := \begin{pmatrix} M_1 & 0 & 0 & 0 \\ 0 & M_1 & 0 & 0 \\ 0 & 0 & \tilde{M}_1 & 0 \\ 0 & 0 & 0 & \tilde{M}_1 \end{pmatrix} , \qquad \Upsilon_2 := \begin{pmatrix} M_2 & 0 & 0 & 0 \\ 0 & \tilde{M}_2 & 0 & 0 \\ 0 & 0 & M_2 & 0 \\ 0 & 0 & 0 & \tilde{M}_2 \end{pmatrix} . \tag{3.6.9}$$

Relations (3.6.8) may be used to define the commutation relations of the generators (entries of Υ) of a quantum group. Standard arguments may be invoked now to require that R satisfies the QYBE

$$R_{12}R_{13}R_{23} = R_{23}R_{13}R_{12} \quad , \tag{3.6.10}$$

where R_{12}, R_{13} and R_{23} are $4^3 \times 4^3$ matrices acting on $\mathbb{C}^4 \otimes \mathbb{C}^4 \otimes \mathbb{C}^4$. Separating in blocks the matrices in both sides of eq. (3.6.10), it follows that the 4×4 matrices $R^{(1)}$, $R^{(4)}$ must satisfy the QYBE and that $R^{(2)}$, $R^{(3)}$ obey the mixed consistency conditions

$$R^{(1)}_{12}R^{(3)}_{13}R^{(3)}_{23} = R^{(3)}_{23}R^{(3)}_{13}R^{(1)}_{12} \ , \qquad R^{(4)}_{12}R^{(2)}_{13}R^{(2)}_{23} = R^{(2)}_{23}R^{(2)}_{13}R^{(4)}_{12} \ . \tag{3.6.11}$$

Notice, that these equations, if considered as an "RTT" equation, tell us that $R^{(2)}, R^{(3)}$ are the representations of the $SL(2)$ quantum group.

Coordinates of a q-Minkowski space can be "packed" into matrix K analogously to that in Section 3.4 (we again use K instead of z used there, to distinct the general consideration from a special case) with q-Lorentz group coaction δ defined by

$$\delta : K \longrightarrow K' = MK\tilde{M}^{-1} \ , \quad K'_{is} = M_{ij}\tilde{M}^{-1}_{ls}K_{jl} \ , \quad \tilde{M}^{-1} = M^{\dagger} \ , \tag{3.6.12}$$

where it is assumed that the matrix elements of K commute with those of M and $\tilde{M}$ but not among themselves. In order to identify the elements of K

$$K = \begin{pmatrix} \alpha & \beta \\ \gamma & \delta \end{pmatrix}$$

with the generators of the q-Minkowski algebra $\mathcal{M}^{(1)}_q$ we require, as in the classical case, that

a) the reality property preserved by (3.6.12);

b) the (real) q-Minkowski length, defined through the q-determinant $det_q K$ of K, invariant under the q-Lorentz transformation (3.6.12);

c) the set of commutation relations for the elements of K is preserved by (3.6.12) for (3.6.4).

The reality condition $K = K^\dagger$ is consistent with (3.6.12), since $\tilde{M}^{-1} = M^\dagger$, as in the classical case.

An elegant way to impose CR for q-Minkowski space coordinates, is to use the so-called reflection equation [152], which in this case has the form ([70, 71], and refs. therein)

$$R^{(1)}K_1R^{(2)}K_2 = K_2R^{(3)}K_1R^{(4)}. \tag{3.6.13}$$

This form provides with the invariance of CR for entries of K (coordinates of q-Minkowski space) with respect to the coaction (3.6.12) if $M, \tilde{M}$ satisfy (3.6.4).

Let us look at some possible solution of (3.6.11). An obvious one is

$$R^{(1)} = R_{12}\ , \quad R^{(2)} = R_{21}\ , \quad R^{(3)} = R_{12}\ , \quad R^{(4)} = R_{21}\ , \tag{3.6.14}$$

with R_{12} given in Section 1.1 (it was the first example of quantum Lorentz group [32]).

For K given by

$$K := \begin{pmatrix} \alpha & \beta \\ \gamma & \delta \end{pmatrix},$$

eq. (3.6.13) is equivalent to the six basic relations

$$\begin{aligned} \alpha\beta &= q^{-2}\beta\alpha\ , & [\delta,\beta] &= q^{-1}(q-q^{-1})\alpha\beta\ , \\ \alpha\gamma &= q^{2}\gamma\alpha\ , & [\beta,\gamma] &= q^{-1}(q-q^{-1})(\delta-\alpha)\alpha\ , \\ [\alpha,\delta] &= 0\ , & [\gamma,\delta] &= q^{-1}(q-q^{-1})\gamma\alpha\ , \end{aligned} \tag{3.6.15}$$

which characterize the algebra of coordinates of the q-Minkowski space denoted as $\mathcal{M}_q^{(1)}$.

The central (commuting) elements of $\mathcal{M}_q^{(1)}$ may be obtained by using the q-trace Tr_q which, for a 2×2 matrix B with elements commuting with those of M (as it is the case of K), is defined by

$$\mathrm{Tr}_q B = \mathrm{Tr}(DB) = q^{-1}b_{11} + qb_{22}\ , \quad D = diag(q^{-1}, q)\,. \tag{3.6.16}$$

The q-trace is invariant under the coaction $B \mapsto MBM^{-1}$,

$$\mathrm{Tr}_q(MBM^{-1}) = \mathrm{Tr}_q(B)\,. \tag{3.6.17}$$

This can be shown easily using the invariance of the q-antisymmetric matrix ϵ^q (which replaces $i\sigma^2$ for $q=1$), by the $SL_q(2)$ matrices,

$$M^{\top}\epsilon^q M = \epsilon^q \quad , \quad \epsilon^q = \begin{bmatrix} 0 & q^{-1/2} \\ -q^{1/2} & 0 \end{bmatrix} = -(\epsilon^q)^{-1} , \tag{3.6.18}$$

since $D = \epsilon^q(\epsilon^q)^{\top}$. The matrix D satisfies

$$M^{\top} D (M^{-1})^{\top} = D \quad , \quad \tilde{M}^{\top} D (\tilde{M}^{-1})^{\top} = D \, . \tag{3.6.19}$$

The centrality of the q-trace can be derived from (3.6.13) with considered solution for the R-matrices. This means, in turn, that

$$K c_n = c_n K \quad , \quad c_n = \mathrm{Tr}_q K^n \quad . \tag{3.6.20}$$

The first two central elements c_1 and c_2 are algebraically independent, but the c_n for $n > 2$ are polynomial functions of them due to the characteristic equation for K

$$qK^2 - c_1 K + \frac{q}{[2]_q}(q^{-1}c_1^2 - c_2)I = 0 \, . \tag{3.6.21}$$

The q-determinant $\det_q K$

$$\det{}_q K = (\alpha\delta - q^2\gamma\beta) \, , \tag{3.6.22}$$

is central, since

$$\det{}_q K = \frac{q^2}{[2]}(q^{-1}c_1^2 - c_2) \, . \tag{3.6.23}$$

Since it can be expressed in terms of q-traces, $det_q K$ is obviously preserved under a similarity transformation $K \mapsto MKM^{-1}$, where M and M^{-1} belong to the *same* quantum group. But, despite the fact that the central elements c_n are not invariant with respect to the q-Lorentz transformation (3.6.12) because it involves M and $\tilde{M}^{-1}$, $det_q K$ is nevertheless preserved under this coaction due to the property [71]

$$\det{}_q(MK\tilde{M}^{-1}) = (\det{}_q M)\det{}_q K(\det{}_q \tilde{M})^{-1} \, ,$$

where the equality $\tilde{M}_1^{-1}\hat{R}M_1 = M_2\hat{R}\tilde{M}_2^{-1}$ have been used. Since

$$\det{}_q M = \det{}_q \tilde{M} = 1 \, ,$$

we may identify this real and central invariant element with the square l_q of the q-Minkowski invariant length

$$l_q = \det{}_q K = \alpha\delta - q^2\gamma\beta \, , \qquad l_q \in \mathcal{M}_q^{(1)} \, . \tag{3.6.24}$$

Other solutions may be found by looking for other consistent sets of matrices $R^{(i)}$ in (3.6.13), (3.6.4). A possible choice is the q-Lorentz group $\mathcal{L}_q^{(2)}$ generated by the same non-commuting entries of M and $\tilde{M}$ as reflected by the equations in (3.6.4), with $R^{(1)} = R^{(4)} = R$, R being of the standard form (1.1.25) and the same $*$-operation, but with $M_1\tilde{M}_2 = \tilde{M}_2 M_1$, so that the elements of M and $\tilde{M}$ commute between themselves. This corresponds to trivial solution for the R-matrices: $R^{(2)} = R^{(3)} = \mathbf{1}$, and leads to the CR

$$R_{12}K_1^{(2)}K_2^{(2)} = K_2^{(2)}K_1^{(2)}R_{21}\,, \tag{3.6.25}$$

$$R_{12}K_1^{(2)}K_2^{(2)} = q^2 K_2^{(2)}K_1^{(2)}R_{12}^{-1}\,, \tag{3.6.26}$$

(the superindex has been added to distinguish $K^{(2)}$ from the previous $K =: K^{(1)}$). It is easy to see that eqs. (3.6.25), (3.6.26) are consistent with the reality condition $K^{(2)} = K^{(2)\dagger}$. Eqs. (3.6.25) (and (3.6.26), which corresponds to $det_q K^{(2)}{=}0$) lead to the following independent commutation relations for the entries of $K^{(2)}$, generating the $\mathcal{M}_q^{(2)}$ algebra (cf. (3.6.15))

$$\begin{aligned}
&\alpha^{(2)}\beta^{(2)} = q^{-1}\beta^{(2)}\alpha^{(2)}\,, && \alpha^{(2)}\gamma^{(2)} = q\gamma^{(2)}\alpha^{(2)}\,,\\
&[\alpha^{(2)},\delta^{(2)}] = 0\,, && [\beta^{(2)},\gamma^{(2)}] = (q-q^{-1})\alpha^{(2)}\delta^{(2)}\,,\\
&\beta^{(2)}\delta^{(2)} = q\delta^{(2)}\beta^{(2)}\,, && \gamma^{(2)}\delta^{(2)} = q^{-1}\delta^{(2)}\gamma^{(2)}\,.
\end{aligned} \tag{3.6.27}$$

The identifications $\alpha \leftrightarrow b, \beta \leftrightarrow a, \gamma \leftrightarrow d, \delta \leftrightarrow c$, make this algebra isomorphic to the $GL_q(2)$ one in eq. (1.1.18). There is an invariant and central element in the $\mathcal{M}_q^{(2)}$ algebra which determines the Minkowski length and metric

$$l_q^{(2)} = \det_q(K^{(2)}) = \alpha^{(2)}\delta^{(2)} - q\gamma^{(2)}\beta^{(2)}\,. \tag{3.6.28}$$

This definition guarantees that $det_q(K'^{(2)}){=}det_q(K^{(2)})$ for the coaction of $\mathcal{L}_q^{(2)}$.

A third possibility $\mathcal{M}_q^{(3)}$, is obtained by setting equal to unity the R-matrices $R^{(1)}$ and $R^{(4)}$ in the (3.6.4). In this case, the matrix elements of M (and $\tilde{M}$) are commuting (they define an $SL(2,C)$ group each), and the non-commutativity of the entries of M and $\tilde{M}$ is described by a certain matrix V through

$$VM_1\tilde{M}_2 = \tilde{M}_2 M_1 V\,. \tag{3.6.29}$$

If $V = diag(q^2, 1, 1q^2)$, this corresponds just to the q-Lorentz group considered in Section 3.4. It is clear that other q-Lorentz groups and corresponding spaces can be constructed in this approach. But as we noted, Minkowski space must be associated with inhomogeneous Poincaré group. Unfortunately, among known solutions within this R-matrix approach only the latter, considered in details in Section 3.4, can be generalized to q-Poincaré symmetry. Others require the addition of dilatation symmetry as we discussed within projective approach in Section 3.3. If to add dilatations, it seems reasonable to consider the complete conformal (four-dimensional) q-group which is mathematically even simpler (as it is *a simple* group) and physically very important and attractive just for high energy phenomena for which the quantum geometry hopefully plays an essential role.

3.6.2 General definition and properties of inhomogeneous quantum groups

The general theory of quantum inhomogeneous groups was developed in [191]. We shall not repeat their involved analysis in details but will present only the basic definitions and results. In the next subsection the classification of a particular class of q-inhomogeneous groups, namely the quantum Poincaré ones, will be described in some more details.

Definition 3.2 *Let H_q be a homogeneous (semisimple) quantum group with completely reducible corepresentations and Λ is a distinguished irreducible corepresentation of H_q. Let a Hopf algebra $Fun_q(G)$ have the properties*

1. *$Fun_q(G)$ is generated (as algebra) by $Fun_q(H)$ and the elements p_s, $s \in \mathcal{I}$ ($\mathcal{I}$ is finite set of indices);*

2. *$Fun_q(H)$ is a sub-bialgebra of $Fun_q(G)$.*

3. $\mathcal{P} = \begin{pmatrix} \Lambda & p \\ 0 & \mathbf{1} \end{pmatrix}$ *is a representation of G;*

4. *There exists $i \in \mathcal{I}$ such that $p_i \notin Fun_q(H)$;*

5. *$\Gamma Fun_q(H) \subset \Gamma$ where $\Gamma = Fun_q(H) \cdot X + Fun_q(H)$, $X = \mathrm{span}\{p_i,\ \ i \in \mathcal{I}\}$.*

Then the quantum group G_q is called inhomogeneous quantum group with homogeneous q-subgroup H_q.

Due to 5., Γ is $Fun_q(H)$-bimodule. By virtue of the above conditions 2.-3.,

$$\Delta Fun_q(H) \subset Fun_q(H) \otimes Fun_q(H) \ ,$$

$$\Delta p = \Lambda_{sk}\otimes p_k + p_s\otimes\mathbf{1}\,, \tag{3.6.30}$$

hence $\Delta\Gamma \subset Fun_q(H)\otimes\Gamma + \Gamma\otimes Fun_q(H)$.

The construction of an antipode and counity is the same as in Section 3.3.

A general form of CR for elements of H_q and p_s are defined by

- a functional $f_{st}: H_q \longrightarrow \mathcal{A}'$ ($\mathcal{A}'$ is some algebra; for simplicity one can take as $\mathcal{A}'$ a usual field $\mathbb{C}$; hereafter we make this choice) with the properties

$$f_{st}(ab) = f_{sm}(a)f_{mt}(b), \quad a,b \in Fun_q(H), \quad f_{st}(\mathbf{1}) = \delta_{st}\,, \tag{3.6.31}$$

$$\Lambda_{st}(f_{tr}\hat{*}a) = (a\hat{*}f_{st})\Lambda_{tr}, \quad a \in Fun_q(H)\,; \tag{3.6.32}$$

here we used the notations

$$a\hat{*}\rho := (\rho\otimes\mathrm{id})\Delta a, \quad \rho\hat{*}a := (\mathrm{id}\otimes\rho)\Delta a, \quad \rho\hat{*}\rho' := (\rho\otimes\rho')\Delta\,,$$

for $a \in Fun_q(H)$, and functionals $\rho, \rho' : H_q \to \mathbb{C}$;

- maps $\phi_s : H_q \longrightarrow H_q$, $s \in \mathcal{I}$ which satisfy

$$\phi_s(ab) = (a\hat{*}f_{st})\phi_t(b) + \phi_s(a)b, \quad a,b \in Fun_q(H), \quad \phi_s(I) = 0\,, \tag{3.6.33}$$

$$\Delta\phi_s(a) = (\Lambda_{st}\otimes I)[(\mathrm{id}\otimes\phi_t)\Delta(a)] + (\phi_s\otimes\mathrm{id})\Delta a, \quad a \in Fun_q(H)\,. \tag{3.6.34}$$

In terms of these maps the CR reads

$$p_s a = (a\hat{*}f_{st})p_t + \phi_s(a), \quad a \in Fun_q(H)\,. \tag{3.6.35}$$

If we take $a = \Lambda_{mn}$, then (3.6.35) becomes

$$p_s\Lambda_{mn} = (f_{st}\otimes id)\Lambda_{mk}\otimes\Lambda_{kn}p_t + \phi_s(\Lambda_{mn}) = f_{st}(\Lambda_{mk})\Lambda_{kn}p_t + \phi_s(\Lambda_{mn})\,.$$

As is shown in [191], a general bialgebra $Fun_q(G)$ satisfying the above conditions 1.-5., is equal to $\tilde{B}/J$, where

- $\tilde{B}$ is the algebra with unity generated by $Fun_q(H)$ and $\tilde{p}_s$ ($s \in \mathcal{I}$) with relations (3.6.35) where ϕ_s is given by

$$\phi_s(a) = a\hat{*}\eta_s - \Lambda_{st}(\eta_t\hat{*}a), \quad a \in Fun_q(H)\,, \quad \eta_s : Fun_q(H) \to \mathbb{C}\,, \tag{3.6.36}$$

for f and η satisfying (3.6.31)-(3.6.34) and such that

$$Fun_q(H) \ni a \longrightarrow \rho(a) = \begin{pmatrix} f(a) & \eta(a) \\ 0 & \varepsilon(a) \end{pmatrix} \in M_{|\mathcal{I}|}(\mathbb{C}) , \qquad (3.6.37)$$

is a unital homomorphism (for f_{st} this is implied by the Eq.(3.6.31); "unital" means just the latter condition in (3.6.31)) Moreover, $\tilde{\mathcal{B}}$ is a bialgebra with comultiplication given by the comultiplication in $Fun_q(H)$ and (3.6.30).

- J is an ideal in $\tilde{\mathcal{B}}$ generated by

$$s_{kl} = (\Xi - \mathbf{1}\otimes\mathbf{1})_{kl,ij}(\tilde{p}_i\tilde{p}_j - \eta_i(\Lambda_{js})\tilde{p}_s + T_{ij} - \Lambda_{im}\Lambda_{jn}T_{mn}) ,$$

for some complex numbers $\{T_{ij}\}_{i,j\in\mathcal{I}}$, satisfying some consistency relations (see below) and

$$\Xi_{ij,sm} := f_{im}(\Lambda_{js}) . \qquad (3.6.38)$$

Consistency relations for T_{ij} are formulated in terms of the functionals $\tau_{ij} : H_q \longrightarrow \mathbb{C}$ of the form

$$\tau_{ij} = \eta_j\hat{*}\eta_i - \eta_i(\Lambda_{js})\eta_s + T_{ij}\varepsilon - (f_{jn}\hat{*}f_{im})T_{mn} .$$

In terms of these functionals the constraints for the T_{ij} read

$$(\Lambda\hat{\oplus}\Lambda)_{kl,ij}(\tau^{ij}\hat{*}b) = b\hat{*}\tau^{kl} , \qquad (3.6.39)$$

where

$$\tau^{kl} = (\Xi - \mathbf{1}\otimes\mathbf{1})_{kl,ij}\tau_{ij}, \qquad (3.6.40)$$

and an element b is from the set generating $Fun_q(H)$ as an algebra with unity. Besides, the following identity must be fulfilled

$$A_3F = 0 , \qquad (3.6.41)$$

where

$$\begin{aligned} A_3 = \mathbf{1}\otimes\mathbf{1}\otimes\mathbf{1} \; &- \; \Xi\otimes\mathbf{1} - \mathbf{1}\otimes\Xi + (\Xi\otimes\mathbf{1})(\mathbf{1}\otimes\Xi) \\ &+ \; (\mathbf{1}\otimes\Xi)(\Xi\otimes\mathbf{1}) - (\Xi\otimes\mathbf{1})(\mathbf{1}\otimes\Xi)(\Xi\otimes\mathbf{1}) , \end{aligned}$$

$$F_{ijk,m} = \tau_{ij}(\Lambda_{km}) ,$$

and with the additional condition that $A_3(Z\otimes\mathbf{1} - \mathbf{1}\otimes Z)T$, where $Z_{ij,m} := \eta_i(\Lambda_{jm})$, is a map from the $Fun_q(H)$-algebra unity to $\Lambda\hat{\oplus}\Lambda\hat{\oplus}\Lambda$.

If $Fun_q(H)$ is a *-Hopf algebra with self-adjoint corepresentation $\bar{\Lambda} = \Lambda$, there exists Hopf *-algebra structure in $Fun_q(G)$ such that its *restriction* on $Fun_q(H)$ coincides with *initial* *-structure on $Fun_q(H)$ and $p_i^* = p_i$ $(i \in \mathcal{I})$. The functionals f, η in this case satisfy

$$f_{ij}(S(a^*)) = \overline{f_{ij}(a)}\ , \qquad a \in Fun_q(H)\ , \tag{3.6.42}$$

$$\eta_i(S(a^*)) = \overline{\eta_i(a)}\ , \qquad a \in Fun_q(H)\ . \tag{3.6.43}$$

Moreover, such a $*$-structure is unique.

3.6.3 Classification of quantum Poincaré groups

An application of general theory of q-inhomogeneous groups developed in [192] leads to the following

Definition 3.3 *A Hopf *-algebra $Fun_q(\mathcal{P})$ is called a quantum Poincaré group if for some q-Lorentz group $\mathcal{L}_q$ it satisfies the conditions*

1. *$Fun_q(\mathcal{P})$ is generated as algebra by $Fun_q(\mathcal{L})$ and the elements p_i, $i \in \mathcal{I}$;*
2. *$Fun_q(\mathcal{L})$ is a Hopf *-subalgebra of $Fun_q(\mathcal{P})$;*
3. *$\mathcal{P} = \begin{pmatrix} \Lambda & p \\ 0 & \mathbf{I} \end{pmatrix}$ is a representation of $Fun_q(\mathcal{P})$;*
4. *There exists $i \in \mathcal{I}$ such that $p_i \notin Fun_q(\mathcal{L})$;*
5. *$\Gamma Fun_q(\mathcal{L}) \subset \Gamma$ where $\Gamma = Fun_q(\mathcal{L})X + Fun_q(\mathcal{L})$, $X = \mathrm{span}\{p_i : i \in \mathcal{I}\}$;*
6. *The left $Fun_q(\mathcal{L})$-module $Fun_q(\mathcal{L}) \cdot \mathrm{span}\{p_ip_j, p_i, \mathbf{I} : i,j \in \mathcal{I}\}$ has a free basis consisting of $10+4+1$ elements.*

Possible q-Lorentz subgroups are generated by $M_{\alpha\beta}$, $\alpha, \beta = 1,2$, satisfying (3.6.1)-(3.6.3), where $X = \sigma Q'$ and the matrices E, E' and Q' have one of the following forms:

1)

$$E = \mathrm{e}_1 \otimes \mathrm{e}_2 - \mathrm{e}_2 \otimes \mathrm{e}_1\ ,$$

$$E' = -\mathrm{e}^1 \otimes \mathrm{e}^2 + \mathrm{e}^2 \otimes \mathrm{e}^1\ ,$$

$$Q' = \begin{pmatrix} t^{-1} & 0 & 0 & 0 \\ 0 & t & 0 & 0 \\ 0 & 0 & t & 0 \\ 0 & 0 & 0 & t^{-1} \end{pmatrix}, \quad 0 < t \leq 1\ ;$$

2) E, E' as above in 1),

$$Q' = \begin{pmatrix} 1 & 0 & 0 & 1 \\ 0 & 1 & 0 & 0 \\ 0 & 0 & 1 & 0 \\ 0 & 0 & 0 & 1 \end{pmatrix} ;$$

3)

$$E = e_1 \otimes e_2 - e_2 \otimes e_1 + e_1 \otimes e_1 ,$$
$$E' = -e^1 \otimes e^2 + e^2 \otimes e^1 + e^2 \otimes e^2 ,$$
$$Q' = \begin{pmatrix} 1 & 0 & 0 & r \\ 0 & 1 & 0 & 0 \\ 0 & 0 & 1 & 0 \\ 0 & 0 & 0 & 1 \end{pmatrix} , \quad r \geq 0 ;$$

4) E, E' as in 3),

$$Q' = \begin{pmatrix} 1 & 1 & 1 & 0 \\ 0 & 1 & 0 & -1 \\ 0 & 0 & 1 & -1 \\ 0 & 0 & 0 & 1 \end{pmatrix} ;$$

5)

$$E = e_1 \otimes e_2 + e_2 \otimes e_1 ,$$
$$E' = e^1 \otimes e^2 + e^2 \otimes e^1 ,$$
$$Q' = i \begin{pmatrix} t^{-1} & 0 & 0 & 0 \\ 0 & -t & 0 & 0 \\ 0 & 0 & -t & 0 \\ 0 & 0 & 0 & t^{-1} \end{pmatrix} , \quad 0 < t \leq 1 ;$$

6) E, E' as in 5),

$$Q' = i \begin{pmatrix} 1 & 0 & 0 & 1 \\ 0 & -1 & 0 & 0 \\ 0 & 0 & -1 & 0 \\ 0 & 0 & 0 & 1 \end{pmatrix} ;$$

7) E, E' as in 5),

$$Q' = i \begin{pmatrix} r & 0 & 0 & l \\ 0 & -r & l & 0 \\ 0 & l & -r & 0 \\ l & 0 & 0 & r \end{pmatrix} ,$$

where

$$r = (t + t^{-1})/2 \ , \quad l = (t - t^{-1})/2 \ , \quad 0 < t < 1 \ .$$

Recall that

$$e_1 = \begin{pmatrix} 1 \\ 0 \end{pmatrix} , \quad e_2 = \begin{pmatrix} 0 \\ 1 \end{pmatrix} , \quad e^1 = \begin{pmatrix} 1 & 0 \end{pmatrix} , \quad e^2 = \begin{pmatrix} 0 & 1 \end{pmatrix} .$$

All the above triples (E, E', Q') give non-isomorphic $\mathcal{L}_q$.

Following [192] we introduce

$$L = sq^{1/2}(\mathbf{1}\otimes\mathbf{1} + q^{-1}EE') \ , \quad \tilde{L} = q\sigma L\sigma \ , \tag{3.6.44}$$

$$G = (V^{-1}\otimes\mathbf{1})(\mathbf{1}\otimes X)(L\otimes\mathbf{1})(\mathbf{1}\otimes V) \ ,$$

$$\tilde{G} = (V^{-1}\otimes\mathbf{1})(\mathbf{1}\otimes\tilde{L})(X^{-1}\otimes\mathbf{1})(\mathbf{1}\otimes V) \ ,$$

$$W = (V^{-1}\otimes V^{-1})(\mathbf{1}\otimes X\otimes\mathbf{1})(L\otimes\tilde{L})(\mathbf{1}\otimes X^{-1}\otimes\mathbf{1})(V\otimes V) \ ,$$

where $V_{\alpha\beta,i} = (\sigma_i)_{\alpha\beta}$, σ_i are the Pauli marices;

(i) $q = q^{1/2} = 1$, in the cases 1)–4);

(ii) $q = -1$, $q^{1/2} = i$, in the cases 5)–7), $s = \pm 1$.

Structure of possible $\mathcal{P}_q$ is defined by the theorem [192] which we just reproduce here

- $Fun_q(\mathcal{P})$ is the universal *-algebra with unity generated by $M_{\lambda\rho}$ and p_i satisfying (3.6.1)- (3.6.3) and

$$p_i a = (a\hat{*}f_{ij})p_j + a\hat{*}\eta_i - \Lambda_{ij}(\eta_j\hat{*}a), \quad a \in Fun_q(\mathcal{L}), \tag{3.6.45}$$

$$(W - \mathbf{1}\otimes\mathbf{1})_{kl,ij}(p_i p_j - \eta_i(\Lambda_{js})p_s + T_{ij} - \Lambda_{im}\Lambda_{jn}T_{mn}) = 0, \tag{3.6.46}$$

$$p_i^* = p_i, \tag{3.6.47}$$

where $f = (f_{ij})_{i,j\in\mathcal{I}}$, $\eta = (\eta_i)_{i\in\mathcal{I}}$ and $T = (T_{ij})_{i,j\in\mathcal{I}}$ are uniquely determined by $s = \pm 1$, $H_{\alpha\beta\gamma\delta}, T_{\alpha\beta\gamma\delta} \in \mathbb{C}$ and the following properties:

a) $Fun_q(\mathcal{L}) \ni a \longrightarrow \rho(a) = \begin{pmatrix} f(a) & \eta(a) \\ 0 & \varepsilon(a) \end{pmatrix} \in M_5(\mathbb{C})$ is a unital homomorphism;

b) $\rho(a^*) = \overline{\rho(S(a))}, \qquad a \in Fun_q(\mathcal{L})$;

c)

$$f_{ij}(M_{\gamma\delta}) = G_{i\gamma,\delta j} , \tag{3.6.48}$$

$$\eta_i(M_{\gamma\delta}) = V^{-1}_{i,\alpha\beta} H_{\alpha\beta\gamma\delta} , \tag{3.6.49}$$

$$T_{ij} = (V^{-1} \otimes V^{-1})_{ij,\alpha\beta\gamma\delta} T_{\alpha\beta\gamma\delta} . \tag{3.6.50}$$

- The $*$-Hopf structure in $\mathcal{P}_q$ is determined by:

$$\Delta M = M \oplus M, \quad \Delta \bar{M} = \bar{M} \oplus \bar{M}, \quad \Delta p = p \oplus I + \Lambda \oplus p,$$

$$\varepsilon(M) = \mathbf{1}, \quad \varepsilon(\bar{M}) = \mathbf{1}, \quad , \varepsilon(p) = 0,$$

$$S(M) = M^{-1}, \quad S(\bar{M}) = \bar{M}^{-1}, \quad S(p) = -\Lambda^{-1} p.$$

Quantum Poincaré groups corresponding to different s are non-isomorphic.

Consistency equations from the general theory of q-inhomogeneous groups reduced in this particular case to a set of polynomial equations for $H_{\lambda\rho\gamma\delta}$, $T_{\lambda\rho\gamma\delta}$ are solved in [192]. The solutions lead to the following classification.

For each case in the classification of q-Lorentz subgroups and each s, the structure constants H and T giving via formulae in (3.6.45)-(3.6.50) all non-isomorphic quantum Poincaré groups $\mathcal{P}_q$, are presented in the following list:

(A) The case 1) in the list of q-Lorentz subgroups, with the parameter $t = 1$ and the parameter s in (3.6.44) of the value $s = -1$:

$$\left. \begin{aligned} H_{\alpha\beta\gamma\delta} &= 0, \\ T_{\alpha\beta\gamma\delta} &= V_{\alpha\beta,i} V_{\gamma\delta,j} T_{ij}, \end{aligned} \right\} \tag{3.6.51}$$

where

a) $T_{03} = -T_{30} = ia$, $T_{12} = -T_{21} = ib$, other $T_{ij} = 0$, $a = \cos\phi$, $b = \sin\phi$ (one parameter family for $0 \leq \phi \leq \pi/2$);

b) $T_{02} = T_{12} = i$, $T_{20} = T_{21} = -i$, other $T_{ij} = 0$;

c) all $T_{ij} = 0$.

(B) 1), $0 < t < 1$, $s = \pm 1$:

$$\left. \begin{aligned} &T_{1122} = ia, \quad T_{1221} = b, \\ &T_{2112} = -b, \quad T_{2211} = -ia, \\ &\text{all } H_{\alpha\beta\gamma\delta} \text{ and other } T_{\alpha\beta\gamma\delta} \text{ equal to } 0 \ , \end{aligned} \right\} \tag{3.6.52}$$

where

a) $a = \cos\phi$, $b = \sin\phi$ (one parameter family for $0 \le \phi < \pi$);

b) $a = b = 0$.

(C) **2)**, $s = 1$:

The first case:

$$\left.\begin{array}{c} H_{1111} = -(a+bi), \quad H_{1122} = a+bi, \quad H_{2112} = -2bi, \\ T_{2111} = c - di, \quad T_{1211} = -c - di, \\ T_{1121} = -c + di, \quad T_{1112} = c + di, \\ \text{other } H_{\alpha\beta\gamma\delta} \text{ and } T_{\alpha\beta\gamma\delta} \text{ equal to } 0\ , \end{array}\right\} \tag{3.6.53}$$

a) $a = 1$, $c = d = 0$ (one parameter family for $b \in \mathbb{R}$);
b) $a = 0$, $b = 1$, $d = 0$ (one parameter family for $c \ge 0$).

The second case:

$$\left.\begin{array}{c} H_{1212} = a+bi, \quad T_{2112} = (a^2+b^2)/2, \quad T_{2111} = c - di, \\ T_{1221} = -(a^2+b^2)/2, \quad T_{1211} = -c - di, \quad T_{1121} = -c + di, \\ T_{1112} = c + di, \quad T_{1111} = -(a^2+b^2)/2, \\ \text{other } H_{\alpha\beta\gamma\delta} \text{ and } T_{\alpha\beta\gamma\delta} \text{ equal to 0 and} \end{array}\right\} \tag{3.6.54}$$

a) $a = 1$, $b = 0$, $c = r\cos\phi$, $d = r\sin\phi$ (two parameter family for $r > 0$, $0 \le \phi < \pi/2$ or $r = \phi = 0$);
b) $a = b = 0$, $c = 1$, $d = 0$;
c) $a = b = c = d = 0$.

(D) **2)**, $s = -1$: corresponds to the solution (3.6.53) and

a) $a = b = 0$, $c = 1$, $d = 0$;

b) $a = b = c = d = 0$.

(E) **3)**, $s = \pm 1$, $r \ge 0$: all $H_{\alpha\beta\gamma\delta}$ and $T_{\alpha\beta\gamma\delta}$ equal to 0.

(F) 4), $s = 1$:

$$\left.\begin{array}{c} H_{2212} = -2bi, \quad H_{2122} = -bi, \quad H_{2112} = a - bi, \\ H_{2111} = bi, \quad H_{1222} = bi, \quad H_{1212} = a, \quad H_{1211} = -bi, \\ H_{1121} = -2bi, \quad H_{1112} = 3bi/4, \quad H_{1111} = -4bi, \\ T_{1112} = 9b^2/8 + 3abi/2, \quad T_{1121} = -9b^2/8 + 3abi/2, \\ T_{1211} = -9b^2/8 - 3abi/2, \quad T_{1221} = 3b^2/2, \\ T_{2111} = 9b^2/8 - 3abi/2, \quad T_{2112} = -3b^2/2, \\ \text{other } H_{\alpha\beta\gamma\delta} \text{ and } T_{\alpha\beta\gamma\delta} \text{ equal to } 0 \end{array}\right\} \tag{3.6.55}$$

a) $a = \cos\phi$, $b = \sin\phi$ (one parameter family for $0 \leq \phi < \pi$);
b) $a = b = 0$.

(G) 4), $s = -1$: all $H_{\alpha\beta\gamma\delta}$ and $T_{\alpha\beta\gamma\delta}$ equal to 0.

(H) 5), $s = \pm 1$, $0 < t < 1$:

$$\left.\begin{array}{c} T_{1122} = ia, \quad T_{1221} = b, \quad T_{2112} = -b, \quad T_{2211} = -ia, \\ \text{all } H_{\alpha\beta\gamma\delta} \text{ and other } T_{\alpha\beta\gamma\delta} \text{ equal to } 0 \end{array}\right\} \tag{3.6.56}$$

a) $a = \cos\phi$, $b = \sin\phi$ (one parameter family for $0 \leq \phi < \pi$);
b) $a = b = 0$.

(I) 6), $s = 1$: all $H_{\alpha\beta\gamma\delta}$ and $T_{\alpha\beta\gamma\delta}$ equal to 0.

(J) 6), $s = -1$:

The first case:

$$\left.\begin{array}{c} H_{1111} = -(a + bi), \quad H_{1122} = a + bi, \quad H_{2112} = -2bi \\ \text{other } H_{\alpha\beta\gamma\delta} \text{ and all } T_{\alpha\beta\gamma\delta} \text{ equal to } 0, \end{array}\right\} \tag{3.6.57}$$

a) $a = \cos\phi$, $b = \sin\phi$ (one parameter family for $0 \leq \phi < \pi$);

b) $a = b = 0$.

The second case:

$$\left.\begin{array}{l} H_{1212} = 1, \quad T_{1111} = -\frac{1}{2}, \\ T_{1221} = -\frac{1}{2}, \quad T_{2112} = \frac{1}{2}, \\ \text{other } H_{\alpha\beta\gamma\delta} \text{ and } T_{\alpha\beta\gamma\delta} \text{ equal to } 0, \end{array}\right\} . \tag{3.6.58}$$

(K) 7), $s = \pm 1$, $0 < t < 1$: all $H_{\alpha\beta\gamma\delta}$ and $T_{\alpha\beta\gamma\delta}$ equal to 0.

In the remaining cases **1)**, $s = 1$, $t = 1$ and **5)**, $s = \pm 1$, $t = 1$, one defines in addition

$$Z_{ij,k} = \eta_i(\Lambda_{jk}) = V^{-1}_{i,\lambda\rho} V^{-1}_{j,\gamma\sigma} (H_{\lambda\rho\gamma\alpha}\delta_{\sigma\beta} - \overline{H_{\rho\lambda\delta\beta}}\delta_{\gamma\alpha}) V_{\alpha\beta,k} \,,$$

(then $H_{\lambda\rho\gamma\alpha} = \frac{1}{2} V_{\lambda\rho,i} V_{\gamma\delta,j} Z_{ij,k} V^{-1}_{k,\alpha\delta}$).

(L) 1), $s = 1$, $t = 1$:

A pair (Z, T) corresponds to a quantum Poincaré group, if and only if

$$T_{mn} = -T_{nm} \in i\mathbb{R}, \qquad Z_{ij,s} g_{sk} = -Z_{ik,s} g_{sj} \in i\mathbb{R} \,,$$

$$\{[(\sigma - \mathbf{1}\otimes\mathbf{1})\otimes\mathbf{1}][(\mathbf{1}\otimes Z)Z - (Z\otimes\mathbf{1})Z]\}_{ijm,n} = -\frac{1}{4} t(\delta_{in} g_{jm} - \delta_{jn} g_{im}) \,, \quad t \in \mathbb{R} \,,$$

$$\lambda_3 (Z\otimes\mathbf{1}) T = 0 \,,$$

where $g_{00} = 1$, $g_{11} = g_{22} = g_{33} = -1$, other $g_{ij} = 0$,

$$\begin{aligned} \lambda_3 \;=\; & \mathbf{1}\otimes\mathbf{1}\otimes\mathbf{1} - \sigma\otimes\mathbf{1} - \mathbf{1}\otimes\sigma + (\sigma\otimes\mathbf{1})(\mathbf{1}\otimes\sigma) \\ & + (\mathbf{1}\otimes\sigma)(\sigma\otimes\mathbf{1}) - (\sigma\otimes\mathbf{1})(\mathbf{1}\otimes\sigma)(\sigma\otimes\mathbf{1}) \,, \end{aligned}$$

is the classical (not normalized) antisymmetrizer.

(M) 5), $s = \pm 1$, $t = 1$:

In addition to the conditions in the preceding item some components of T_{mn}, $Z_{ij,k}$ must vanish (see [192] for the details).

This list gives *all the quantum Poincaré groups (up to isomorphism but not necessarily non-isomorphic)*.

In particular,

- The classical Poincaré group is obtained in the case 1), $s = 1$, $t = 1$, $H = 0$, $T = 0$;
- The quantum Poincaré group of Section 3.4 corresponds to 1), $s = 1$, $t > 0$, $H = 0$, $T = 0$ (t is denoted by q there);
- The quantum Poincaré group obtained in the preceding Section 3.5.3 via the contraction of anti-de Sitter group corresponds to 1), $s = 1$, $t = 1$,

$$H_{1111} = -H_{1122} = \frac{1}{2}H_{1221} = \frac{1}{2}H_{2112} = -H_{2211} = H_{2222} = ih/2, \quad h \in \mathbb{R},$$

  other $H_{\alpha\beta\gamma\delta}$ and all $T_{\alpha\beta\gamma\delta}$ equal to 0.

Let

$$\mathcal{B}^N = Fun_q(\mathcal{L}) \cdot \text{span}\{p_{i_1} \cdot \ldots \cdot p_{i_n} : i_1, \ldots, i_n \in \mathcal{I}, \quad n = 0, 1, \ldots, N\}.$$

Then $\mathcal{B}^N$ is a free left $Fun_q(\mathcal{L})$-module and number of monomials in p_s of degrees less than a given number is the same as in the case of classical Poincaré group. It is clear that because of inhomogeneous CR for p_s, this is not correct for a number of monomials of fixed degree.

Taking into account properties of the usual Minkowski space and the Poincaré group action on it, the q-Minkowski space $\mathcal{M}_q$ can be defined as follows

Definition 3.4 $(\mathcal{M}_q, \delta)$ *describes a quantum Minkowski space associated with a quantum Poincaré group $\mathcal{P}_q$, if the following conditions are satisfied*

1. *$\mathcal{M}_q$ is a unital *-algebra generated by x_i, $i \in \mathcal{I}$, $x_i^* = x_i$ and $\delta : \mathcal{M}_q \longrightarrow Fun_q(\mathcal{P}) \otimes \mathcal{M}_q$ is the coaction*

$$\delta x_i = \Lambda_{ij} \otimes x_j + p_i \otimes \mathbf{I} \; ; \tag{3.6.59}$$

2. *if $\delta W \subset Fun_q(\mathcal{L}) \otimes W$ for a linear subspace $W \subset \mathcal{M}_q$, then $W \subset \mathbb{C}\mathbf{I}$ (analog of translation homogeneity of usual Minkowski space);*
3. *if $(\mathcal{M}'_q, \delta')$ also satisfies 1)–2) for some $x_i' \in \mathcal{M}'_q$ then there exists a unital *-homomorphism $\rho : \mathcal{M}_q \longrightarrow \mathcal{M}'_q$ such that $\rho(x_i) = x_i'$ and $(\mathrm{id} \otimes \rho)\delta = \delta' \rho$ (universality of $(\mathcal{M}_q, \delta)$).*

The structure of $\mathcal{M}_q$ is defined by the following result:

$\mathcal{M}_q$ is the universal unital *-algebra generated by x_i, $i = 0,1,2,3$, satisfying $x_i^* = x_i$ and

$$(W - \mathbf{1}\otimes\mathbf{1})_{ij,kl}(x_k x_l - \eta_k(\Lambda_{lm})x_m + T_{kl}) = 0, \tag{3.6.60}$$

and δ is given by (3.6.59). These CR are very close to (3.6.46) and again (as for $\mathcal{B}^N$) the number of monomials in x^i of degrees less than a given number is the same as in the case of classical Minkowski space.

Podles and Woronowicz have constructed explicitly R-matrices for all the $\mathcal{P}_q$.

Let

$$w := (V^{-1}\otimes V^{-1})(\mathbf{1}\otimes X\otimes\mathbf{1})(E\otimes\sigma E)\,,$$

$$Z_{ij,k} = \eta_i(\Lambda_{jk})\,,$$

$$W_1 = \begin{pmatrix} W & Z & -W\cdot Z & (W - \mathbf{1}\otimes\mathbf{1})T \\ 0 & 0 & \mathbf{1} & 0 \\ 0 & \mathbf{1} & 0 & 0 \\ 0 & 0 & 0 & \mathbf{1} \end{pmatrix},$$

$$W_2 = \begin{pmatrix} 0 & 0 & 0 & w \\ 0 & 0 & 0 & 0 \\ 0 & 0 & 0 & 0 \\ 0 & 0 & 0 & 0 \end{pmatrix}.$$

Then for all deformations $\mathcal{P}_q$, the $\hat{\mathrm{R}}$-matrix for corepresentation $\mathcal{P}$ (see Definition 3.3) satisfies QYBE (1.5.5), if and only if

a) $\hat{R} = x\mathbf{1}$ ($x \in \mathbb{C}\setminus\{0\}$), or

b) $\hat{R} = yW_1 + zW_2$ ($y, z \in \mathbb{C}$; in the case 4), $s = 1$, $b \neq 0$, one must put $y = 0$).

Those $\hat{R}$ are invertible if and only if we have the case a) or b) with $y \neq 0$.

Let us shortly discuss the list of the q-groups and the corresponding q-Minkowski spaces.

First of all, we note that the matrices E, E' which define part of CR for possible q-Lorentz subgroups contain no continuous parameters of deformation. So it is easy to see that *only the case 1)* of possible q-Lorentz subgroups corresponds to *smooth* deformation of Poincaré symmetry. It is a quite interesting question: what is the physical meaning (if any) of the other *discrete* deformations? But it is definitely outside the present point of view on q-symmetry as the genuine symmetry of the nature (which reveal itself at high energy and small distances) and on classical symmetry as an approximate one (at large distances). Such a point of view requires *a smooth* deformation and, hence, we have to restrict ourselves only to the case 1) in the list of q-Lorentz subgroups and to the cases (B), s=1 or (L) in the list of q-Poincaré groups. These deformations contain the smooth parameter t which defines CR for elements of $Fun_q(\mathcal{L})$ and *homogeneous* part of CR for q-translations (see (3.6.47)) and coordinates (3.6.60). Another set of smooth parameters are packed in $H_{\alpha\beta\gamma\delta}$ and $T_{\alpha\beta\gamma\delta}$. These parameters control *inhomogeneous* parts of CR for q-translations and q-coordinates (cf. (3.6.47) and (3.6.60)).

In Section 3.4 we considered in details the deformation with nontrivial homogeneous part of the CR ($t \neq 1$; recall that this parameter was denoted in this particular case by q) and trivial inhomogeneous part $H = T = 0$. Contrary to this, the deformation discussed in the preceding Section 3.5 (q-contraction of simple q-groups) corresponds to trivial homogeneous and nontrivial inhomogeneous parts of the CR.

Thus, we have considered in this book both possible essentially different deformations and any other smooth deformation of Poincaré group must be some composition of these limiting cases.

Chapter 4

Non-commutative Geometry and Internal Symmetries of Field Theoretical Models

In the preceding Chapter we discussed the attempts to deform the geometry of Minkowski space and the corresponding groups of space-time symmetries. But besides the latter, the modern realistic field theoretical models of fundamental interactions exhibit a rich structure of the so-called internal or gauge symmetries and patterns of their spontaneous breaking (see e.g. [36]). Gauge symmetry together with the requirements of renormalizability and anomaly cancellations put strong restrictions on the possible choice of models in the sector of gauge and matter fields, leaving, in fact, a freedom in the choice of the very gauge group only. At present, experimental results give confidence that elementary particle interactions at energies $\sim 100\ GeV$ are defined by the $SU(3)\otimes SU(2)\otimes U(1)$ gauge symmetry. The situation with the so-called *Higgs sector* which is responsible for spontaneous symmetry breaking is quite different. Theoretically, there are many different ways to construct it and this sector is very complicated for experimental testing. So there exist different modifications of the Standard Model (SM) of fundamental interactions (see e.g.[203] and refs. therein).

In any case, it is clear that for *theoretical* explanation of the special choice of gauge symmetry in the SM and its spontaneous breaking, it is necessary to incorporate some additional and more fundamental ideas. Historically, the first and a proved-to-be very fruitful approach to theoretical explanations of internal symmetries was the Kaluza-Klein idea about multidimensionality of the space-time, its spontaneous compactification and dimensional reduction

of corresponding field theoretical models[5]. Unfortunately, this theory itself, even in its modern supergravity version has a number of essential shortcomings including non-renormalizability[96].

It is a common belief that these problems can be solved in a more complicated and fundamental theory of superstrings[112]. The latter, being very promising and ambitious, has serious problems (hopefully technical) in attempts to link string phenomena which reveal itself at extremely high Plank scale of energy ($\sim 10^{19}$ GeV) with the present experimental results and facilities ($\sim 10^2 \div 10^3$ GeV).

Another way of modification of original Kaluza-Klein idea, which probably can (or must) be incorporated into the superstring theory, is provided by non-commutative geometry. A pioneer work in this direction had been made by Connes and Lott[65] on the basis of Connes' formulation of non-commutative geometry[63, 64]. The main idea of the approach is to formulate the model in the extended space-time in the spirit of Kaluza-Klein approach but to use for this extension the internal space described in terms of non-commutative geometry. After this work, there appeared variety of different modifications and interpretations of the original idea. At present, these attempts are still far from their ultimate aim - to give *natural* explanation and description of a Higgs sector of the SM and its possible generalizations (Grand Unified Models). Being close to each other from conceptual point of view, the different models are based on different concrete postulates and lead to different predictions. Besides they have some specifically field theoretical problems which are definitely outside the scope of this book (for example, the problem of stability of predictions with respect to renormalization). So we will not review all of them. Instead, after introduction of basic facts about the Connes-Lott model (Section 4.1) we will discuss in Section 4.2 its essential properties on the example of simplest internal spaces and in the form maximally close to the usual Kaluza-Klein theory. This will allow us to understand better the origin of problems in models of this kind and provide a link with space-time deformation discussed in the previous Chapter. Actually, we will argue that a consistent theory of this type must include, along with quantum internal geometry, the space-time deformation as well and moreover we shall raise the question about the dynamical nature of the deformation parameters. In the concluding Section of this Chapter (and the book), we will introduce basic definitions of q-deformed fibre bundle geometry which may provide appropriate geometrical framework for further development of the ideas.

4.1 Non-commutative geometry of Yang-Mills-Higgs models

In this section we will give basic facts about general construction of the Connes-Lott model [63, 65, 64, 126], skipping many peculiar mathematical details which do not influence essentially the very idea of the approach.

Let us start from the list of building blocks of usual Yang-Mills-Higgs theory. Any gauge model of fundamental interactions is defined by the following basic structures:

1. A finite dimensional, real, compact Lie group G with a positive definite, bilinear invariant form on the Lie algebra L of G, parameterized by a few (if the group is *semisimple*) positive numbers g_i, the coupling constants;
2. A unitary representation ρ_L on a Hilbert space $\mathcal{H}_L$ of the left handed fermions ψ_L, and a representation ρ_R on $\mathcal{H}_R$ of the right handed fermions ψ_R;
3. A representation ρ_S on $\mathcal{H}_S$ of (Higgs) scalars φ and an invariant, positive polynomial $V(\varphi)$, $\varphi \in \mathcal{H}_S$ of order 4 (the Higgs potential);
4. One complex number or Yukawa coupling g_Y for *every* trilinear invariant, i.e. singlet in the decomposition of the representation associated to $\left(\bar{\mathcal{H}}_L \hat{\otimes} \mathcal{H}_R \hat{\otimes} \mathcal{H}_S\right) \oplus \left(\bar{\mathcal{H}}_L \hat{\otimes} \mathcal{H}_R \hat{\otimes} \bar{\mathcal{H}}_S\right)$.

The gauge symmetry of a model with invariant Lagrangian is spontaneously broken if minimum (vacuum) value $v \in \mathcal{H}_S$ of the Higgs potential is not gauge invariant: $\rho_S(g)v \neq v$, for some $g \in G$.

Recall that the Standard Model is defined by the following inputs:

1.
$$G = SU(3) \times SU(2) \times U(1) ,$$
with three coupling constants g_3, g_2, g_1;

2.
$$\mathcal{H}_L = \bigoplus_1^3 \left[(1,2,-1) \oplus (3,2,\frac{1}{3})\right], \qquad (4.1.1)$$
$$\mathcal{H}_R = \bigoplus_1^3 \left[(1,1,-2) \oplus (3,1,\frac{4}{3}) \oplus (3,1,-\frac{2}{3})\right], \qquad (4.1.2)$$

where (n_3, n_2, y) denotes the tensor product of an n_3 dimensional representation of $SU(3)$, an n_2 dimensional representation of $SU(2)$ and the one dimensional representation of $U(1)$ with hypercharge y. The three terms in the direct sums (4.1.1),(4.1.2) correspond to three generations of elementary particles[203];

3. The Higgs potential has the form

$$V(\varphi) = \lambda(\varphi^\dagger\varphi)^2 - \frac{\mu^2}{2}\varphi^\dagger\varphi, \qquad \varphi \in \mathcal{H}_S, \quad \lambda, \mu^2 > 0.$$

where

$$\mathcal{H}_S = (1, 2, -1),$$

and any doublet φ of the Higgs fields subjected to the constraint

$$\sqrt{\varphi^\dagger\varphi} = |\mu|/(2\sqrt{\lambda})$$

is a minimum and *the unbroken* subgroup is $G_\ell = SU(3) \times U(1)_{em}$;

4. There are 27 Yukawa couplings in the Standard Model.

The gauge fields are 1-forms with values in the Lie algebra L of G: $\mathcal{A} \in \Omega^1(\mathcal{M}, L)$. The mass matrix for the gauge bosons A appears from the gauge invariant Lagrangian

$$\mathcal{L}_\varphi = \eta^{\mu\nu}\gamma_{ij}\nabla_\mu\varphi^{\dagger i}\nabla_\nu\varphi^j \ ,$$

where

$$\nabla\varphi := d\varphi + \tilde{\rho}_S(\mathcal{A})\varphi \ ,$$

is covariant differential ($\tilde{\rho}_S$ denotes Lie algebra representation on $\mathcal{H}_S$), $\eta^{\mu\nu}$ is the usual Minkowski metric and γ_{ij} is the invariant metric on $\mathcal{H}_S$. The mass matrix

$$(\tilde{\rho}_S(\mathcal{A})v)^* \tilde{\rho}_S(\mathcal{A})v \ ,$$

contains the masses of the gauge bosons and vanishes on the generators of unbroken subgroup. In the example of the Standard Model, the unbroken subgroup is generated by the gluons and the photon which remain massless.

A fermionic mass matrix μ defines a linear map $\mu : \mathcal{H}_R \longrightarrow \mathcal{H}_L$. To produce it, one adds *by hand* to the Dirac Lagrangian Yukawa terms of the form

$$\sum_{j=1}^{n} g_{Yj}\left(\psi_L^\dagger, \psi_R, \varphi\right)_j + \sum_{j=n+1}^{m} g_{Yj}\left(\psi_L^\dagger, \psi_R, \varphi^\dagger\right)_j \ + \text{ complex conjugate,} \quad (4.1.3)$$

n is the number of singlets in $\left(\bar{\mathcal{H}}_L \hat{\otimes} \mathcal{H}_R \hat{\otimes} \mathcal{H}_S\right)$, and $m-n$ the number of singlets in $\left(\bar{\mathcal{H}}_L \hat{\otimes} \mathcal{H}_R \hat{\otimes} \bar{\mathcal{H}}_S\right)$. This gives the fermionic mass matrix μ as a function of the Yukawa couplings $g_{Y\,j}$ and the vacuum value v

$$\psi_L^* \mu \psi_R := \sum_{j=1}^{n} g_{Y\,j} \left(\psi_L^*, \psi_R, v\right)_j + \sum_{j=n+1}^{m} g_{Y\,j} \left(\psi_L^*, \psi_R, v^*\right)_j .$$

Now we turn to Connes-Lott models which is defined by the following structures:

1. A finite dimensional, associative algebra A over the field $\mathbb{R}$ or $\mathbb{C}$ with unit $\mathbb{I}$ and involution *;
2. Two *-representations of A, ρ_L and ρ_R on Hilbert spaces $\mathcal{H}_L$ and $\mathcal{H}_R$, such that $\rho := \rho_L \oplus \rho_R$ is faithful;
3. A mass matrix μ i.e. a linear map $\mu : \mathcal{H}_R \longrightarrow \mathcal{H}_L$;
4. A certain number of coupling constants depending on the degree of reducibility of $\rho_L \oplus \rho_R$;
5. A linear map $\mathcal{D}$ from $\mathcal{H} = \mathcal{H}_R \oplus \mathcal{H}_L$ into itself

$$\mathcal{D} := \begin{pmatrix} 0 & \mu \\ \mu^* & 0 \end{pmatrix},$$

which plays the role of the Dirac operator and is called internal Dirac operator.

The general Connes' formalism involves three steps.

(1) One starts with the above listed data. Note that the original Connes-Lott formalism takes the space-time as Euclidean and the Lorentz signature is recovered only at the end, through the Wick rotation. In the case of a $U(1) \times U(1)$ theory with symmetry breaking, A is equal to the direct sum of two algebras of complex functions on space-time (one is labelled by "left" and the other by "right"); this algebra is still commutative and can be written as the space of 2×2 diagonal matrices with entries which are functions of space-time. We will analyze this case in more details in slightly different but equivalent (in this special case) formalism. In a more complicated setting, where the gauge group is not abelian, one has just to replace algebra of functions on space-time A by its tensor product with an appropriate matrix algebra. This case will be also discussed in a simplified formalism in the next Section. The description

of realistic Standard Model in Connes' approach is based on the algebra $A = Fun(\mathcal{M}) \otimes \Big(\mathbb{C} \oplus \mathbb{H} \oplus M_3(\mathbb{C}) \Big)$, where $\mathbb{H}$ is the algebra of quaternions and $M_3(\mathbb{C})$ are 3×3 complex matrices.

(2) At the second step one has to construct a differential algebra Ω_D (whose definition relies on the choice of $\mathcal{D}$), out of which one defines the generalized connections and curvatures.

(3) The third step is the construction of the Yang-Mills (or generalized Yang-Mills) Lagrangian itself and involves the so-called Dixmier trace as a substitution for integration. The triple $(A, \mathcal{H}, D)$ is called a *K-cycle* or *a spectral triple*.

There are other variants of the formalism which are based either on more simple algebras and pure bosonic (gauge and Higgs) sector of models [92, 93, 94, 11] (see below in this section) or start with the same Dirac-Yukawa operator but do not require the construction of the algebra Ω_D and the use of the Dixmier trace [67]. Their clear advantage is simplicity but they have not the generality of Connes' approach. The only quantity to be computed is the expression for the generalized curvature $\mathcal{F}$ in terms of the generalized connection $\mathcal{A}$.

An auxiliary differential algebra ΩA of the Connes and Lott construction is the so-called *universal differential envelope of A*. It is defined iteratively:

1. $\Omega^0 A := A$;

2. $\Omega^1 A$ is generated by symbols da, $a \in A$, with usual defining relations

$$d\mathbf{I} = 0, \quad d(ab) = (da)b + adb.$$

Therefore $\Omega^1 A$ consists of finite sums of terms of the form $a_0 da_1$

$$\Omega^1 A = \left\{ \sum_j a_0^j da_1^j, \quad a_0^j, a_1^j \in A \right\} ;$$

3. Likewise for higher p (we omit the wedge signs)

$$\Omega^p A = \left\{ \sum_j a_0^j da_1^j ... da_p^j, \quad a_q^j \in A \right\} .$$

The differential d is defined by

$$d(a_0da_1...da_p) := da_0da_1...da_p \ .$$

This implies the usual property: $d^2 = 0$. Notice that the universal differential envelope ΩA of a commutative algebra A is not necessarily a graded commutative. The universal differential envelope of any algebra has no cohomology. This means that every closed form ω : $d\omega = 0$ of degree $p \geq 1$, is exact, i.e. $\omega = d\kappa$ for some $(p-1)$form κ.

The involution * is extended from the algebra A to $\Omega^1 A$ by putting

$$(da)^* := d(a^*) =: da^*.$$

With the definition $(\omega\omega')^* = (\omega')^*\omega^*$, the involution is extended to the whole differential envelope. This means that one uses the second possibility (1.1.4) for involution of differential forms on an algebra mentioned in Section 1.1.

The next step is to extend the representation $\rho := \rho_L \oplus \rho_R$ on $\mathcal{H} := \mathcal{H}_L \oplus \mathcal{H}_R$ from the algebra A to its universal differential envelope ΩA. This extension is the central piece of Connes' algorithm and here the internal Dirac operator (see item 5. on page 285) comes to the play

$$\pi : \Omega A \longrightarrow \mathrm{End}(\mathcal{H}) \ , \tag{4.1.4}$$

$$\pi(a_0da_1...da_p) := (-i)^p \rho(a_0)[\mathcal{D}, \rho(a_1)]...[\mathcal{D}, \rho(a_p)] \ . \tag{4.1.5}$$

A straightforward calculation shows that π is in fact a representation of ΩA as an algebra with involution, and one can define also a differential, again denoted by d, on $\pi(\Omega A)$ by

$$d\pi(\omega) := \pi(d\omega). \tag{4.1.6}$$

However, this definition is not meaningful if there are forms $\omega \in \Omega A$ with $\pi(\omega) = 0$ and $\pi(d\omega) \neq 0$. This requires the construction of a new quotient differential algebra $\Omega_{\mathcal{D}} A$

$$\Omega_{\mathcal{D}} A := \pi(\Omega A)/J \ ,$$

where J is the ideal

$$J := \pi(d \ker \pi) =: \bigoplus_p J^p \ .$$

Now on the quotient, the differential (4.1.6) is well defined. Degree by degree one has

$$\Omega^0_{\mathcal{D}} A = \rho(A) \ ,$$

since $J^0 = 0$,

$$\Omega^1_{\mathcal{D}} A = \pi(\Omega^1 A) \ ,$$

since ρ is faithful, and in degree $p \geq 2$,

$$\Omega^p_{\mathcal{D}} A = \pi(\Omega^p A)/\pi(d(\ker \pi)^{p-1}) \ .$$

Consider the vector space of anti-hermitian 1-forms $\{H \in \Omega^1_{\mathcal{D}} A,\ H^* = -H\}$. A general element H is of the form

$$H = i \begin{pmatrix} 0 & h \\ h^* & 0 \end{pmatrix}$$

with $h : \mathcal{H}_R \to \mathcal{H}_L$ a finite sum of terms

$$\rho_L(a_0)[\rho_L(a_1)\mu - \mu\rho_R(a_1)] \ ,$$

$a_0, a_1 \in A$. These elements are called Higgs or gauge potentials. In fact, the space of gauge potentials carries an affine representation of the group G of unitaries

$$G := \{g \in A,\ gg^* = g^*g = 1\}$$

defined by

$$H^g := \rho(g)H\rho(g^{-1}) + \rho(g)d(\rho(g^{-1})) \tag{4.1.7}$$

$$= \rho(g)H\rho(g^{-1}) + (-i)\rho(g)[\mathcal{D}, \rho(g^{-1})] \tag{4.1.8}$$

$$= \rho(g)[H - i\mathcal{D}]\rho(g^{-1}) + i\mathcal{D}$$

$$= i \begin{pmatrix} 0 & h^g \\ (h^g)^* & 0 \end{pmatrix} ,$$

with $h^g - \mu := \rho_L(g)[h - \mu]\rho_R(g^{-1})$. H^g is the "gauge transformed of H". As usual every gauge potential H defines a covariant derivative $d + H$. This means that if a form $\omega \in \Omega_{\mathcal{D}} A$ is covariant under the left action of G on $\Omega_{\mathcal{D}} A$

$${}^g\omega := \rho(g)\omega \ ,$$

then

$$(d+H^g)\,{}^g\omega = {}^g\,[(d+H)\omega]\,.$$

Also we define the curvature $\mathcal{F}$ of H by

$$\mathcal{F} := dH + H^2 \;\in \Omega^2_{\mathcal{D}} A.$$

Note that H^2 is considered as element of $\Omega^2_{\mathcal{D}} A$ which means its projection P from $\pi(\Omega^2 A)$. The curvature $\mathcal{F}$ is a hermitian 2-form with *homogeneous* gauge transformations

$$\mathcal{F}^g := d(H^g) + (H^g)^2 = \rho(g)\mathcal{F}\rho(g^{-1}).$$

To construct a dynamical action one has to define invariant functional $\mathcal{F}(H^g) = \mathcal{F}(H)$. To avoid misunderstanding, recall that we work at fixed space-time point and the algebra A corresponds to a (non-commutative) geometry of an internal space.

As the elements of $\Omega_{\mathcal{D}} A$ are operators on the Hilbert space $\mathcal{H}$, i.e. they are the concrete matrices, there is a natural scalar product defined by

$$<\hat{\omega}, \hat{\kappa}> := \mathrm{Tr}(\hat{\omega}^* \hat{\kappa}), \quad \hat{\omega}, \hat{\kappa} \in \pi(\Omega^p A)\,, \tag{4.1.9}$$

for elements of equal degree and by zero for two elements of different degrees. With this scalar product $\Omega_{\mathcal{D}} A$ is a subspace of $\pi(\Omega A)$, by definition orthogonal to the ideal J. As a subspace, $\Omega_{\mathcal{D}} A$ inherits a scalar product

$$(\omega, \kappa) = <\hat{\omega}, P_J \hat{\kappa}>, \quad \omega, \kappa \in \Omega^p_{\mathcal{D}} A\,,$$

where P_J is the orthogonal projector in $\pi(\Omega A)$ onto the orthogonal completion of J and $\hat{\omega}$ and $\hat{\kappa}$ are any representatives in the classes ω and κ. Again the scalar product vanishes for forms with different degrees.

The Higgs potential $V_0(H)$ (at fixed space-time point) is defined as a functional

$$V_0(H) := (\mathcal{F}, \mathcal{F}) = \mathrm{Tr}[(dH + H^2)P_J(dH + H^2)].$$

It is a polynomial of degree 4 in H with real, non-negative values. Furthermore it is gauge invariant, $V_0(H^g) = V_0(H)$, because of the homogeneous transformation property of the curvature $\mathcal{F}$ and because the orthogonal projector P_J commutes with all gauge transformations, $\rho(g)P_J = P_J\rho(g)$. The most remarkable property of the preliminary Higgs potential is that, in most cases, its vacuum spontaneously breaks the group G.

We will try to understand the origin of this symmetry breaking in the next section using the geometrical formulation of a simple model of this type. Here we only note that in *sharp contrast* with the case of usual Yang-Mills theory, it is possible to construct using the connection H another 1-form

$$\Phi := H - i\mathcal{D}_G =: i\begin{pmatrix} 0 & \varphi \\ \varphi^* & 0 \end{pmatrix} \in \Omega^1_{\mathcal{D}} A , \tag{4.1.10}$$

where

$$\mathcal{D}_G := -i\int_G \pi(g^{-1}dg)\,(d\mu_g) ,$$

$(d\mu_g)$ is the Haar measure of the compact Lie group G and $\varphi = h - \mu_G$, with μ_G defined as

$$\mu_G = \mu - \int_G \rho_L(g^{-1})\mu\rho_R(g)(d\mu_g) ,$$

(do not confuse the mass operator μ with the Haar measure $(d\mu_g)$). This 1-form is transformed *homogeneously*,

$$\Phi^g = H^g - i\mathcal{D}^g_G = \rho(g)\Phi\rho(g^{-1}) . \tag{4.1.11}$$

Actually it is this fact that permits, to construct Higgs fields out of gauge ones. We will consider the meaning of this object in the next section also.

Finally, the vectors ψ_L, ψ_R, and H become space-time dependent by combining with functions from $Fun(\mathcal{M})$ (i.e. forming tensor product of the algebras). Note that in the case of the algebra of complex functions on the manifold $\mathcal{M}$ the differential algebra $\Omega_{\mathcal{D}}A$ constructed with the use of *ordinary* Dirac operator coincides with de Rham's differential algebra of differential forms on $\mathcal{M}$. Consider the algebra $A_t := Fun\mathcal{M}\otimes A$. The group of unitaries of the tensor algebra A_t is the gauged version of the group of unitaries of the internal algebra A, i.e. the group of functions from space-time into the group G (i.e. the usual local gauge group). Consider the representation $\rho_t := \cdot\otimes\rho$ of the tensor algebra on the tensor product $\mathcal{H}_t := \mathcal{H}_S\otimes\mathcal{H}$, where $\mathcal{H}_S$ is the Hilbert space of square integrable spinors on which functions act by multiplication: $(f\psi)(x) := f(x)\psi(x)$, $f \in Fun(\mathcal{M})$, $\psi \in \mathcal{H}_S$. The Dirac operator in this generalized case is defined as follows

$$\mathcal{D}_t = \mathcal{D}d\otimes 1 + \gamma_5\otimes\mathcal{D} ,$$

(the differential d in this definition acts on the first factor in the tensor product $\mathcal{H}_S\hat{\otimes}\mathcal{H}$ and, hence, this is a *usual* differential, acting on a function). One can repeat the above construction for the infinite dimensional algebra A_t [64].

The definition of the Higgs potential in the infinite dimensional space

$$V_t(H_t) = (\mathcal{F}_t, \mathcal{F}_t) ,$$

requires a suitable regularization of the sum of eigenvalues over the space of spinors $\mathcal{H}_S$. Here, we have to assume space-time to be compact and Euclidean. Then the regularization is achieved by the Dixmier trace which allows an explicit computation of V_t. It consists of three pieces: the Yang-Mills action, covariant Klein-Gordon action and the Higgs potential.

To illustrate the main ideas of Connes-type models, let us consider briefly the model developed in [92, 93, 94]. It does not reproduce the complete structure of the Standard Model but carries its basic properties (gauge invariance and spontaneous symmetry breaking). The pleasant property of the simplified models is that their basic objects are well familiar from the theory of ordinary Lie algebras.

The model is based on the algebra

$$A = Fun(\mathcal{M}) \otimes M_n ,$$

where M_n is the algebra of $n \times n$ complex matrices serving as the non-commutative analog of internal space. Notice that such algebra is close to the Connes-Lott's choice. Let $\{\lambda_i\}_{i=1}^{n^2-1}$ be a basis of $su(n)$, chosen so that the structure constants C^i_{jk} are real and the Killing metric is equal to Kronecker symbol

$$-\mathrm{Tr}(\lambda_i \lambda_j) = \delta_{ij} .$$

To construct differential calculus on the algebra, one considers set of derivatives of M_n (see, e.g., [13]). It is well known that any derivative of M_n is an inner derivative and

$$e_i := m\, \mathrm{ad}_{\lambda_i} ,$$

form their basis with CR

$$[e_i, e_j] = m C^k_{ij} e_k ,$$

m is constant with dimension of mass. The set $\{x^\mu, \lambda^i\}$ generates the algebra A. Exterior differential for any $t \in M_n$ is defined by the equality

$$dt(e_i) = e_i t .$$

This gives, in particular

$$d\lambda^i(e_j) = m[\lambda_j, \lambda^i] = m C^i_{kj} \lambda^k .$$

The set $\Omega^p M_n$ is constructed in the same way as in general Connes' approach for non-commutative algebras (see preceding Section) and $\Omega^p A$ is a straightforward combination of $\Omega^p M_n$ and usual p-forms on $\mathcal{M}$. Notice, that at first sight A can be viewed as an algebra of matrix-valued functions. However, in this case the exterior differential defined above would not be appropriate since its kernel, the elements $t \in A$ with $dt = 0$, is the set of t which are not only *constant* matrix-valued functions on $\mathcal{M}$ but any multiples of the identity matrix.

The set of $d\lambda^i$ forms a system of generators of $\Omega^1(M_n)$ as a left or right M_n-module and as usual in non-commutative geometry $\lambda^i d\lambda^j \neq (d\lambda^j)\lambda^i$. However, for matrix algebra there is another set of generators of 1-forms characterised by the equality

$$\theta^i(e_j) = \delta^i_j \ .$$

These 1-forms commute with elements of M_n and anticommute with each other

$$\theta^i\theta^j = -\theta^j\theta^i \ .$$

They are related to the $d\lambda^i$ by the equation

$$d\lambda^i = mC^i_{jk}\lambda^j\theta^k \ , \qquad \theta^i = m^{-1}\lambda_j\lambda^i d\lambda^j \ ,$$

and satisfy the same equation as the components of the Maurer-Cartan form [145, 95] on the group $SU(n)$

$$d\theta^i = -\frac{1}{2}mC^i_{jk}\theta^j\theta^k \ .$$

The set of 1-forms $\Omega^1(A)$ can be written as the direct sum

$$\Omega^1(A) = \Omega^1_H \oplus \Omega^1_V \ ,$$

where

$$\Omega^1_H := M_n \otimes \Omega^1(Fun(\mathcal{M})) \ ,$$

$$\Omega^1_V := Fun(\mathcal{M}) \otimes \Omega^1(M_n) \ ,$$

reminiscent the splitting of a cotangent space in Kaluza-Klein type models. Another object similar to that in usual multidimensional approach, is the quadratic form

$$ds^2 := \eta_{\mu\nu}dx^\mu dx^\nu + g_{ij}\theta^i\theta^j \quad \mu,\nu = 0,\ldots,3;\ i,j = 1,\ldots,(n^2-1) \ .$$

Here g_{ij} are the components of the unique metric on M_n annihilated by appropriately defined Lie derivative: $L_X g = 0$, where $X = X^i e_i$, is an arbitrary derivative. From the generators θ^i one can construct the 1-form in Ω^1_V

$$\theta := -m\lambda_i\theta^i ,$$

and define

$$\omega := \mathcal{A} + \theta + \phi\,, \quad \mathcal{A} \in \Omega^1_H\,, \quad \phi \in \Omega^1_V\,,$$

which is non-commutative generalization of connection 1-form. A gauge transformation defines a mapping of $\Omega^1(A)$ into itself of the form

$$\omega' = g^{-1}\omega g + g^{-1}dg\,,$$

with the elements g being from the group $\mathcal{U}_n$ of unitaries of A (invertible elements which satisfy $g^\dagger g = 1$). If one defines

$$\theta' = g^{-1}\theta g + g^{-1}d_V g\,,$$

$$\mathcal{A}' = g^{-1}\mathcal{A}g + g^{-1}d_H g\,,$$

the form ϕ transforms homogeneously

$$\phi' = g^{-1}\phi g\,,$$

and can describe the Higgs field. The curvature 2-form Ω and the field strength $\mathcal{F}$ are defined as usual

$$\Omega := d\omega + \omega^2\,, \quad \mathcal{F} := d_H\mathcal{A} + \mathcal{A}^2\,.$$

The gauge invariant action for the gauge potential $\mathcal{A}$ and the Higgs field ϕ is given by

$$S = \frac{1}{4}\mathrm{Tr}\int\left(\Omega_{MN}\Omega^{MN}\right)\,, \quad M,N = 0,\ldots,(n^2-1)+3\,.$$

Written explicitly, the corresponding Lagrangian becomes

$$\mathcal{L} = \frac{1}{4}\mathrm{Tr}\,(\mathcal{F}_{\mu\nu}\mathcal{F}^{\mu\nu}) + \frac{1}{2}\,(\nabla_\mu\phi_\mu\nabla^\mu\phi^\mu) - V(\phi)\,,$$

where ∇_μ is the usual covariant derivative with respect to the connection form $\mathcal{A}$ and the Higgs potential $V(\phi)$ is given by

$$V(\phi) = -\frac{1}{4}\mathrm{Tr}\left([\phi_i,\phi_j] - mC^k_{ij}\phi_k\right)^2\,,$$

which is quartic polynomial in ϕ and has two minima at

$$\phi_i = 0 \qquad \text{and} \qquad \phi_i = m\lambda_i \ .$$

Thus in the model there exist two classical vacua separated by a potential barrier: the first one corresponds to symmetric phase with zero mass gauge bosons; the second minima corresponds to spontaneously broken $SU(n)$ symmetry with gauge bosons of the mass $m_A^2 = 2nm^2$.

More details of the calculations can be found in the cited original papers and we will proceed in the following section to the geometrical analysis of a simple model of this type.

4.2 Posets, discrete differential calculus and Connes-Lott-like models

We know already that the simplest and geometrically and physically the most transparent situation in frame of non-commutative geometry is connected with the commutative coordinate algebra, the non-commutativity being related to differential calculus only. In this Section we will try to understand the origin of a gauge symmetry breaking in the Connes-Lott model in this simplest case. For this aim we start from reminding the construction of Yang-Mills-Higgs models in frame of the so-called *coset space dimensional reduction* (for extensive review see, e.g., [150, 136]). This is a theory of Kaluza-Klein type [5] in multidimensional space with compact additional subspace. For definiteness, we will consider Manton's construction [179] where the additional subspace is two dimensional sphere (which is a particular case of *coset spaces*, hence the name of the approach). In Subsection 4.2.1 a finite approximation of manifolds by the so-called posets will be given. In the final part of the section we will discuss gauge symmetry on finite sets of points and the origin of spontaneous symmetry breaking in gauge models based on non-commutative geometry.

4.2.1 Yang-Mills-Higgs theory from dimensional reduction

Manton considered a 6-dimensional space which is direct product of 4-dimensional flat Minkowski space and a small 2-dimensional sphere of fixed radius R. Convenient coordinates are $y^M = (x^\mu, \psi, \phi)$, where x^μ $(\mu = 0, \ldots 3)$ are the usual 4-dimensional Minkowski coordinates and ψ, ϕ are the angles of the sphere.

The Yang-Mills action in the 6-dimensional space is

$$\mathcal{L} = \frac{1}{g^2}\int d^6y\sqrt{-\det h}\ \mathrm{Tr}\,(F_{MN}F_{kl})\,h^{MK}h^{NL}\ , \tag{4.2.1}$$

where $h^{MN} = diag\{1,-1,-1,-1,-1/R^2,-1/(R^2\sin^2\psi)\}$ is the metric of the space and F_{MN} is the gauge curvature tensor $F_{MN} = \partial_M A_N - \partial_N A_M - [A_M, A_N]$ corresponding to a simple gauge group G. To perform dimensional reduction of (4.2.1), Manton put the condition, that A_M must be spherically symmetric in the fifth and sixth dimensions. This requirement gives the equation

$$\left(\partial_m \xi_i^M\right) A_N + \xi_i^N \partial_N A_M = \partial_M W_i - [A_M, W_i]\ ,$$

where ξ_i $(i = 1,2,3)$ are vector fields generating the symmetry and W_i are some fields (in the lhs of this equation one can recognize the Lie derivative of A_M with respect to ξ_i). This equation has the general solution

$$A_\mu = A_\mu(x)\ , \qquad A_\psi = -\Phi_1(x)\ ,$$

$$A_\phi = \Phi_2(x)\sin\psi - \Phi_3\cos\psi\ ,$$

if the following constraints are satisfied

$$[\Phi_3, \Phi_1(x)] = -\Phi_2(x)\ , \tag{4.2.2}$$

$$[\Phi_3, \Phi_2(x)] = \Phi_1(x)\ , \tag{4.2.3}$$

$$[\Phi_3, A_i(x)] = 0\ , \tag{4.2.4}$$

where A_i, Φ_1, Φ_2 are Lie algebra valued fields in four dimensions and Φ_3 is an arbitrary *constant* element of Cartan subalgebra of Lie algebra of the gauge group G. The general form of solution for A_N can be inserted into (4.2.1). The spherical symmetry permits an integration over the angles ψ, ϕ, giving the action for a 4-dimensional gauge theory with scalar fields

$$\mathcal{L} = \frac{4\pi R^2}{g^2}\int d^4x\ \mathrm{Tr}\left\{F_{\mu\nu}F_{\mu\nu} - \frac{2}{R^2}\nabla_\mu\Phi_i\nabla_\mu\Phi_i + \frac{1}{R^4}\left(\varepsilon_{\mu\nu\rho}\Phi_\rho + [\Phi_\mu, \Phi_\nu]\right)^2\right\}\ ,$$

where $\nabla_\mu\Phi_i = \partial_\mu\Phi_i - [A_\mu, \Phi_i]$. We see that this Lagrangian contains quartic term for the scalar fields and detailed computations with use of the constraints (4.2.2)-(4.2.4), allow to rewrite it in the usual Yang-Mills-Higgs form. The residual unbroken gauge subgroup proves to be the group of stability of Φ_3, i.e. the subgroup of G leaving Φ_3 fixed. We will not go into details of these calculations. The only important lesson for us now is that peculiar properties of gauge fields in multidimensional space (actually in an additional subspace) can lead to spontaneous breaking of gauge symmetry.

A potentially dangerous property of Manton-like models is that their consistent consideration have to include other modes of gauge (and other if matter fields are added) fields besides the spherically symmetric ones. This leads to infinite number of heavy particles which may strongly contradict phenomenology (if one tries to link such a model to the Standard one, then the characteristic scale of those masses must be of the order of the Standard Model scale and hence not very heavy).

Thus it is tempting to substitute the additional manifold by a discrete set of points which does not lead to such infinite tower of undesirable states (particles).

A simple preliminary model of this type is given in this section after some additional necessary mathematics.

We hope that clear understanding of the gauge theory on spaces with discrete additional subspaces may be useful for the development of Yang-Mills-Higgs models in the general frame of non-commutative geometry.

4.2.2 Finite approximation of topological spaces

A well known fact in quantum field theory is that the higher the dimension of an underlying space-time, the worse its ultraviolet properties. In particular, complete six-dimensional Yang-Mills theory is non-renormalizable and has contradictions with phenomenology mentioned in the preceding subsection. One can try to avoid this problem either via a sharp transition to some non-commutative algebra instead of internal space as we discussed in Section 4.1, or by substitution of continuous sphere S^2 by its discrete approximation. In what follows we will consider the latter possibility. It is desirable that such approximation would not be too abrupt. In particular, it must carry a trace of the topology of initial space. The general formalism for such approximations was developed in [224, 10] (see also refs. therein).

Let $\mathcal{M}$ be a continuous topological space, like for example the sphere S^N or the Euclidean space $\mathbf{R}^N$. Choose the sets O_λ which cover $\mathcal{M}$

$$\mathcal{M} = \bigcup_\lambda O_\lambda \,, \tag{4.2.5}$$

and identify any two points x, y of M, if every set O_λ containing either point contains the other too. Now one can define an equivalence relation

$$x \sim y \text{ means } x \in O_\lambda \iff y \in O_\lambda \quad \text{for every } \; O_\lambda \,. \tag{4.2.6}$$

It is reasonable to replace $\mathcal{M}$ *by* $\mathcal{M}/\sim := P(\mathcal{M})$ *as the approximation for* $\mathcal{M}$.

We assume that the number of sets O_λ is finite when $\mathcal{M}$ is compact so that $P(\mathcal{M})$ is an approximation to $\mathcal{M}$ by a finite set in this case and that each O_λ is open and that

$$\mathcal{U} = \{O_\lambda\} , \tag{4.2.7}$$

is a topology for $\mathcal{M}$. This implies that $O_\lambda \cup O_\mu$ and $O_\lambda \cap O_\mu \in \mathcal{U}$, if $O_{\lambda,\mu} \in \mathcal{U}$. In general, points in $P(\mathcal{M})$ come from open sets and sets of the form $O_\lambda \setminus [O_\lambda \cap O_\mu]$, i.e. sets which are not open in $\mathcal{M}$. Consider this for a cover of $\mathcal{M} = S^1$. Let O_2, O_4 be two open sets around "north" and "south" poles of S^1, correspondingly, which cover the circle. The overlap $O_2 \cap O_4$ is clearly disconnected and consists of two open sets O_1, O_3. One can write

$$S^1 = O_1 \cup (O_2 \backslash O_2 \cap O_4) \cup O_3 \cup (O_4 \backslash O_2 \cap O_4) .$$

This shows that the map $S^1 \to P_4(S^1)$ ($P_N(\mathcal{M})$ denotes $P(\mathcal{M})$ with N points) is given by

$$\begin{aligned} O_1 &\to x_1 , \quad & O_2 \setminus [O_2 \cap O_4] &\to x_2 , \\ O_3 &\to x_3, \quad & O_4 \setminus [O_2 \cap O_4] &\to x_4 . \end{aligned} \tag{4.2.8}$$

$P(\mathcal{M})$ inherits the quotient topology (for an introduction to topology see, e.g. [214]) from $\mathcal{M}$ which is defined as follows. Let Φ be the map from $\mathcal{M}$ to $P(\mathcal{M})$ obtained by identifying equivalent points. An example of Φ is given by (4.2.8). In the quotient topology, a set in $P(\mathcal{M})$ is declared to be open if its inverse image for Φ is open in $\mathcal{M}$. It is the finest topology compatible with the continuity of Φ.

For $P_4(S^1)$, this means that the open sets are

$$\{x_1\}, \ \{x_3\}, \ \{x_1, x_2, x_3\}, \ \{x_1, x_4, x_3\}, \tag{4.2.9}$$

and their unions and intersections (an arbitrary number of the latter being allowed as $P_4(S^1)$ is finite).

A partial order $\preceq$ can be introduced in $P(M)$ by declaring that $x \preceq y$ if every open set containing y contains also x. This turns $P(M)$ to a *partially ordered set* or a *poset.*

For $P_4(S^1)$, this order is

$$x_1 \preceq x_2, \quad x_1 \preceq x_4; \quad x_3 \preceq x_2, \quad x_3 \preceq x_4. \tag{4.2.10}$$

Speaking rigorously one has to add the relations $x_j \preceq x_j$. They used to be omitted and one writes $x \prec y$ to indicate that $x \preceq y$ and $x \neq y$.

Any poset can be represented by a Hasse graph constructed by arranging its points at different levels and connecting them using the following rules:

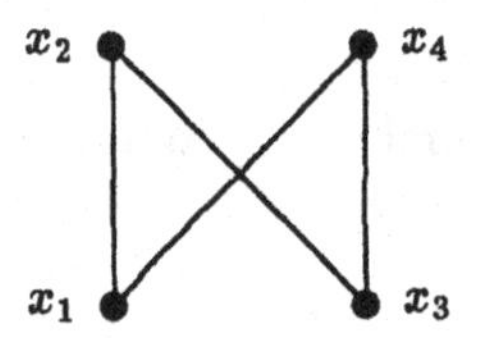

Figure 4.1: The Hasse graph for the circle poset $P_4(S^1)$

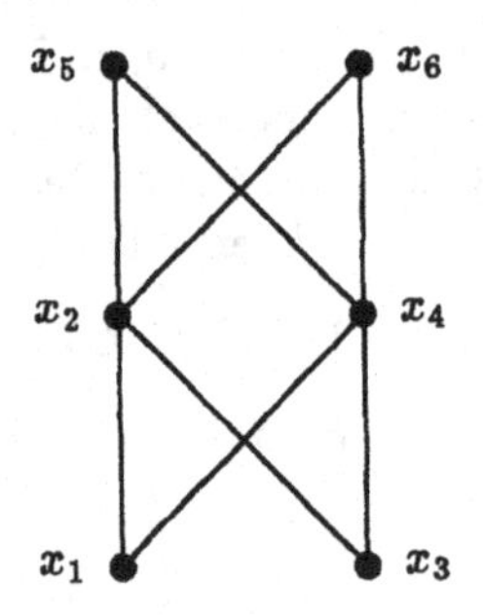

Figure 4.2: The Hasse graph for the two-sphere poset $P_6(S^2)$

1) If $x \prec y$, then y is higher than x.

2) If $x \prec y$ and there is no z such that $x \prec z \prec y$, then x and y are connected by a line called a link.

In case 2), y is said to cover x.

The Hasse graph for $P_4(S^1)$ is shown in Fig.4.1.

The smallest open set O_x containing x consists of all y preceding x ($y \preceq x$). In the Hasse graph, it consists of x and all points we encounter as we travel along links from x to the bottom. In Fig.4.1, this rule gives $\{x_1, x_2, x_3\}$ as the smallest open set containing x_2, just as in (4.2.9).

As another example, consider the Hasse graph of Fig.4.2 for a two-sphere poset $P_6(S^2)$. Its open sets are generated by

$$\begin{gathered}\{x_1\}, \quad \{x_3\}, \quad \{x_1, x_2, x_3\}, \quad \{x_1, x_4, x_3\}, \\ \{x_1, x_2, x_5, x_4, x_3\}, \quad \{x_1, x_2, x_6, x_4, x_3\},\end{gathered} \tag{4.2.11}$$

by taking unions and intersections.

A point x of a poset P can be assigned a rank $r(x)$ as follows. A point of P is regarded as of rank 0, if it is linked to no point, or is a highest point. Let

P' be the poset obtained from P by removing all rank zero points and their links. The highest points of P' are assigned rank 1. We continue in this way to rank all points. The rank of a poset is just the maximum rank occurring among the points.

A remarkable property of the poset approximation is its ability to accurately reproduce the fundamental group of the manifold and hence to carry important topological properties of initial manifolds.

More detailed topological structure of an arbitrary poset can be investigated with the help of boundary operator δ, the basic tool of homology theory [122, 214]. To define it on a poset, one needs to introduce the so-called incidence numbers $I(x_i, x_j)$ attached to each link x_k—x_l, so that $x_k \prec x_l$, with the following properties [122]:

$$I(x_i, x_j) = 0\ , \quad \text{unless } x_j \text{ covers } x_i\ ;$$

$$\sum_{x_j} I(x_i, x_j) I(x_j, x_k) = 0\,. \tag{4.2.12}$$

The *boundary operator* δ is then defined by

$$\delta x_i = \sum_{x_j} I(x_i, x_j) x_j\ ,$$

$$\delta x_i = 0\ , \quad \text{if no point covers } x_i\ , \tag{4.2.13}$$

and fulfills the basic property

$$\delta^2 = 0\,. \tag{4.2.14}$$

Until now we considered the approximations for manifolds in standard geometrical terms: points, sets etc. To make a link with non-commutative geometry and gauge theory constructed in the frame which we discussed in the preceding section, we have to do the standard step and to pass to the dual description: in terms of functions and differential forms on discrete sets approximating the manifold.

Such a dual pattern was developed in [79]. For this aim one starts, as usual, from the algebra of $\mathbb{C}$-valued functions on a discrete set of N points K_N. Multiplication is defined pointwise, i.e. $(fh)(x_i) = f(X_i)\, h(x_i)$. Functions, of course, depend on the discrete index only and it is reasonable to omit the character x in the argument: $f(i) := f(x_i)$. There is a distinguished set of characteristic functions e_i on K_N, defined by $e_i(j) = \delta_{ij}$. They satisfy the relations

$$e_i\, e_j = \delta_{ij}\, e_i\ , \qquad \sum_i e_i = \mathbf{1}\ , \tag{4.2.15}$$

where $\mathbf{I}$ denotes the constant function $\mathbf{I}(i) = 1$. Each $f \in A$, can then be written as

$$f = \sum_i f(i)\, e_i\,. \tag{4.2.16}$$

The algebra A can be extended to a $\mathbb{Z}$-graded *differential algebra* $\Omega(A) = \bigoplus_{r=0}^{\infty} \Omega^r(A)$ (where $\Omega^0(A) = A$), via the action of a linear operator $d : \Omega^r(A) \to \Omega^{r+1}(A)$, satisfying

$$d\mathbf{I} = 0 \quad , \quad d^2 = 0 \quad , \quad d(\omega_r\, \omega') = (d\omega_r)\, \omega' + (-1)^r\, \omega_r\, d\omega' \tag{4.2.17}$$

where $\omega_r \in \Omega^r(A)$. The spaces $\Omega^r(A)$ of *r-forms* are A-bimodules. $\mathbf{I}$ is taken to be the unity in $\Omega(A)$. From the above properties of the set of functions e_i, one obtains

$$e_i\, de_j = -(de_i)\, e_j + \delta_{ij}\, de_i\,, \tag{4.2.18}$$

and

$$\sum_i de_i = 0\,. \tag{4.2.19}$$

It is convenient to define

$$e_{ij} := e_i\, de_j \quad (i \neq j) \quad , \quad e_{ii} := 0\,, \tag{4.2.20}$$

(note that $e_i\, de_i \neq 0$) and

$$e_{i_1 \ldots i_r} := e_{i_1 i_2} e_{i_2 i_3} \cdots e_{i_{r-1} i_r}\,, \tag{4.2.21}$$

which for $i_k \neq i_{k+1}$ equals to $e_{i_1}\, de_{i_2} \cdots de_{i_r}$. The differential of the characteristic functions can be expressed in terms of e_{ij}

$$de_i = \sum_j (e_{ji} - e_{ij})\,, \tag{4.2.22}$$

and more generally

$$de_{i_1 \ldots i_r} = \sum_j \sum_{k=1}^{r+1} (-1)^{k+1} e_{i_1 \ldots i_{k-1} j i_k \ldots i_r}\,. \tag{4.2.23}$$

The forms $e_{i_1 \ldots i_r}$ provide a basis for $\Omega^{r-1}(A)$ and any $\psi \in \Omega^{r-1}(A)$ can be written as

$$\psi = \sum_{i_1, \ldots, i_r} e_{i_1 \ldots i_r}\, \psi_{i_1 \ldots i_r}\,, \tag{4.2.24}$$

with $\psi_{i_1 \ldots i_r} \in \mathbb{C}$, $\psi_{i_1 \ldots i_r} = 0$, if $i_s = i_{s+1}$ for some s. From (4.2.23) one has

$$d\psi = \sum_{i_1,\ldots,i_{r+1}} e_{i_1 \ldots i_{r+1}} \sum_{k=1}^{r+1} (-1)^{k+1} \psi_{i_1 \ldots \hat{i}_k \ldots i_{r+1}} , \tag{4.2.25}$$

(the hat "$\hat{}$" means that the corresponding index must be omitted). In particular, for $f \in \Omega^0$

$$df = \sum_{i,j} e_{ij} (f(j) - f(i)) \qquad (\forall f \in A) , \tag{4.2.26}$$

so that $d(const) = 0$.

If no further relations are imposed, the differential algebra corresponds to the *universal* differential algebra (or *universal differential envelope*, cf. Section 4.1) of A. Again, as in the preceding Section, one can define an *inner product* on a differential algebra $\Omega(A)$ which should have the properties $(\psi_r, \phi_s) = 0$, for $r \neq s$, and

$$(\psi, c\phi) = c(\psi, \phi) , \qquad \forall c \in \mathbb{C} , \qquad \overline{(\psi, \phi)} = (\phi, \psi) . \tag{4.2.27}$$

Furthermore, we should require that $(\psi, \phi) = 0$, $\forall \phi$, implies $\psi = 0$.

An *adjoint* d^* : $\Omega^r(A) \to \Omega^{r-1}(A)$ of d with respect to an inner product is then defined by

$$(\psi_{r-1}, d^* \phi_r) := (d\psi_{r-1}, \phi_r) . \tag{4.2.28}$$

This allows us to construct a *Laplace-Beltrami operator* as follows

$$\Delta := -d\, d^* - d^*\, d . \tag{4.2.29}$$

In the case of the universal differential algebra on A, an inner product has the form

$$(\psi, \phi) = \mu_r \sum_{i_1,\ldots,i_r} \bar{\psi}_{i_1 \ldots i_r} \phi_{i_1 \ldots i_r} . \tag{4.2.30}$$

If we are interested to approximate a differentiable manifold by a discrete set, the universal differential algebra is too large to provide us with a corresponding analog of the algebra of differential forms on the manifold. We need "smaller" differential algebras. Relations (4.2.21),(4.2.21) permit to achieve such a reduction by setting some of e_{ij} to zero. This does not generate any relations for the remaining (nonvanishing) $e_{k\ell}$ and is consistent with the rules of differential calculus. It generates relations for the forms of higher grades,

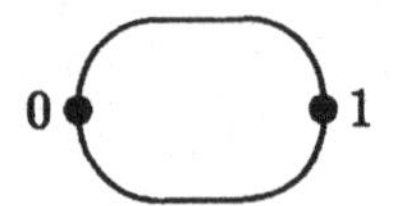

Figure 4.3: The graph associated with the universal differential algebra on a two-point set.

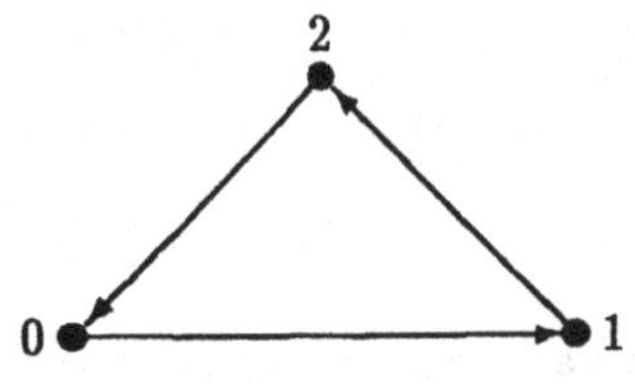

Figure 4.4: The graph associated with a differential algebra on a set of three elements.

however. In particular, it may require that some of those $e_{i_1 \ldots i_r}$ with $r > 2$, which do not contain as a factor any of the e_{ij} which are set to zero, have to vanish.

The reductions of the universal differential algebra obtained in this way, are conveniently represented by graphs as follows. The elements of K_N is represented by vertices and to $e_{ij} \neq 0$ one associates an arrow from i to j. The universal differential algebra then corresponds to the graph where *all* the vertices are connected pairwise by arrows in *both* directions. Of course in this special case, arrows do not carry any information and one can link vertices just by lines. For example, the graph in Fig.4.3 corresponds to the universal differential algebra on a set of two elements.

Deletion of some of the arrows leads to a graph which represents a reduction of the universal differential algebra. To illustrate, let us consider a set of three elements with the differential algebra determined by the graph in Fig.4.4. The nonvanishing basic 1-forms are then e_{01}, e_{12}, e_{20} only, and $e_{10} = e_{21} = e_{02} = 0$. From these we can only build the basic 2-forms e_{012}, e_{120} and e_{201}. However, (4.2.23) yields

$$0 = de_{10} = \sum_{k=1}^{3}(e_{k10} - e_{1k0} + e_{10k}) = -e_{120} \ , \tag{4.2.31}$$

and similarly $e_{012} = 0 = e_{201}$. Hence there are no 2-forms (and thus also no higher forms). This may be interpreted in such a way that the differential algebra assigns a *one*-dimensional structure to the three-point set. Using (4.2.22),

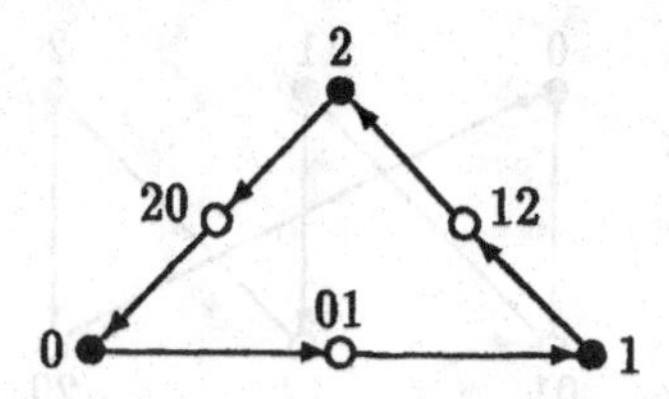

Figure 4.5: The extension of the graph associated with a differential algebra on a set of three elements by addition of new vertices corresponding to the nonvanishing basic 1-forms.

we have

$$de_0 = e_{20} - e_{01} \quad , \quad de_1 = e_{01} - e_{12} \quad , \quad de_2 = e_{12} - e_{20} \, . \qquad (4.2.32)$$

These graphs can be extended in the following way. One adds *new* vertices for all forms survived after the reduction and connects by arrows a vertex representing a given p-form with vertices representing $(p+1)$-form which appear in rhs of Eq.(4.2.23) for this p-form. The direction of the arrows depends on the sign with which $(p+1)$-form appears in (4.2.23): for plus sign the arrow is directed to p-form, for minus sign it is directed to the corresponding $(p+1)$-form. One can check that in this way one obtains exactly the Hasse graphs introduced above which determine a finite topology. In particular, for the example of a 2-point set in Fig.4.3 one obtains the Hasse graph Fig.4.1, for $P_2(S^1)$ or the graph Fig.4.2, for $P_2(S^2)$, depending on the chosen reduction: from the basic 1-forms e_{01} and e_{10} we can construct forms $e_{010101\ldots}$ and $e_{101010\ldots}$ of arbitrary grade. If we set the 2-forms e_{010} and e_{101} equal to zero, the Hasse graph determines a topology which approximates the topology of the circle S^1. If, however, we set the 3-forms e_{0101} and e_{1010} equal to zero, an approximation of S^2 is obtained, and so forth. A slightly more complicated graph in Fig.4.4, extended as the graph described above, can be drawn as in Fig.4.5 or in the form of Hasse graph Fig.4.6.

For reduced algebras, one can introduce the useful notion of *dimension* of a differential algebra as the grade of its highest nonvanishing forms. Applying it to subgraphs, leads to a *local* notion of dimension.

Consider a field Ψ on K_N as a cross section of a vector bundle over K_N, e.g., a cross section of the trivial bundle $K_N \times \mathbb{C}^n$. In the algebraic language, the latter corresponds to the *left* $\mathcal{A}$-module with an action $\Psi \mapsto G\Psi$ of a (local) gauge group, a subgroup of $GL(n, \mathcal{A})$ with elements $G = \sum_i G(i)\, e_i$, with $G(i) \in GL(n, \mathbb{C})$. Let us introduce *the covariant exterior derivatives*

$$D\Psi = d\Psi + \mathcal{A}\Psi \, , \qquad (4.2.33)$$

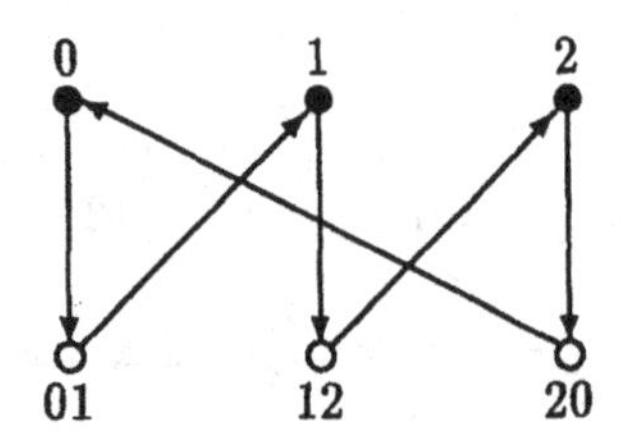

Figure 4.6: The (oriented) Hasse graph derived from the preceding graph

where $\mathcal{A}$ is a 1-form. These expressions are indeed covariant if $\mathcal{A}$ obeys the usual transformation law of a *connection 1-form*,

$$\mathcal{A}' = G\,\mathcal{A}\,G^{-1} - dG\,G^{-1}\,. \tag{4.2.34}$$

Since dG is a discrete derivative, $\mathcal{A}$ cannot be Lie algebra valued. It is rather an element of $\Omega^1(A) \otimes_A M_n(A)$, where $M_n(A)$ is the space of $n \times n$ matrices with entries in A.

An element $U \in \Omega^1(A) \otimes_A GL(n, A)$, is called a *transport operator* (the reason will become clear in the following) if it transforms as $U \mapsto G\,U\,G^{-1}$ under a gauge transformation. Since $U = \sum_{i,j} e_{ij}\,U_{ij}$, with $U_{ij} \in GL(n, \mathbb{C})$, we find

$$U'_{ij} = G(i)\,U_{ij}\,G(j)^{-1}\,. \tag{4.2.35}$$

Using (4.2.34), (4.2.21) and (4.2.26), it can be shown that such a transport operator is given by

$$U := \sum_{i,j} e_{ij}\,(\mathbf{I} + A_{ij})\,, \tag{4.2.36}$$

where $\mathbf{I}$ is the identity in the group.

We see again, as in the preceding section, that in contrast to the case of gauge theory on smooth manifolds, one can construct from the connection *another 1-form*, which is transformed homogeneously under gauge group. But now it does not look as some miracle and we can easily recognize in U the discrete version of Wilson parallel shifting operator(see, e.g.[206]). The actual difference, taking its origin in the very discretization, is that for a continuous theory, U is *a nonlocal* operator while A is *a local* one and they have *different status*. On a discrete set, they *both* are defined on links connecting the points i and j, and so can be used *equally*.

The covariant derivatives introduced above can now be written as follows

$$D\Psi = \sum_{i,j} e_{ij}\,\nabla_j\Psi(i) \quad , \quad \nabla_j\Psi(i) := U_{ij}\,\Psi(j) - \Psi(i)\,.$$

Analogous construction exists for right A-moduli Ξ

$$D\Xi = \sum_{i,j} e_{ij}\, \nabla_j \Xi(i)\, U_{ij} \quad , \quad \nabla_j \Xi(i) := \Xi(j)\, U_{ij}^{-1} - \Xi(i) \; ,$$

where $\Xi = \sum_i e_i\, \Xi(i)$, but one needs the additional assumption that U_{ij} is *invertible.*

The curvature form is defined as usual

$$\mathcal{F} := d\mathcal{A} + \mathcal{A}^2 \; ,$$

and the Yang-Mills action

$$S_{YM} := \mathrm{tr}\,(\mathcal{F}, \mathcal{F}) = \sum_{i,j,k} \mathrm{Tr}\,(\mathcal{F}^{\dagger}_{ijk}\, \mathcal{F}_{ijk})$$

$$= \mathrm{Tr} \sum_{i,j,k} (U^{\dagger}_{jk}\, U^{\dagger}_{ij}\, U_{ij}\, U_{jk} - U^{\dagger}_{jk}\, U^{\dagger}_{ij}\, U_{ik} - U^{\dagger}_{ik}\, U_{ij}\, U_{jk} + U^{\dagger}_{ik}\, U_{ik}) \; , \tag{4.2.37}$$

is then gauge-invariant, if $G^{\dagger} = G^{-1}$.

The usual compatibility condition for parallel transport and conjugation is

$$(D\Psi)^{\dagger} = D(\Psi^{\dagger}) \; , \tag{4.2.38}$$

which is equivalent to

$$A^{\dagger} = -A \; , \tag{4.2.39}$$

and implies $F^{\dagger} = F$. Let $\bar{}$ denote an involutive mapping of K_N, ($\bar{\bar{k}} = k$, for all $k \in K_N$). Define an *involution* * on $\Omega(A)$ by requiring $(\psi\phi)^* = \phi^*\psi^*$, $(d\omega_r)^* = (-1)^r d\omega_r^*$ and $(f^*)(k) = \overline{f(\bar{k})}$. Using the Leibniz rule, one finds

$$e^*_{k\ell} = -e_{\bar{\ell}\bar{k}} \; , \tag{4.2.40}$$

and thus $e^*_{i_1 \dots i_r} = (-1)^{r+1}\, e_{\bar{i}_r \dots \bar{i}_1}$.

Using (4.2.40), (4.2.39) becomes

$$U^{\dagger}_{ij} = U_{\bar{j}\bar{i}} \; , \tag{4.2.41}$$

which implies $F^{\dagger}_{ijk} = F_{\bar{k}\bar{j}\bar{i}}$. Evaluation of the Yang-Mills action (4.2.37) with (4.2.41) gives

$$S_{YM} = \mathrm{tr} \sum_{i,j,k} (U_{\bar{k}\bar{j}}\, U_{\bar{j}\bar{i}}\, U_{ij}\, U_{jk} - U_{\bar{k}\bar{j}}\, U_{\bar{j}\bar{i}}\, U_{ik} - U_{\bar{k}\bar{i}}\, U_{ij}\, U_{jk} + U_{\bar{k}\bar{i}}\, U_{ik}) \; . \tag{4.2.42}$$

For $P(S^2)$ with the universal differential algebra one finds

$$\begin{aligned} F &= e_{010}\left(U_{01}U_{10} - \mathbf{1}\right) + e_{101}\left(U_{10}U_{01} - \mathbf{1}\right) \\ &= e_{010}\left(U_{10}^{\dagger}U_{10} - \mathbf{1}\right) + e_{101}\left(U_{10}U_{10}^{\dagger} - \mathbf{1}\right), \end{aligned} \tag{4.2.43}$$

and

$$S_{YM} = 2\,\mathrm{Tr}\left(U_{10}^{\dagger}U_{10} - \mathbf{1}\right)^2 \tag{4.2.44}$$

which has the form of a Higgs potential.

We are ready now to give the interpretation of "miraculous" spontaneous symmetry breaking in gauge models with non-commutative geometry, at least in the simple situation with discrete internal space and non-commutative differential calculus on it. Considering for definiteness the example of two-point internal space, we see that being combined with (usual, continuous) space-time geometry, the model has the gauge symmetry group $\prod^{\otimes}_{x,i} G(x^{\mu}, i)$; $x \in \mathcal{M}$, $i = 0, 1$. In more standard notations this can be expressed as $\mathcal{M}$-space local invariance with respect to the group $G \otimes G$.

As is clear from the expression for the Lagrangian (4.2.44), the vacuum of the model corresponds to

$$U_{01}(x) = \mathbf{1}\,. \tag{4.2.45}$$

The transformation rules of the transport operator (4.2.35) immediately give the result that the subgroup of stability of this field configuration is

$$\tilde{G} := diag(G \otimes G) \subset G \otimes G\,.$$

Thus "chiral" gauge symmetry $G \otimes G$ of this model is indeed broken spontaneously. But in terms of gauge field (connection) $A(x, i)$, the solution (4.2.45) corresponds to usual and natural vacuum field configuration

$$A_{\mu}(x, i) = 0\,,$$

which itself does not lead to any spontaneous breaking. Hence, the origin of the symmetry breaking is in the appearance (due to special, namely, discrete geometry of the internal space) of the operator U_{01} which has gauge transformation rule (4.2.35), and because of the assumed invertibility has vacuum expansion around unity $U_{01} \sim \mathbf{1} + ...$ This introduces a constant field analogous to Φ_3 in Manton's construction of Standard Model briefly discussed in the first subsection. Just as Φ_3, the operator $U_{01} \sim \mathbf{1}$ breaks the initial gauge symmetry $G \otimes G$ down to subgroup $\tilde{G}$ of its stability.

An interesting problem is to generalize the description of gauge models with discrete internal spaces to more complicated and realistic cases of

Standard-like models using more rich posets and additional constraints on gauge fields analogous to the requirement of spherical symmetry imposed in the Manton's model.

Analogy with multidimensional Kaluza-Klein like models leads to one more conclusion. It is well known that original Kaluza-Klein idea of multidimensional space-time gets powerful predictability when included in a more fundamental theory of multidimensional gravity with *spontaneous compactification* (see [5]). This means that geometrically all dimensions are considered equivalently, and the distinguished internal space (compact manifold of a small size) appears dynamically as a result of evolution of multidimensional Universe. Being transformed to the case of non-commutative geometry, the analogous approach would mean that both 4-dimensional space-time and internal space must be treated in the same frame of non-commutative (deformed) geometry with *dynamical* parameters of deformations. An appropriate evolution of the latter could explain visible distinction of space-time and internal space properties (different scales of discretization).

We are still far from formulation of a theory of such type. The geometrical basis of Kaluza-Klein theory with spontaneous compactification (as well as gauge field theory in general) is the differential geometry of fibre bundles. Thus a natural and necessary step in the direction of construction of "multidimensional non-commutative field theory" seems to be the development of non-commutative analog of fibre bundles geometry. Next Section is devoted to a short introduction to basic notions of such a theory. Interesting enough, the self-consistent definition of a non-commutative fibre bundle requires itself that *both* a base (space-time from physical point of view) and a fibre (internal space or gauge group) be described in terms of non-commutative geometry *simultaneously*.

4.3 Basic elements of quantum fibre bundle theory

Classical differential geometry of fibre bundles provides mathematical foundation for gauge theory of particle interactions and Kaluza-Klein-type models. Up to now there are no indications that gauge symmetry *must* be deformed, and *physically* meaningful and complete q-deformed gauge models do not yet exist. Nevertheless, the consideration in the last two Sections shows that this quantum geometry can be useful for the natural incorporation of Higgs fields in the gauge theory. Reformulation of the Connes' idea in terms of quantum fibre bundles may have many advantages. Among them:

- better understanding of relations with "classical" counterpart (standard Yang-Mills-Higgs theory);
- more general or complimentary formalism for the Yang-Mills-Higgs theory in frame of non-commutative geometry;
- matching the Connes approach with well developed and fruitful theory of quantum groups etc.

As one more preliminary remark, we must add that usual (classical) geometry of fibre bundles is rich enough to describe spontaneous symmetry breaking. Namely, this is provided by the so-called reduction of corresponding fibre bundle [145, 95]. A possible application of this geometrical construction to the Standard Model was elaborated in [37] (note that in this particular approach, fibre bundle and dimensional reduction are achieved simultaneously but in general these are quite different notions!). The problem of classical geometry is that it suggests *too many* possibilities for the Higgs sector. We may hope that quantum geometry somehow restricts these possibilities and gives natural explanation of particular structure of Higgs Lagrangian of the Standard Model.

Now we will give basic definitions of quantum fibre bundle geometry following [29].

In general, the procedure of fibre bundle deformation is usual for quantization of geometrical objects: one passes from manifolds to algebras of functions on them, reverses all "structural arrows" of the construction because of this dual form of description and after substitution of the specific commutative algebra by a more general non-commutative algebra, one obtains an object of non-commutative geometry.

Recall the standard definition of principal fibre bundle.

Definition 4.1 *Let M be a smooth manifold and G a Lie group. A principal bundle $P(M,G)$ over the base manifold M consists of a smooth manifold P and a smooth action of G on P such that G acts freely on P from the right, i.e. $P \times G \ni (u,a) \mapsto ua = R_a u \in P$ is an action and $R_a u = u$, for some $u \in P$ means $a = e$ (e is group unity). Moreover, $M \cong P/G$ and the canonical projection $\pi : P \to M$ is a smooth map.*

Another way to express the free character of the action, convenient for dual reformulation, is to say that the map

$$P \times G \to P \times P, \qquad (u,a) \mapsto (u,ua)\,, \tag{4.3.1}$$

is an inclusion.

Locally $P \cong U \times G,\ U \subset M$. This means that if $U \subset M$ is an open set covered by one chart, then there exists a map $\phi_U : \pi^{-1}(U) \to G$ such that $\phi_U(ua) = \phi_U(u)a$ and such that the map $\pi^{-1}(U) \to U \times G$, defined by $u \mapsto (\pi(u), \phi_U(u))$ is an isomorphism.

The quantum counterpart of this construction is

Definition 4.2 *The space $P = P(B, A)$ is called a quantum principal bundle with universal differential calculus, structure quantum group A and base B, if:*

1. *A is a Hopf algebra;*
2. *P is a right A-comodule algebra with a coaction δ_R: $P \to P \otimes A$;*
3. *$B = P^A = \{u \in P : \delta_R u = u \otimes 1\}$;*
4. *$(m \otimes \mathrm{id})(\mathrm{id} \otimes \delta_R) : P \otimes P \to P \otimes A$ is a surjection (co-freeness condition); m is a multiplication map in P.*

Remark: this definition must be completed by one more condition which reflects the fact that in the classical case smoothness and dimensional considerations combine with freeness of the action to ensure that the quotient is a manifold and the fiber through a point u is a copy of a Lie group G. At the differential level, the Lie algebra L of G is included in the vertical part of the tangent space $T_u P$ by the map which generates fundamental vector fields. Dimensional arguments then imply that this map is an isomorphism of L with the vertical part of each $T_u P$. We refer the reader to [29] for the exact algebraic formulation (in dual form) which replaces this complex of consequences arising from the smoothness and dimensional considerations. Roughly speaking, in place of dimensional arguments one supposes directly that the image of the fundamental vector fields through each point span all the vertical vectors through the point.

Note that B is a subalgebra of P. For if $u, v \in B$ then

$$\delta_R(uv) = \delta_R(u)\delta_R(v) = (u \otimes 1)(v \otimes 1) = (uv) \otimes 1 \ .$$

Hence $uv \in B$. There is a natural inclusion $j : B \hookrightarrow P$, which corresponds to the canonical projection π in the classical case.

As well as many other constructions of quantum geometry in this book, we illustrate the definition of a quantum fibre bundle using the q-group $SU_q(2)$. Now the algebra of functions on it, $Fun_q(SU(2))$ with the standard CR (1.1.17) and the involution (1.6.22), can be considered as the algebra P in the Definition

4.2. If one takes as an algebra A (q-fibre) the algebra $Fun_q(U(1))$, generated by the elements t, t^{-1} with the constraints $tt^{-1} = t^{-1}t = 1$, the right coaction $\delta_R :\longrightarrow P\otimes A$ is defined as follows

$$\delta_R\left(\begin{pmatrix} a & b \\ -b^*/q & a^* \end{pmatrix}\right) = \begin{pmatrix} a & b \\ -b^*/q & a^* \end{pmatrix}\begin{pmatrix} t & 0 \\ 0 & t^{-1} \end{pmatrix} = \begin{pmatrix} a\otimes t & b\otimes t^{-1} \\ -b^*/q\otimes t & a^*\otimes t^{-1} \end{pmatrix} .$$

The subalgebra B satisfying the requirement 3. in the Definitin 4.2 is generated by the quadratic combinations

$$aa^* , \quad bb^* , \quad ab^* , \quad ba^* ,$$

modulo $SU_q(2)$ commutator relations and the determinant constraint $aa^* + bb^* = 1$. Let us choose as the independent combinations $X_- := ab^*$, $X_+ := ba^* = X_-^*$, $X_3 := \frac{1}{2}(aa^* - bb^*)$. This is analog of the usual map $S^{(3)} \longrightarrow S^{(2)}$: $X_i = \xi^\dagger \sigma_i \xi$, $i = 1, 2, 3$; σ_i are the Pauli matrices, $\xi = \begin{pmatrix} a \\ b \end{pmatrix}$ and $\sum_i X_i^2 = 1$, if $\xi^\dagger \xi = 1$. In the q-deformed case one finds that

$$X_- X_+ + q^2 X_3^2 = \frac{q^2}{4}(aa^* + bb^*)^2 = \frac{q^2}{4} .$$

This means that the algebra B (quantum analog of a base space) in this example is the particular variant (cf. Section 1.7.3) of a quantum sphere $S_q^{(2)}$. As a result, we have obtained the quantum fibre bundle $P(S_q^{(2)}, U_q(1))$.

In *classical* geometry the simplest case of the so-called *trivial* fibre bundle can be described with one chart $U \equiv M$, and thus it has *globally* the structure of the cross product $P \cong M \times G$.

To introduce an analogous non-commutative object, one needs the notion of the *convolution product* of linear maps on a Hopf algebra (or coalgebra) A which is defined as follows.

Definition 4.3 *Let B be an algebra and $f_1, f_2 : A \rightarrow B$ two linear maps. The convolution product of f_1 and f_2 (denoted by $g = f_1 * f_2$) is the linear map $g : A \rightarrow B$ given by $g(a) := m_b(f_1 \otimes f_2)\Delta(a) = \sum f_1(a_{(1)}) f_2(a_{(2)})$ for any $a \in A$.*

The convolution product is associative and converts the set of linear maps $\mathrm{Lin}(A, B)$ into an algebra. Note that if B has a unity η_B (viewed as a map), then $f * (\eta_B \circ \varepsilon) = (\eta_B \circ \varepsilon) * f = f$, so that $\eta_B \circ \varepsilon$ is the identity in the convolution algebra $\mathrm{Lin}(A, B)$.

Definition 4.4 *A linear map $f : A \to B$ is convolution invertible, if there exists a map $f^{-1} : A \to B$ such that $f^{-1} * f = f * f^{-1} = \eta_B \circ \varepsilon$.*

Definition 4.5 *Let A be a Hopf algebra and P an A-comodule algebra with invariant subalgebra B. Assume that there exists a convolution invertible map $\Phi : A \hookrightarrow P$ such that*

$$\delta_R \circ \Phi = (\Phi \otimes id) \circ \Delta, \qquad \Phi(\mathbf{I}_A) = \mathbf{I}_P \ , \tag{4.3.2}$$

(Δ is a comultiplication in A). Then P is a quantum principal bundle. $P(B, A, \Phi)$ is called a trivial bundle with trivialization Φ.

Below we will restrict our discussion to this case (i.e. we shall not discuss topologically nontrivial fibre bundles, see [29]) and now give important for applications (at least in the classical case) the definition of trivial *associated (vector)* bundle $E = P \otimes_A V$.

Definition 4.6 *Let $(A, \Delta, \varepsilon, S)$ be a Hopf algebra. We say that $E(B, V, A)$ is a trivial quantum vector bundle with base B, fibre V and structure quantum group A if:*

1. *B is an algebra with unity;*
2. *(V, ρ_L) is a left A-comodule algebra, i.e. there exists a map $\rho_L : V \to A \otimes V$, such that*

$$(\Delta \otimes \mathrm{id})\rho_L = (\mathrm{id} \otimes \rho_L)\rho_L, \qquad (\varepsilon \otimes \mathrm{id})\rho_L = \mathrm{id} \ ;$$

3. *$E = B \otimes V$.*

Let us note that E is a left A-comodule algebra. The coaction $\delta_L : E \to A \otimes E$, is given by $\delta_L = \rho_L \otimes id$ and the multiplication $(v_1 \otimes b_1)(v_2 \otimes b_2) = v_1 v_2 \otimes b_1 b_2$ is the tensor product one.

A *quantum gauge transformation* of a trivial vector bundle $E(B, V, A)$ is then a convolution invertible map $\gamma : A \to B$, such that $\gamma(\mathbf{I}_A) = \mathbf{I}_B$. It is easy to see that a set of convolution invertible maps form a group (in the ordinary classical sense!) with the convolution as a multiplication and $\eta_B \circ \varepsilon$ as the group unity.

Similarly, if V is a left A-comodule and $f_1 : A \to B$, $f_2 : V \to B$, then $(f_1 * f_2)(v) = \sum f_1(v^{(\bar{1})}) f_2(v^{(\bar{2})})$, for any $v \in V$. Finally if Γ is any bimodule of B and $f_1 : A \to B$, $f_2 : V \to \Gamma$, we define $(f_1 * f_2)(v) = \sum f_1(v^{(\bar{1})}) f_2(v^{(\bar{2})})$.

A map $\sigma : V \to B$ is a *section* of E, if it transforms under the action of gauge transformation γ according to the law

$$\sigma \overset{\gamma}{\longmapsto} \sigma^\gamma = \gamma * \sigma \, .$$

A acts on V according to the left coaction ρ_L. In the classical case, these maps are induced by the usual section $s : M \to v$, where v is a typical fibre (such definition of a section is correct for a trivial bundle) and gauge transformation $g : M \to G$. Indeed, they induce the algebra maps:

$$\sigma : Fun(v) \ni f(s) \longrightarrow f(s(x)) =: f' \in B = Fun(M), \; s \in v, \; x \in M \, ,$$

$$\gamma : Fun(G) \ni h(g) \longrightarrow h(g(x)) =: h'(x) \in B = Fun(M), \; g \in G \, .$$

Moreover the gauge transformation g acting pointwise induces a transformation of sections $s \mapsto s^g$, which in components reads

$$(s^g)^i(x) = g^i_j(x) s^j(x) \, ,$$

for all $x \in U$. This in turn gives rise to the transformation of σ,

$$\sigma^\gamma(v^i) = \gamma(g^i_j)\sigma(v^j) \, .$$

This explains the definitions of quantum gauge transformations and sections of quantum vector bundles.

Let $\Gamma(E)$ be a set of sections of E and $\Gamma^1(E)$ be a set of maps $V \to \Omega^1(B)$ (from V to a space of 1-forms on B). A linear map $\nabla : \Gamma(E) \to \Gamma^1(E)$ is a *covariant exterior differential* on the trivial quantum vector bundle E, if for any quantum gauge transformation γ on E, there exists map $\nabla^\gamma : \Gamma(E) \to \Gamma^1(E)$ such that for any section $\sigma \in \Gamma(E)$,

$$\nabla^\gamma \sigma^\gamma = \gamma * (\nabla\sigma) \, . \tag{4.3.3}$$

The covariant differential has the form well familiar from the classical case

$$\nabla = d + \beta * \, , \tag{4.3.4}$$

where a map $\beta : A \to \Gamma_B$, Γ_B being a space of 1-forms on B, transforms by the quantum gauge transformation γ of E as

$$\beta \overset{\gamma}{\longmapsto} \beta^\gamma = \gamma * \beta * \gamma^{-1} + \gamma * d(\gamma^{-1}) \, . \tag{4.3.5}$$

Indeed,

$$\nabla^\gamma \sigma^\gamma = d\sigma^\gamma + \beta^\gamma * \sigma^\gamma = d(\gamma * \sigma) + (\gamma * \beta * \gamma^{-1} + \gamma * d(\gamma^{-1})) * \gamma * \sigma$$
$$= d\gamma * \sigma + \gamma * d\sigma + \gamma * \beta * \sigma - d\gamma * \sigma = \gamma * (\nabla\sigma).$$

Hence ∇ transforms as a covariant derivative. A map $\beta : A \to \Gamma_B$ is called a *connection one-form* on E.

To any connection β on a trivial quantum vector bundle $E(B, V, A)$, one can associate its *curvature* $F : A \to \Omega^2(B)$ defined as

$$F = d\beta + \beta * \beta\,. \tag{4.3.6}$$

The properties of this map are very similar to the classical ones except that the usual product of functions is replaced by the convolution product:

1. For any section $\sigma \in \Gamma(E)$

$$\nabla^2\sigma = F * \sigma\,; \tag{4.3.7}$$

2. For any quantum gauge transformation γ of E

$$F^\gamma = \gamma * F * \gamma^{-1}\,; \tag{4.3.8}$$

3. The Bianchi identity

$$dF + \beta * F - F * \beta = 0\,. \tag{4.3.9}$$

The definitions above are based on the left $\mathcal{A}$-comodule. There is a well established symmetry between left and right constructions [29].

If G is a matrix quantum group (see Section 1.5) and V is a quantum space of its corepresentation, all the expressions can be written in a more explicit form. Introducing the shorthand notations

$$\sigma^i := \sigma(v^i), \quad (\sigma^\gamma)^i := \sigma^\gamma(v^i)\,,$$

$$\beta^i_j := \beta(t^i{}_j), \qquad (\beta^\gamma)^i{}_j := \beta^\gamma(t^i{}_j), \qquad F^i{}_j := F(t^i{}_j), \quad \gamma^i{}_j := \gamma(t^i{}_j)\,,$$

one has the following formulae:

$$(\sigma^\gamma)^i = \gamma^i{}_j\sigma^j\,,$$

$$(\beta^\gamma)^i{}_j = \gamma^i{}_k\beta^k{}_l(\gamma^{-1})^l{}_j + \gamma^i{}_k d(\gamma^{-1})^k{}_j\,,$$

$$\nabla\sigma^i = d\sigma^i + \beta^i{}_j\sigma^j\,,$$

$$F^i{}_j = d\beta^i{}_j + \beta^i{}_k\beta^k{}_j\,,$$

$$\nabla^2\sigma^i = F^i{}_j\sigma^j\,,$$

$$dF^i{}_j + \beta^i{}_k F^k{}_j - F^i{}_k \beta^k{}_j = 0 \; .$$

Further development of the formalism [29] gives generalization to non-commutative geometry of all basic objects of classical differential fibre bundle geometry including the case of topologically nontrivial bundles. Note that an unusual feature encountered here, even at the level of trivial bundles, is that the group of gauge transformations (which remains an *ordinary* group) does not consist only of algebra maps from A (the quantum group) to B (the base quantum space) as one might naively expect, but needs to be enlarged, as soon as B is non-commutative, to the more general class of "convolution-invertible" maps. It contains all algebra maps but also other "nonlocal" maps. The point is that the convolution (composition of gauge transformations) does not close on algebra maps.

In general, this framework does not need any kind of Lie algebra and covers in particular, the case when A is functions on a discrete group. This opens the way to the unification of (quantum) fibre bundle approach to generalized gauge theory with the models of Connes-Lott type.

Until now there has been no achievement on the use of q-fibre bundles for construction of a realistic gauge field theory. However, based on our experience from the usual groups, where the language of fibre bundle is very useful and is, in fact, equivalent to a Lagrangian formulation (see, e.g. [95]), we believe that in the q-deformed case it may appear to be even a more fruitful approach. The essence of all this is that a naive and straightforward substitution of an ordinary gauge group by a q-deformed analog does not lead to a meaningful theory, in particular because one has only the dual space description and no exponential map between the algebra and group transformations exsits. The q-fibre bundle formalism considered in this chapter gives an adequate mathematical background for the generalization of the gauge invariance principle to spaces described by non-commutative geometry.

Short Glossary of Selected Notions from the Theory of Classical Groups

In this Glossary we recall some basic definitions from the theory of Lie groups and algebras which are essentially used in the main text. Of course, rigorous definitions can be given only in the appropriate context of a complete exposition of the theory and we refer the reader to the books [13, 31, 95, 124, 142, 145, 219, 233, 236, 250] for further details and clarifications.

A-Module

An abelian group M with additively written composition law $\gamma = \alpha + \beta$, $\alpha, \beta, \gamma \in M$ is called *left A-module* with respect to an algebra A if a multiplication of elements of M by elements of A from the left is defined, with the properties

$$
\begin{aligned}
a(\alpha + \beta) &= a\alpha + a\beta \,, \\
(a + b)\alpha &= a\alpha + b\alpha \,, \\
(ab)\alpha &= a(b\alpha) \,, \qquad a, b \in A; \;\; \alpha, \beta \in M \,.
\end{aligned}
$$

Right A-module and *A-bimodule* are defined quite analogously.

Notice that one usually defines a module with respect to a ring (see, e.g. [236]) which is not necessary an algebra. In our book we deal with algebras (any algebra is the ring) and thus sligtly modify the standard definition. Another remark is that the second binary operation (multiplication) can exist also in M which converts it into a ring or an algebra (in general, non-commutative).

Complex and boundary operator

An abstract *complex* is a sequence of abelian groups G_n, $n \in \mathbb{Z}_+$

$$
\cdots \longrightarrow G_n \xrightarrow{\delta_n} G_{n-1} \xrightarrow{\delta_{n-1}} \cdots \longrightarrow G_1 \xrightarrow{\delta_1} G_0 \,,
$$

where the *boundary homomorphisms* (*boundary operators*) δ_n have the defining property

$$\delta_n \circ \delta_{n+1} = 0 \ .$$

Haar and invariant measures on a Lie group

Positive definite linear form $\mu(f)$ on $Fun(G)$, G being a Lie group, ($f \in Fun(G)$; $\mu(f) > 0$ if f is positively definite function) is called *left* or *right Haar measure* if it is left invariant

$$\mu(T_g^L f) = \mu(f) \ , \qquad T_g^L f := f(g^{-1}x) \ , \quad x, g \in G \ ,$$

or right invariant

$$\mu(T_g^R f) = \mu(f) \ , \qquad T_g^R f := f(xg) \ , \quad x, g \in G \ ,$$

correspondingly.

Left *and* right invariant (bi-invariant) Haar measure is called *invariant measure.*

Ideal and quotient algebra

A subalgebra $I \subset A$ of an algebra A is called the *left (right, two-sided) ideal* in A if for any $u \in I$ and any $a \in A$, $au \in I$ (correspondingly, $ua \in I$, both $au \in I$, $ua \in I$).

Any two-sided ideal $i \subset A$ defines the decomposition of the algebra A into equivalence classes (*cosets*) with the equivalence relation:

$$a \sim b, \ \text{ iff } \ a - b \in I; \quad a, b \in A \ .$$

In the set of the cosets one can define the natural multiplication

$$(a + I)(b + I) = (ab + I) \ ,$$

i.e. product of the cosets of the elements a and b is equal to the coset of ab. This algebra of cosets is called *quotient algebra.*

Isomorphism, automorphism, homomorphism, endomorphism

Let $\mathcal{X}$ and $\mathcal{X}'$ be two sets with some relations among elements of the each set.

For example, $\mathcal{X}$ and $\mathcal{X}'$ can be (Lie) groups, and the corresponding relations can be the group multiplications : $ab = c$ for $a, b, c \in \mathcal{X}$ and $a'b' = c'$

for $a', b', c' \in \mathcal{X}'$. Another example is ordered sets with defined inequalities $a > b$, $a, b \in \mathcal{X}$ and $a' > b'$, $a', b' \in \mathcal{X}'$.

Let there exists a one-to-one map $\rho : \mathcal{X} \longrightarrow \mathcal{X}'$ preserving the relations among elements of $\mathcal{X}, \mathcal{X}'$, i.e. if some relation is fulfilled for $a, b \in \mathcal{X}$ then the corresponding relation is fulfilled for $\rho(a), \rho(b) \in \mathcal{X}'$ and vice versa. In this case the sets $\mathcal{X}$ and $\mathcal{X}'$ are called *isomorphic* ones: $\mathcal{X} \cong \mathcal{X}'$, and the map ρ is called *isomorphism*.

In particular, if the sets coincide $\mathcal{X} = \mathcal{X}'$, a one-to-one map ρ, preserving some relations, is called *automorphism*.

If each element $a \in \mathcal{X}$ is mapped into a unique image, a single element $a' \in \mathcal{X}'$, but reverse is not in general true (e.g. a' may be the image of several elements of $\mathcal{X}$ or not be the image of any elements of $\mathcal{X}$) and the map preserves structure relations in $\mathcal{X}$ and $\mathcal{X}'$, then this map is called *homomorphism*.

Homomorphic map of a set into itself is called *endomorphism*.

Real forms of Lie algebras and groups

The *complex extension* L^c of arbitrary real Lie algebra L (Lie algebra over $\mathbb{R}$) is the Lie algebra which consists of elements of the form $X = X + iY$; $X, Y \in L$ (as the vector space) and with the Lie multiplication

$$\begin{aligned} Z &= [Z_1, Z_2] = ([X_1, X_2] - [Y_1, Y_2]) + i([X_1, Y_2] + [Y_1, X_2]) \\ &=: X + iY , \\ Z_1 &:= X_1 + iY_1 , \quad Z_2 := X_2 + iY_2 . \end{aligned}$$

The *real form* of a complex Lie algebra L^c (Lie algebra over $\mathbb{C}$) is the real Lie algebra L^r, such that its complex extension coincides with L^c.

Let G be the complex Lie group generated by a complex Lie algebra L^c. The subgroup G^r which corresponds to (i.e. generated by) the real form L^r of the Lie algebra L^c is called *real form* of the complex Lie group G.

Representations: faithful, irreducible, reducible, completely reducible (decomposable), indecomposable, adjoint

A representation of an algebra A (group G) is a homomorphism of A (or G) into an algebra (group) of linear transformations of some vector space V.

A representation is termed *faithful* if the homomorphism is an isomorphism.

A subspace $V_1 \subset V$ of a representation space V is called *invariant subspace* with respect to an algebra A (group G) if $Tv \in V_1$ for all $v \in V_1$ and all $T \in A$ (or $T \in G$).

A representation is called *irreducible* if the representation space V has not invariant subspaces (exept the whole space V and zero space $\{0\}$). Otherwise the representation is called *reducible*.

A representation is called *completely reducible* or *decomposable* if all linear transformations of the representation can be presented in the form of block-*diagonal* matrices, each block acting in the corresponding invariant subspace. Otherwise the representation is called *indecomposable*.

The simplest example of indecomposable representation is provided by two-dimensional representation of the abelian group $G = \mathbb{R}$ (the group multiplication corresponds to the addition in $\mathbb{R}$)

$$\mathbb{R} \ni x \longrightarrow T_x := \begin{pmatrix} 1 & x \\ 0 & 1 \end{pmatrix} ,$$

which acts in the linear space $V^{(2)}$, i.e.

$$T_x \begin{pmatrix} u_1 \\ u_2 \end{pmatrix} = \begin{pmatrix} u_1 + xu_2 \\ u_2 \end{pmatrix} .$$

The subspace $V^{(2)}_{(1)}$ of vectors $u = \begin{pmatrix} u_1 \\ 0 \end{pmatrix}$ is invariant with respect to T_x $\forall x \in \mathbb{R}$, the orthogonal subspace $V^{(2)}_{(2)}$ consisting of vectors $u = \begin{pmatrix} 0 \\ u_2 \end{pmatrix}$ is not invariant. It is impossible to achieve the decomposition into invariant subspaces by any (linear) transformations of bases in $V^{(2)}$.

The representation of a Lie group G (Lie algebra L) in the vector space of the Lie algebra L itself is called *adjoint representation* and the corresponding transformations are denoted by Ad_g, $g \in G$ (ad_X, $X \in L$). In the case of Lie algebra, the adjoint representation is defined by the commutator in L

$$\mathrm{ad}_X Y = [X, Y] , \quad X, Y \in L .$$

Root system of a semisimple Lie algebra, positive and simple roots

Let L be a semisimple Lie algebra with the Cartan subalgebra $H \subset L$ and α be a linear function on H. If the linear subspace $L^\alpha \subset L$ defined by the condition

$$L^\alpha := \{Y \in L \mid [X, Y] = \alpha(X) Y \ \ \forall X \in H\} ,$$

does exist (i.e. $L^\alpha \neq 0$), the function α is called *the root* of L, and L^α is called *the root subspace*. System of non-zero roots is denoted by Δ. Actually all root subspaces are one-dimensional, so that L^α is *root vector*.

The Cartan subalgebra and root vectors give the very convenient basis for an arbitrary semisimple Lie algebra L and provide their classification. In particular, *Cartan-Weyl basis* of a semisimple Lie algebra L consists of a basis $\{H_i\}$ of the Cartan subalgebra H and root vectors $E_\alpha \in L^\alpha$. In this basis, for any $\alpha, \beta \in \Delta$ the defining CR have the form

$$\begin{aligned} [H_i, E_\alpha] &= \alpha(H_i)E_\alpha \,, \qquad H_i \in H \,, \\ [E_\alpha, E_\beta] &= \begin{cases} 0 & \text{if } \alpha+\beta \neq 0 \text{ and } \alpha+\beta \notin \Delta \,, \\ H_\alpha & \text{if } \alpha+\beta = 0 \,, \\ N_{\alpha,\beta}E_{\alpha+\beta} & \text{if } \alpha+\beta \in \Delta \,, \end{cases} \end{aligned}$$

where the constants $N_{\alpha,\beta}$ satisfy the identity $N_{\alpha,\beta} = -N_{-\beta,-\alpha}$.

Thus the problem of the classification is reduced to the study of possible sets of the constants $N_{\alpha,\beta}$.

A root α is called *positive* if the first coordinate of the corresponding H_α is positive. Subsystem of positive roots is denoted by Δ_+.

A positive root is called *simple* if it cannot be expressed as sum of two other positive roots.

There is one-to-one correspondence between roots $\alpha \in \Delta$ and elements $H_\alpha \in H$ of the Cartan subalgebra defined by the equality

$$\langle X, H_\alpha \rangle = \alpha(X) \,, \qquad \forall X \in H \,.$$

Here $\langle \cdot, \cdot \rangle$ denotes the *Killing form* (scalar product) on L

$$\langle X, Y \rangle = \mathrm{Tr}(\mathrm{ad}_X \mathrm{ad}_Y) \,, \qquad \forall X, Y \in L \,.$$

Usually, the scalar product $\langle H_\alpha, H_\beta \rangle$ is denoted simply as (α, β).

Semidirect sum of Lie algebras and semidirect product of Lie groups (inhomogeneous Lie algebras and groups)

Let M and T are Lie algebras and $D : X \to D(X)$, $X \in M$ is a homomorphism of M into the set of linear operators in the vector space T, such that every D is a differentiation of T (i.e. D satisfies the Leibniz rule). Define in the direct space of M and T the Lie algebra structure using Lie brackets in M and T and

$$[X, Y] = D(X)Y \,, \qquad X \in M, \; Y \in T \,.$$

One can check that all the Lie axioms are satisfied. The obtained Lie algebra L is called *semidirect sum* of M and T

$$L := T \niplus M \ .$$

Such an algebra generates *semidirect product* of the Lie groups G_M and G_T, which can be defined independently as follows. Let G_T be arbitrary group, G_T^A be the group of all automorphisms of G_T, $G_M \subset G_T^A$ be some subgroup and $\Lambda(g)$ be the image of $g \in G_T$ under an automorphism $\Lambda \in G_M$.

The semidirect product $G = G_T \otimes G_M$ of the groups G_T and G_M is the group of all ordered pairs (g, Λ), where $g \in G_T$, $\Lambda \in G_M$, with the group multiplication

$$(g, \Lambda)(g', \Lambda') = (g\Lambda(g'), \Lambda\Lambda') \ ,$$

unity (e, id), e being the unity in G_T, and the inverse elements

$$(g, \Lambda)^{-1} = (\Lambda^{-1}(g^{-1}), \Lambda^{-1}) \ .$$

Solvable, nilpotent, semisimple and simple Lie algebras; Cartan and Borel subalgebras

A Lie algebra L is called *solvable* if the recursive definition

$$L^{(0)} := L, \ L^{(1)} := [L^{(0)}, L^{(0)}], \ldots, L^{(n+1)} := [L^{(n)}, L^{(n)}], \ldots \qquad n = 0, 1, 2, \ldots$$

yields the zero subalgebra after a finite number of steps, i.e. $L^{(n)} = 0$ for some $n < \infty$.

A Lie algebra is called *nilpotent* if the recursive definition

$$L_{(0)} := L, \ L_{(1)} := [L_{(0)}, L], \ldots, L_{(n+1)} := [L_{(n)}, L], \ldots \qquad n = 0, 1, 2, \ldots$$

yields the zero subalgebra after a finite number of steps, i.e. $L_{(n)} = 0$ for some $n < \infty$.

A Lie algebra L is *semisimple* if it has no non-zero abelian ideals.

A Lie algebra L is *simple* if it has no ideals (except zero ideal and the very L).

The Cartan subalgebra H of a semisimple Lie algebra L is the maximal abelian subalgebra in L with completely reducible adjoint representation.

The Borel subalgebra $\mathcal{B}$ of a simple Lie algebra L is the maximal solvable subalgebra in L.

Tensor operators

Let $g \to M(g)$ be a matrix representation of a group G in a finite dimensional vector space V and $g \to U_g$ is a unitary representation of G in a Hilbert space $\mathcal{H}$. A set $\{T^a\}_{a=1}^{\dim V}$ of operators in $\mathcal{H}$ is called *tensor operator* if

$$U_g^{-1} T^a U_g = M^a_{\ b}(g) T^b \ .$$

Universal enveloping algebra

Universal enveloping algebra of a Lie algebra L is a quotient algebra

$$\mathcal{U}(L) := A_L / J_{[\cdot,\cdot]} \ ,$$

where A_L is a (free) associative algebra generated by all $X_i \in L$ $(i = 1, \ldots, \dim L)$, $J_{[\cdot,\cdot]}$ is the two-sided ideal generated by elements of the form $XY - YX - [X,Y]$, $\forall X, Y \in L$.

Bibliography

[1] E. Abe, *Hopf Algebras* (Cambridge Tracts in Math., No.74, Cambridge Univ. Press, Cambridge-New York, 1980).

[2] F.C. Alcaraz and V. Rittenberg, *Phys. Lett.* **B314** (1993) 377.

[3] L. Alvarez-Gaume, C. Gomes and G. Sierra, *Nucl. Phys.* **B330** (1990) 347.
J.-L. Gervais, *Comm. Math. Phys.* **130** (1990) 257.

[4] G.E. Andrew, *q-Series: Their Development and Application in Analysis, Number Theory, Combinatorics, Physics and Computer Algebra* (Rhidel, Providence, 1986)

[5] T. Applequist, A.Chodos and P.G.O.Freund, eds., *Modern Kaluza-Klein Theory*, (Addison-Wesley, New York, 1987).

[6] M. Arik and D.D. Coon, *J. Math. Phys.* **16** (1975) 1776.

[7] M. Arik and D.D. Coon, *J. Math. Phys.* **17** (1976) 524.

[8] V.I. Arnold, *Mathematical Methods in Classical Mechanics* (Springer, Berlin, 1978).

[9] W. Arveson, *An Invitation to C^*-Algebras* (Springer-Verlag, New York, 1976).

[10] A.P. Balachandran, G. Bimonte, E. Ercolessi, G. Landi, F. Lizzi, G. Sparano and P. Teotonio-Sobrinho, *Nucl. Phys. Proc. Suppl.* **B37** (1995) 20.

[11] B.S. Balakrishna, F. Gürsey and K.C. Wali, *Phys. Lett.* **B254** (1991) 430.

[12] V.V. Bazhanov, *Comm. Math. Phys.* **113** (1987) 471.

[13] A.O. Barut and R. Raczka, *Theory of Group Representations and Applications* (Polish Sci.Publishers, Warsaw, 1977).

[14] F. Bayen, M. Flato, C. Fronsdal, A. Lichnerowicz and D. Sternheimer, *Ann. Phys.* **111** (1978) 61; *Ann. Phys.* **111** (1978) 111.

[15] L. Baulieu and E.G. Floratos, *Phys. Lett.* **B258** (1991) 171.

[16] A.A. Belavin and V.G. Drinfel'd, *Funkt. Anal. Appl.* **16** (1982) 159; *Soviet Sci. Rev.* **C4** (1984) 93.

[17] F.A. Berezin, *The method of second quantization* (Academic Press, New York, 1966).

[18] F.A. Berezin, *Teor. Mat. Fiz.* **6** (1971) 194.

[19] F.A. Berezin, *Comm. Math. Phys.* **40** (1975) 153.

[20] F.A. Berezin, *Sov. Phys. Usp.* **23** (1980) 1981.

[21] F.A. Berezin and M.S. Marinov, *Ann. Phys.* **104** (1977) 336.

[22] D. Bernard, *Lett. Math. Phys.* **17** (1989) 239.

[23] L.C. Biederharn, *J. Phys.* **A22** (1989) L873.

[24] L.C. Biedenharn and M. Tarlini, *Lett. Math. Phys.* **20** (1990) 271.

[25] L.C. Biedenharn and M.A. Lohe, in *From Symmetries to Strings, Forty Years of Rochester Conferences* ed. A. Das (World Scientific, Singapore, 1990).

[26] D. Blokhintsev, *Space and Time in Microworld* (Riedel, Dortrecht, 1973).

[27] F. Bonechi, E. Celeghini, R. Giachetti, E. Sorace and M. Tarlini, *Phys. Rev. Lett.* **68** (1992) 3718.

[28] P. Bonneau, M. Flato, M. Gerstenhaber and G. Pinczon, *Comm. Math. Phys.* **161** (1994) 125.

[29] T. Brzeziński and S. Majid, *Comm. Math. Phys.* **157** (1993) 591 and **167** (1995) 235.

[30] N. Burroughs, *Comm. Math. Phys.* **127** (1990) 109.

[31] R.N. Cahn, *Semi-Simple Lie Algebras and Their Representations* (Benjamin/Reading, Massachusetes, 1984).

[32] U. Carow-Watamura, M. Schlieker, M. Scholl and S. Watamura, *Z. Phys.* **C48**, (1990) 159.

[33] U. Carow-Watamura, M. Schlieker, S. Watamura and W. Weich, *Comm. Math. Phys.* **142** (1991) 605.

[34] U. Carow-Watamura, M. Schlieker and S. Watamura, *Z. Phys.* **C49** (1991) 439.

[35] E. Celeghini, R. Giachetti, E. Sorace and M. Tarlini *J. Math. Phys.* **31** (1990) 2548; **32** (1991) 1155; **32** (1991) 1159.

[36] M. Chaichian and N.F. Nelipa, *Introduction to Gauge Field Theories* (Springer-Verlag, Berlin, Heidelberg, 1984).

[37] M. Chaichian, A. Demichev, A.Ya. Rodionov and N.F. Nelipa, *Nucl. Phys.* **B279** (1987) 452.

[38] M. Chaichian and P. Kulish, *Phys. Lett.* **B234** (1990) 72.

[39] M. Chaichian, D. Ellinas and P. Kulish, *Phys. Rev. Lett.* **65** (1990) 980.

[40] M. Chaichian and D. Elinas, *J. Phys.* **A23** (1990) L291.

[41] M. Chaichian and P. Kulish, in *Proc. of 14th John Hopkins Workshop on Current Problems in Particle Theory, Debrecen, Hungary, August 1990* eds. G. Domokos et al. (World Scientific, Singapore, 1991).

[42] M. Chaichian, P. P. Kulish and J. Lukierski, *Phys. Lett.* **B243** (1990) 401.

[43] M. Chaichian, P. Kulish and J. Lukierski, *Phys. Lett.* **B262** (1991) 43.

[44] M. Chaichian, A. P. Isaev, J. Lukierski, Z. Popowicz and P. Prešnajder, *Phys. Lett.* **B262** (1991) 32.

[45] M. Chaichian, J. A. De Azcárraga, P.Prešnajder and F. Rodenas, *Phys. Lett.* **B291** (1992) 411.

[46] M. Chaichian and P. Prešnajder, *Phys. Lett.* **B277** (1992) 109.

[47] M. Chaichian, J.F. Gomes and P. Kulish, *Phys. Lett.* **B311** (1993) 93.

[48] M. Chaichian, J.F. Gomes and R. Gonzalez Felipe *Phys. Lett.* **B341** (1994) 147.

[49] M. Chaichian, R. Gonzalez Felipe and P. Prešnajder *J. Phys.* **A28** (1995) 2247.

[50] M. Chaichian, R. Gonzalez Felipe and C. Montonen *J. Phys.* **A26** (1994) 4017.

[51] M. Chaichian, H. Grosse and P. Prešnajder, *J. Phys.* **A27** (1994) 2045.

[52] M. Chaichian, J.A. De Azcárraga and F. Rodenas, in *Symmetries in Science VI*, ed. B.Gruber (Plenum Press, New York, 1994), p. 157.

[53] M. Chaichian and A.P. Demichev, *Phys. Lett.* **B320** (1994) 273.

[54] M. Chaichian, M. Mnatsakanova and Yu. Vernov, *J. Phys.* **A27** (1994) 2053.

[55] M. Chaichian and A. Demichev, in *Proceedings of the Clausthal Symposium on "Nonlinear, Dissipative, Irreversible Quantum Systems"*, eds. H.-D. Doebner, V.K. Dobrev and P. Nattermann (World Scientific, Singapore,1995) p.401.

[56] M. Chaichian and A. Demichev, *Phys. Lett.* **A207** (1995) 23.

[57] M. Chaichian and A. Demichev, *J. Math. Phys.* **36** (1995) 398.

[58] M. Chaichian and A.P. Demichev, *Polynomial algebras and higher spins* (Preprint, Helsinki University HU-SEFT R 96-02, Helsinki, 1996; to appear in *Phys. Lett.* **A**).

[59] M. Chaichian and P. Prešnajder, *q-Virasoro algebra, q-conformal dimensions and free q-superstring* (Preprint, Helsinki University HU-SEFT R 96-04, Helsinki, 1996; to appear in *Nucl. Phys.* **B**).

[60] Z. Chang, *Phys. Rep.* **262** (1995) 137.

[61] V. Chari and A. Pressley, *A Guide to Quantum Groups* (Cambridge Univ. Press, Cambridge, 1994).

[62] G.-H. Chen, L.-M. Kuang and M.-L. Ge, *Phys. Lett.* **A213** (1996) 231.

[63] A. Connes, in *The Interface of Mathematics and Particle Physics*, eds. D. Quillen, G.B. Segal and S.T. Tson (Claredon Press, Oxford, 1990).

[64] A. Connes, *Non-Commutative Geometry* (Academic Press, New York, 1994).

[65] A. Connes and J. Lott, *Nucl. Phys. Proc. Suppl.* **B18** (1990) 29.

[66] D.D. Coon, S. Yu and M.M. Baker, *Phys. Rev.* **D5** (1972) 1429.

[67] R. Coquereaux, G. Esposito-Farese, G. Vaillant, *Nucl. Phys.* **B353** (1991) 689.

[68] C. Cryssomalakos, B. Drabant, M. Schliker, W. Weich and B. Zumino, *Comm. Math. Phys.* **147** (1992) 635.

[69] T.L. Curtright and C.K. Zachos, *Phys. Lett.* **B 243** (1990) 237.

[70] J.A. de Azcárraga, P.P. Kulish and F. Rodenas, *Lett. Math. Phys.* **32** (1994) 173.

[71] J.A. de Azcárraga, P.P. Kulish and F. Rodenas, *Fortschr. Phys.* **44** (1996) 1.

[72] J.A. de Azcárraga, P.P. Kulish and F. Rodenas, *Phys. Lett.* **B351** (1995) 123.

[73] C.De Concini and V. Kac, *Prog. in Math.* **92** (1990) p.471; *Int. J. Mod. Phys.* **A7** (1992) 141.

[74] A. Demichev, *J. Phys.* **A29** (1996) 2737.

[75] E. Demidov, Yu. Manin, E. Mukhin and D. Zhdanovich, *Prog. Theor. Phys. Suppl.* **102** (1990) 203.

[76] J. Dixmier, *C^*-algebras* (North-Holland, Amsterdam, 1977).

[77] A. Dimakis and F. Müller-Hoissen, *Phys. Lett.* **B295** (1992) 242.

[78] A. Dimakis and F. Müller-Hoissen, *J. Phys.* **A25** (1992) 5625.

[79] A. Dimakis and F. Müller-Hoissen, *J. Math. Phys.* **35** (1994) 6703.

[80] V.K. Dobrev, *J. Phys.* **A26** (1993) 1317.

[81] V.K. Dobrev, *J. Phys.* **A27** (1994) 4841 (& Note Added 6633).

[82] V.K. Dobrev, in *Proceedings of the Clausthal Symposium on "Nonlinear, Dissipative, Irreversible Quantum Systems"*, eds. H.-D. Doebner, V.K. Dobrev and P. Nattermann (World Scientific, Singapore,1995) p.407.

[83] B. Drabant, B. Jurčo, M. Schlieker and B. Zumino, *Lett. Math. Phys.* **26** (1992) 91.

[84] B. Drabant, M. Schlieker, W. Weich and B. Zumino, *Comm. Math. Phys.* **147** (1992) 605.

[85] B. Drabant and W. Weich, *J. Math. Phys.* **36** (1995) 891.

[86] V.G. Drinfel'd, in *Proc. of the International Congress of Mathematicians (Berkley, 1986)* (American Mathematical Society, 1987), p.798.

[87] V.G. Drinfel'd, *Sov. Math. Dokl.* **27** (1983) 68.

[88] V.G. Drinfel'd, *Sov. Math. Dokl.* **32** (1985) 254.

[89] V. G. Drinfel'd, *Sov. Math. Dokl.* **36** (1988) 212.

[90] V.G. Drinfel'd, *Leningrad. Math. J.* **1** (1990) 1419.

[91] V.G. Drinfel'd and V.V. Smirnov, in *Itogi Nauki i Tekhniki, Seriya Sovremennye Problemy Matematiki* **24** (1984) 81 (Translated: Plenum, New York, 1985) p. 1975.

[92] M. Dubois Violette, R. Kerner and J. Madore, *J. Math. Phys.* **31** (1990) 316; 323.

[93] M. Dubois Violette, R. Kerner and J. Madore, *Class. Quant. Grav.* **6** (1991) 1709.

[94] M. Dubois Violette, J. Madore and R. Kerner, *Class. Quant. Grav.* **8** (1991) 1077.

[95] B.A. Dubrovin, A.T. Fomenko and S.P. Novikov, *Modern Geometry* Vol. 1,2 (Springer-Verlag, Berlin, 1984 (v.1), 1985 (v.2)).

[96] M.J. Duff, B.E.W. Nillson and C. Pope, *Phys. Rep.* **130** (1986) 1.

[97] B. Enriquez, *Lett. Math. Phys.* **25** (1992) 111.

[98] H. Ewen, O. Ogievetsky and J. Wess, *Lett. Math. Phys.* **22** (1991) 297.

[99] H. Exton, *q-Hypergeometric functions and applications* (Ellis Horwood, Chichester, 1983).

[100] L.D. Faddeev, in *Mathematical Physics Review, Sect.C: Math. Phys. Rev.* (Academic, Harwood, 1980) Vol. 1, p.107.

[101] L.D. Faddeev, in *Proceed. Les Houches Summer School, 1982* (Elsevier Sci. Publ., Amsterdam, 1984), p. 563.
L.D. Faddeev, in *Field and Particles*, eds. H. Mitter and W. Schweiger (Springer, Berlin, 1990), p.89.

[102] L.D. Faddeev and L.A. Takhtajan, *Lecture Notes in Physics* **246** (1986) 166.

[103] L. Faddeev, N. Reshetikhin and L. Takhtajan, in *Braid group, knot theory and statistical mechanics*, eds. C.N. Yang and M.L. Ge (World Scientific, Singapore, 1989).

[104] L.D. Faddeev, N.Yu. Reshetikhin and L.A. Takhtadjan, *Algebra i Analiz* 1 (1989) 178 (Transl.: *Leningrad Math. J.* **1** (1990) 193).

[105] L. Faddeev, in *Integrable Systems, Quantum Groups and Quantum Field Theory*, eds. L.A. Ibort and M.A. Rodríguez (Kluwer Academic Publ., Dordrecht, 1993) p.1.

[106] B. L. Feigin and D. B. Fuchs, *Func. Anal. App.* **16** (1982) 114.

[107] I. B. Frenkel and N. Jing, *Proc. Nat. Acad. Sci. USA* **85** (1988) 9373.

[108] E. Frenkel and N. Reshetikin, *Quantum affine algebras and deformations of the Virasoro and W-algebras* (Preprint, May 1995).

[109] R. Gilmore, *Lie Groups, Lie Algebras and Some of Their Applications* (Willey-Interscience, New York, 1974).

[110] G. Gasper and M. Rahman, *Basic Hypergeometric Series* (Cambridge Univ. Press, Cambridge, 1990).

[111] S.J. Gates, M.T. Grisaru, M. Roček and W. Siegel, *Superspace* (Benjamin/Cummings, Massachusets, 1983).

[112] M. Green, D. Schwarz and E. Witten, *Superstring Theory*, Vol. 1 (Cambridge Univ. Press, Cambridge, 1987).

[113] O.W. Greenberg, *Phys. Rev. Lett.* **64** (1990) 705; *Phys. Rev.* **D43** (1991) 4111.

[114] H. Grosse and J. Madore, *Phys. Lett.* **B283** (1992) 218.

[115] H. Grosse, C. Klimčik and P. Prešnajder, *Simple field theoretical models on noncommutative manifolds* (Preprint IHES/P/95/90, Bures-sur-Yvette, 1995).

[116] G. Gurevich and S. Majid, *Pac. J. Math.* **162** (1994) 27.

[117] F.D.M. Haldane and C. Rezayi, *Phys. Rev.* **B31** (1985) 2529.

[118] T. Hayashi, *Comm. Math. Phys.* **127** (1990) 129.

[119] A. Hebecker, S. Schreckenberg, J. Schwenk, W. Weich and J. Wess, *Lett. Math. Phys.* **31** (1994) 279.

[120] W. Heisenberg (1954), as quoted in
H.P. Dürr, *Werner Heisenberg und die Physik unserer Zeit*, (S.299, Fr.Vieweg u. Sohn, Braunschweig, 1961) and
H. Rampacher, H. Stumpf and F. Wagner, *Fortsch. Phys.* **13** (1965) 385.

[121] J. Hietarinta, *Phys. Lett.* **A165** (1992) 245.

[122] P.J. Hilton and S. Wylie, *Homology Theory* (Cambridge Univ. Press, Cambridge, 1966).

[123] W.V.D. Hodge and D. Pedoe, *Methods of Algebraic Geometry*, Vol. I, (Cambridge Univ. Press, Cambridge, 1947).

[124] J.E. Humphrey, *Introduction to Lie Algebras and Representation Theory*, (Springer, New York, 1972).

[125] E. Inönü and E.P. Wigner, *Proc. Nat. Acad. Sci. USA* **39** (1956) 510.

[126] B. Iochum and T. Schücker, *Yang-Mills-Higgs Versus Connes-Lott* (Preprint CPT-94/P.3090, Centre de Physique Theorique, Marseille, 1995)

[127] F.N. Jackson, *Quart. J. Pure and Appl. Math.* **41** (1910) 193; *Am. J. Math.* **32** (1910) 305.

[128] A. Jannussis et al. , *Hadronic J.* **3** (1980) 1622.
A. Jannussis et al. , *Lett. Nuovo Cim.* **38** (1983) 155.

[129] M. Jimbo, *Lett. Math. Phys.* **10** (1985) 63; *Lett. Math. Phys.* **11** (1986) 247.

[130] M. Jimbo, *Comm. Math. Phys.* **102** (1986) 537.

[131] B. Jurčo, *Lett. Math. Phys.* **22** (1991)177.

[132] V.G. Kac, *Advances in Math.* **26** (1977) 8.

[133] V.G. Kac, *Lecture Notes in Math.* **676** (1978) 597.

[134] V. G. Kac, *Lecture Notes in Phys.* **94** (1979) 441.

[135] V.G. Kac, *Infinite dimensional Lie algebras* (Birkhäuser, Boston-Basel-Stuttgart, 1983).

[136] D. Kapetanakis and G. Zoupanos, *Phys. Rep.* **219** (1992) 1.

[137] C. Kassel, *Quantum Groups* (Springer, New York, 1995).

[138] G. Keller, *Lett. Math. Phys.* **21** (1991) 273.

[139] A. Kempf, *Czech. J. Phys.* **44** (1994) 1041.

[140] A. Kempf and S. Majid, *J. Math. Phys.* **35** (1994) 6802.

[141] S.M. Khoroshkin and V.N. Tolstoy, *Comm. Math. Phys.* **141** (1991) 599.

[142] A.A. Kirillov, *Elements of the Theory of Representations* (Springer, Berlin, 1976).

[143] A.N. Kirillov and N. Reshetikhin, *Representations of the algebra $\mathcal{U}_q(sl_2)$, q-orthogonal polynomials and invariants of links* (LOMI Preprint E-9-88, Leningrad, 1988).

[144] A.N. Kirillov and N. Reshetikhin, *Comm. Math. Phys.* **134** (1990) 421.

[145] S. Kobayashi and K. Nomizu. *Foundations of Differential Geometry* Vol. 1,2 (Interscience Publishers, New York, 1969).

[146] T. Kobayashi and T. Uematsu, *Z. Phys.* **C56** (1992) 193.

[147] T.H. Koornwinder, in *Orthogonal Polynomials*, ed. P. Nevai (Kluwer Acad.Pub., 1990), p.257.

[148] T.H. Koornwinder, in *Representations of Lie Groups and Quantum Groups*, eds. V.Baldoni and M.A.Picardello, Pitman Research Notes in Mathematics, Ser. 311 (Longman Scientific and Technical, 1994).

[149] K. Kowalski and J. Rembielinski, *J. Math. Phys.* **34** (1993) 2153.

[150] Y.A. Kubyshin, J.M. Mourão, G. Rudolf and I.P. Volobuev, *Lecture Notes in Physics* **349** (Springer, Heidelberg, 1989).

[151] P. Kulish and Yu. Reshetikhin, *J. Sov. Math.* **23** (1983) 2435 (translation from: *Zapiski Nauch. Seminarov LOMI* **101** (1981) 101.

[152] P. Kulish and E. Sklyanin, *Lecture Notes Phys.* **151** (1982) 61.

[153] P. Kulish and E. Sklyanin, *J. Phys.* **A25** (1992) 5963.

[154] P. Kulish and Sasaki, *Prog. Theor. Phys.* (1993) .

[155] P. Kulish and E. Damaskinsky, *J. Phys.* **A23** (1990) L 415.

[156] P. Kulish and N. Reshetikhin, *Lett. Math. Phys.* **18** (1989) 143.

[157] P.P. Kulish, *Theor. Math. Phys.* **86** (1991) 108.

[158] V.V. Kuryshkin, *Ann. Found. L. de Broglie* **5** (1980) 111.

[159] L.D. Landau and E.M. Lifshitz, Quantum Mechanics (Pergamon Press, New York, 1981).

[160] A. Lerda, *Anyons: Quantum Mechanics of Particles with Fractional Statistics* (Springer Verlag, Berlin, 1992).

[161] S. Levendorskii and Ya.S. Soibelman, *J. Geom. Phys.* **7** (1990) 241.

[162] S. Levendorskii and Ya.S. Soibelman, *Comm. Math. Phys.* **139** (1991) 141.

[163] J. Lukierski, A. Nowicki, H. Ruegg and V. Tolstoy, *Phys. Lett.* **B264** (1991) 331.

[164] J. Lukierski, and A. Nowicki, *Phys. Lett.* **B279** (1992) 299.

[165] J. Lukierski and H. Ruegg, *Phys. Lett.* **B329** (1994) 189.

[166] J. Lukierski, A. Nowicki, H. Ruegg and V.N. Tolstoy, *J. Phys.* **A27** (1994) 2389.

[167] G. Lusztig, *Adv. in Math.* **70** (1988) 237.

[168] G. Lusztig, *Contemp. Math.* **82** (1989) 59.

[169] G. Lusztig, *J. Amer. Math. Soc.* **30** (1990) 257; **30** (1990) 447.

[170] G. Lusztig, *Geom. Dedicata* **35** (1990) 89.

[171] G. Lusztig, *Introduction to Quantum Groups* (Progress in Math., 110; Birkhäuser, Boston, 1993).

[172] A. Macfarlane, *J. Phys.* **A22** (1989) 4581.

[173] J. Madore, *Noncommutative Geometry and Applications* (Cambridge Univ. Press, Cambridge, 1995).

[174] S. Majid, *Foundations of Quantum Group Theory* (Cambridge Univ. Press, Cambridge, 1995).

[175] S. Majid, in *Quantum Groups, Integrable Statistical Models and Knot Theory*, eds. M-L. Ge and H.J. de Vega (World Scientific, Singapore, 1993) p. 231.

[176] S. Majid, *Some Remarks on the q-Poincare Algebra in R-Matrix Form* (DAMPT/95-08, Cambridge, 1995).

[177] Y. Manin, *Quantum groups and non-commutative geometry* (Center des Recherches Mathématiques, Montréal, 1988).

[178] Yu.I. Manin, *Comm. Math. Phys.* **123** (1989) 163.

[179] N.S. Manton, *Nucl. Phys.* **B158** (1979) 141.

[180] M.A. Martin-Delgado, *J. Phys. A: Math. Gen.* **24** (1991) 1285; P.V. Neškovic and B.V. Uroševic, *Int. J. Mod. Phys.* **A7** (1992) 3379.

[181] T. Masuda, K. Mimachi, Y. Nakagami, N. Noumi and K. Ueno, *C.R. Acad. Si. Paris, Ser.I. Math.* **307** (1988) 559.

[182] J.E. Moyal, *Proc. Cambridge Phil. Soc.* **45** (1949) 99.

[183] M.A. Naimark, *Linear Representations of the Lorentz Group* (Pergamon, Oxford, 1964).

[184] M. Nomura, *J. Math. Phys.* **30** (1989) 2397.

[185] O. Ogievetsky, W.B. Schmidke, J. Wess and B. Zumino, *Comm. Math. Phys.* **150** (1992) 495.

[186] A. Pais and V. Rittenberg, *J. Math. Phys.* **16** (1975) 2062; **17** (1975) 598.

[187] V. Pasquier, *Comm. Math. Phys.* **118** (1988) 355.

[188] V. Pasquier and H. Saleur, *Nucl. Phys.* **B330** (1990) 523.

[189] R. Penrose, *Rep. Math. Phys.* **12** (1977) 65.

[190] P. Podlés, *Lett. Math. Phys.* **14** (1987) 193.

[191] P. Podleś and S. L. Woronowicz, *On the structure of inhomogeneous quantum groups* (Preprint UC Berkly, 1995, hep-th 9412058).

[192] P. Podleś and S. L. Woronowicz, *Comm. Math. Phys.* **178** (1996) 61.

[193] A.P. Polychronakos, *Phys. Lett.* **B256** (1991) 35.

[194] W. Pusz and S.L. Woronowicz, *Rep. Math. Phys.* **27** (1989) 231.

[195] C.R. Putnam, *Commutation properties of Hilbert space operators and related topics* (Springer, Berlin, 1967).

[196] G. Rideau, *Lett. Math. Phys.* **24** (1992) 147.

[197] N.Yu. Reshetikhin, *Quntized universal enveloping algebra, the Yang-Baxter equation and invariants of links I,II* (preprints LOMI E-4-87, E-17-87, Leningrad, LOMI, 1988)

[198] N.Yu. Reshetikhin, *Lett. Math. Phys.* **20** (1990) 331.

[199] N. Reshetikhin and F. Smirnov, *Comm. Math. Phys.* **131** (1990) 157.

[200] V. Rittenberg and M. Scheunert, *J. Math. Phys.* **33** (1992) 436.

[201] P. Roche and D. Ardaudon, *Lett. Math. Phys.* **17** (1989) 295.

[202] L.J. Rogers, *Proc. London Math. Soc.* **24** (1893) 337.

[203] G.G. Ross, *Grand Unified Theories* (Addison Wesley, Reading, MA, 1984).

[204] M. Rosso, *Comm. Math. Phys.* **117** (1988) 581.

[205] M. Rosso, *Comm. Math. Phys.* **124** (1989) 307.

[206] H.J. Rothe, *Lattice Gauge Theories. An Introduction* (World Scientific, Singapore, 1992).

[207] H. Ruegg, *J. Math. Phys.* **31** (1990) 1085.

[208] H. Saleur, *Nucl. Phys.* **B336** (1990) 363.

[209] A. Schirrmacher, *Z. Phys.* **C50** (1991) 321.

[210] A. Schirrmacher, *J. of Phys.* **A24** (1991) L1249.

[211] A. Schirrmacher, J. Wess and B. Zumino *Z. Phys.* **C49** (1991) 317.

[212] M. Schliecer, W. Weich and R. Weixler, *Z. Phys.* **C53** (1992) 79.

[213] W.B. Schmidke, J. Wess and B. Zumino, *Z. Phys.* **C52** (1991) 471.

[214] A. Schwarz, *Topology for Physicists*, (Springer, Berlin, 1994).

[215] J. Schwenk and J. Wess, *Phys. Lett.* **B291** (1992) 273.

[216] J. Schwinger *Phys. Rev.* **82** (1951) 664; *Phys. Rev. Lett.* **3** (1959) 296.

[217] J. Schwinger in *Quantum Theory of Angular Momentum*, eds. L.C. Biedenharn and H. van Dam (New York, Academic, 1958).

[218] M.A. Semenov-Tian-Shansky, *Funkt. Anal. Appl.* **17** (1983) 259.

[219] J-P. Serre, *Complex Semisimple Lie Algebras* (Springer-Verlag, New York, 1987).

[220] S. Shabanov, *Phys. Lett.* **B293** (1992) 117.

[221] H. Snyder, *Phys. Rev.* **71** (1947) 38.

[222] Y.S. Soibelman, *Algebra i Analiz* **2** (1990) 190.

[223] Y.S. Soibelman, *Selected Topics in Quantum Groups* (Publ.RIMS -804, Kyoto, 1991).

[224] R.D. Sorkin, *Int. J. Theor. Phys.* 30 (1991) 923.

[225] A. Sudbery, *J. Phys.* **A23** (1990) L697.

[226] L.A. Takhtajan, *Adv. Stud. Pure Math.* **19** (1989) 435.

[227] L.A. Takhtajan, in *Introduction to Quantum Groups and Integrable Massive Models of Quantum Field Theory*, eds. M.L. Ge and B.H. Zhao (World Scientific, Singapure, 1990).

[228] T. Tjin, *Int. J. Mod. Phys.* **A7** (1992) 6175.

[229] V.N. Tolstoy, *Lecture Notes in Physics* **370** (1990) 188.

[230] L.L. Vaksman and L.I. Korogodsky, *Dokl. Akad. Nauk SSSR* **304** (1989) 1036.

[231] L.L. Vaksman and Ya.S. Soibelman, *Funct. Anal. Appl.* **22** (1988) 170.

[232] G. Veneziano, *Nuovo Cim.* **57A** (1968) 190;
S. Fubini, D. Gordon and G. Veneziano, *Phys. Lett.* **B29** (1969) 679;
S. Fubini and G. Veneziano, *Nuovo Cim.* **A64** (1969) 811; *Ann. Phys.* **63** (1971) 12.

[233] N.Ya. Vilenkin, *Special Functions and the Theory of Group Representations* (Amer.Math.Soc.Transl. of Math. Monographs, vol.22, 1968).

[234] M. Virasoro, *Phys. Rev.* **D1** (1970) 2933.

[235] S.P. Vokos, B. Zumino and J. Wess, *Z. Phys* **C48** (1990) 65.

[236] B.L. van der Waerden, *Algebra I,II* (Springer, Berlin, 1971,1967).

[237] B.L. van der Waerden, *Group Theory and Quantum Mechanics* (Springer, Berlin, 1974).

[238] A. Weinstein, *J. Diff. Geom.* **18** (1983) 523.

[239] J. Wess and B. Zumino, *Nucl. Phys. Suppl.* **B18** (1990) 302.

[240] P.B. Wiegmann and A.Z. Zabrodin, *Phys. Rev. Lett.* **72** (1994) 1890.

[241] E.P. Wigner, *Phys. Rev.* **77** (1950) 711.

[242] E. Witten, *Comm. Math. Phys.* **121** (1989) 351.

[243] E. Witten, *Nucl. Phys.* **B330** (1990) 285.

[244] S.L. Woronowicz, *Publ. RIMS* **23** (1987) 117.

[245] S.L. Woronowicz, *Comm. Math. Phys.* **111** (1987) 613.

[246] S.L. Woronowicz, *Invent. Math.* **93** (1988) 35.

[247] S.L. Woronowicz, *Comm. Math. Phys.* **122** (1989) 125.

[248] S.L. Woronowicz, *Comm. Math. Phys.* **136** (1991) 399.

[249] S.L. Woronowicz and S. Zakrzewski, *Compositio Math.* **90** (1994) 211.

[250] B.G. Wybourn, *Classical Groups for Physicists* (Wiley-Interscience, New York, 1974).

[251] C. Zachos, in *Quantum Groups*, eds. T. Curtroght, D. Fairlie and C. Zachos, (World Scientific, Singapore, 1991);
C. Zachos, in *Deformation Theory and Quantum Groups with Applications Mathematical Physics*, eds. M. Gerstenhaber and J. Stasheff (Contemporary Mathematics 134, American Math. Society, Providence, RI, 1992) p.351.

[252] S. Zakrzewski, *J. Phys.* **A27** (1994) 2075.

[253] P. Zaugg, *J. Phys.* **A28** (1995) 2589.

[254] P. Zaugg, *J. Math. Phys.* **36** (1995) 1547.

[255] B. Zumino, in Proceedings of X*th* IAMP Conf., Leipzig (1991), ed. K. Schmüdgen (Springer-Verlag, Berlin, 1992), p.20.

Index

adjoint action q-deformed 48
algebra unital associative 23
anti-de Sitter q-algebra 250
anyonic statistics 176
antipode (coinverse) 27
automorphism 65, 317

bialgebra 27
bosonization of braided groups 172
boundary operator 316
on posets 299
braided group 166
braid-statistics relations 172

$\mathbb{C}^*$-algebra 3
Cartan-Weyl basis 47, 319
Chevalley basis 47
Clebsch-Gordan coefficients 155
q-deformed 155
coaction of quantum group 61
coassociativity 26
coinverse (antipode) 27
co-Jacobi identity 35
co-Leibniz rule 35
comodule 62
complex 37, 315
comultiplication (coproduct) 20, 26
cocommutative 20
non-cocommutative 21
connection quantum 313
Connes-Lott model 283
contraction of q-algebra 121
contraction of quantum groups 257
convolution invertible map 311
convolution product of linear maps 310
co-Poisson homomorphism 35
co-Poisson Lie algebra 41
coproduct (comultiplication) 20
corepresentation of Hopf algebra 93
irreducible 94
unitary 94
cosets 316
coset space dimensional reduction 294
counity 26
curvature quantum 313
cyclic representation 107

derivatives q-deformed 82
determinant quantum 63
differential calculus
on algebra 10
on quantum group 80
on quantum space 81
dilatation 210, 215, 224
dimension of a differential algebra 303
local 303
duality 26

endomorphism 52, 317
Euclidean length quantum 69
Euclidean space quantum 68
exponent q-deformed 60, 129
product representation 144

second 131
summation theorem 135
exterior differential operator
on algebra 10
quantum covariant 312
exterior quantum algebra 64
on Euclidean quantum space 69

fibre bundle
associated (vector) quantum 311
principal 308
quantum 309
trivial 310
trivial quantum 311
field theory on two-dimensional q-space 222
finite difference operators on equidistant lattice 210
free particle on two-dimensional q-space 221
fuzzy sphere 260

gauge transformation quantum 311
general linear groups quantum 15
Grassmann variables 147, 150
generalized 147

Haar measure 36, 316
Hamiltonian curve 33
Hasse graph 297
Hecke condition 64
Heisenberg algebra 9
homomorphism 18, 317
Hopf algebra 2, 23, 28
dual 28
quasi-triangular 53
structure maps 28
structure constants 29
triangular 53
Hopf subalgebra (q-subalgebra) 56, 58

ideal in algebra 316
ideal q-gas 193
integral calculus
on quantum group 85
on quantum space 87
internal Dirac operator 285
invariant measure 316
invariant subspace 317
*-involution in Hopf algebra 65
isomorphism 3, 317

Jackson derivative 83
multidimensional 215
symmetric 154
Jackson integral 89
Jacobi identity 33
Jaynes-Cummings model q-deformed 182
Jordan-Schwinger construction 151
q-deformed 152

Killing form 319

Laplace-Beltrami operator on poset 301
Leibniz rule 32
Lie algebra
complex extension 317
nilpotent 320
real form 317
semisimple 47, 320
simple 44, 320
solvable 110, 320
Lie bialgebra 40
Lie-Kirillov-Kostant bracket 33
Lorentz algebra quantum 248
Lorentz group quantum 233
general definition 262

magnetic translations 184
Manin triple 40

matrix group quantum 60
Maurer-Cartan equation quantum 81
Minkowski length q-deformed 235
Minkowski space 230
  quantum 234, 264
module (A-module) 10, 315
  bimodule 10, 315
  left, right 315
Moyal bracket 19

n-bein q-deformed 74
nilpotency 82, 119, 147

operator kernel 132, 133

path integral
  q-deformed 143, 146
  generalized Grassmann 148
Poincaré algebra
  κ-deformed 255, 259
  twisted 239
Poincaré group
  classification 274
  general definition 271
  twisted 233
Poisson algebra 32
Poisson bracket 19
Poisson-Hopf algebra 34
  deformation 43
Poisson-Lie group 33
Poisson manifold 33
Poisson map 32
poset 297
principal fibre bundle 308
  quantum 309
  trivial 310
    quantum 311

q-binomial coefficients 50, 97, 155
q-commutator 50
q-dimension of representation 105
q-integral
  definite 89
  indefinite 87
q-Jacobi polynomials 98
q-oscillator algebra 112
  central element 113
  quasi-classical limit 132
  representations 112, 117
    Bargmann-Fock 122, 125
    cyclic 118
    Macfarlane realization 123
  $SU_q(n)$-covariant 159
q-square bracket 21
q-subalgebra (Hopf subalgebra) 56, 58
q-subgroup 77
q-tetrade 237
q-tensor operator 156
q-trace 81
quantum double 55
quantum group 1
  compact 94
  Euclidean 126
  inhomogeneous 224
    general definition 268
    projective construction 225
  linear general $GL_q(n)$ 15, 63
  linear unimodular $SL_q(n)$ 63
    $SL_q(2)$ 17
    real $SL_q(n,\mathbb{R})$ 65
  matrix 60
  orthogonal $SO_q(N)$ 67
  symplectic $Sp_q(n)$ 67
  unitary
    $SU_q(n)$ 66
    $SU_q(2,2)$ 232
quantum plane 126
quantum space 61
  Euclidean 68
  vector 61

quantum universal enveloping algebra 21
quotient algebra 60, 316

real forms of quantum groups 65
real form of Lie group 317
reflection equation 172
Regge trajectory
linear 198
nonlinear 202
representation 317
adjoint 48, 318
cyclic 107
decomposable 105, 318
faithful 94, 317
indecomposable 103, 318
irreducible 3, 317
of $Fun_q(G)$ 109
reducible 101, 318
completely 318
representations of QUEA
for generic q 99
for root of unity q 102
indecomposable 103
r-matrix classical 38
R-matrix 15, 54
multiparametric 72
universal 53
root system of Lie algebra 51, 318
root positive 51, 319
root simple 47, 319

Schouten bracket 38
Schur relation quantum 96
Serre relations 47
section of q-fibre bundle 312
semidirect sum of Lie algebras (inhomogeneous Lie algebras)319
semidirect product of Lie groups (inhomogeneous Lie groups) 320
smooth deformation 19
Standard model 283
star-involution (*-involution) in Hopf algebra 65
star-product 18
string theory 198
q-deformation 203
structure space 3
superalgebra quantum 166
supergroup quantum 167
$GL_q(1|1)$ 167
$SL_q(1|1)$ 168
$U_q(1|1)$ 168
$SU_q(1|1)$ 168
superdeterminant q-deformed 168
superoscillator q-deformed 169
symplectic leaves 33, 107
symplectic manifold 33

tensor operator 320
q-deformed (q-tensor operator) 156
transmutation of quantum group 172, 175
TT-relations 16
twist of Hopf algebra 70

unity map 23
universal differential envelope of algebra 286, 301
universal enveloping algebra 20, 321
quantum 21

vector space quantum 61
Veneziano model 198
q-deformed 199
vertex operator q-deformed 200
Virasoro algebra 203
q-deformed 204

Weyl relation 12
Weyl symbol 18

Wigner-Eckart theorem 156
    q-generalization 157

XXX-model 177
XXZ-model 180

Yang-Baxter equation
    quantum 15, 54
    classical 38
    modified classical 38
Yang-Mills-Higgs theory 283